JOURNAL OF CHROMATOGRAPHY LIBRARY — volume 72

advances in LC–MS instrumentation

JOURNAL OF CHROMATOGRAPHY LIBRARY — volume 72

advances in LC–MS instrumentation

edited by

Achille Cappiello

Istituto di Scienze Chimiche "Fabrizio Bruner"
Università degli Studi di Urbino "Carlo Bo"
Urbino, Italy

ELSEVIER

Amsterdam • Boston • Heidelberg • London • New York • Oxford
Paris • San Diego • San Francisco • Singapore • Sydney • Tokyo

Elsevier
Radarweg 29, PO Box 211, 1000 AE Amsterdam, The Netherlands
The Boulevard, Langford Lane, Kidlington, Oxford OX5 1GB, UK

First edition 2007

Library of Congress Cataloging-in-Publication Data
A catalog record for this book is available from the Library of Congress

British Library Cataloguing in Publication Data
A catalogue record for this book is available from the British Library

ISBN-13: 978-0-444-52773-8
ISBN-10: 0-444-52773-7
ISSN: 0301-4770

For information on all Elsevier publications
visit our website at books.elsevier.com

Printed and bound in The Netherlands

07 08 09 10 11 10 9 8 7 6 5 4 3 2 1

Table of Contents

List of Contributors

Aviv Amirav, School of Chemistry, Sackler Faculty of Exact Sciences, Tel Aviv University, Tel Aviv 69978, Israel

Michael P. Balogh, Waters Corp. Milford, MA, USA

Thorsten Benter, Division of Physical Chemistry, University of Wuppertal, Gauss Str. 20, 42119 Wuppertal, Germany

Achille Cappiello, Istituto di Scienze Chimiche "F. Bruner", Università degli Studi di Urbino "Carlo Bo", Piazza Rinascimento, 6, 61029 Urbino, Italy

Joseph A. Caruso, Department of Chemistry, University of Cincinnati, P.O. Box 45221-0172, Cincinnati, OH 45221, USA

Tom A. van de Goor, Agilent Technologies, Life Science & Chemical Analysis, 5301 Stevens Creek Boulevard, Santa Clara, CA 95052, USA

Ori Granot, School of Chemistry, Sackler Faculty of Exact Sciences, Tel Aviv University, Tel Aviv 69978, Israel

Daren Levin, Department of Chemistry and Chemical Biology and Barnett Institute of Chemical and Biological Analysis, Northeastern University, Boston, MA 02115, USA

Raanan A. Miller, V.P. Technology & CTO, Sionex Corporation, 8-A Preston Court, Bedford, MA 01730, USA

Laura Molin, CNR, Area della Ricerca di Padova, ISTM, Corso Stati Uniti, 4, 35127 Padova, Italy

Erkinjon G. Nazarov, Chief Scientist, Sionex Corporation, 8-A Preston Court, Bedford, MA 01730, USA

Pierangela Palma, Istituto di Scienze Chimiche "F. Bruner", Università degli Studi di Urbino "Carlo Bo", Piazza Rinascimento, 6, 61029 Urbino, Italy

Andrea Raffaelli, CNR-ICCOM, c/o Dipartimento di Chimica e Chimica Industriale, Istituto di Chimica dei Composti Organo Metallici, Sezione di Pisa, Università di Pisa, Via Risorgimento, 35, 56126 Pisa, Italy

Oliver J. Schmitz, Division of Analytical Chemistry, University of Wuppertal, Gauss Str. 20, 42119 Wuppertal, Germany

Roberta Seraglia, CNR, Area della Ricerca di Padova, ISTM, Corso Stati Uniti, 4, 35127 Padova, Italy

Loris Tonidandel, CNR, Area della Ricerca di Padova, ISTM, Corso Stati Uniti, 4, 35127 Padova, Italy

Pietro Traldi, CNR, Area della Ricerca di Padova, ISTM, Corso Stati Uniti, 4, 35127 Padova, Italy

Katarzyna Wrobel, Instituto de Investigaciones Científicas, Universidad de Guanajuato, L de Retana N° 5, 36000 Guanajuato, Mexico

Kazimierz Wrobel, Instituto de Investigaciones Científicas, Universidad de Guanajuato, L de Retana N° 5, 36000 Guanajuato, Mexico

Introduction

Mass spectrometry (MS) is the most specific and flexible technique for the detection and identification of organic and inorganic compounds. MS can provide not only molecular weight information but also a wealth of structural details that together give a unique fingerprint for each analyte. MS is by far the detector with the highest information output for unit of sample weight. The dynamic combination of gas chromatography and mass spectrometry (GC–MS) has yielded the most specific and sensitive method for the characterization of the components of complex volatile mixtures.

Under this circumstance, the use of mass spectrometer as a sophisticated HPLC detector has been deeply investigated and brought as one of the major scientific advancement in the world of chemistry and biochemistry. Since the majority of all known compounds are amenable by HPLC, the integration of HPLC and MS greatly extends the range of compound classes that can benefit the peculiar MS detection. An uninterrupted evolution, observed in the last 25 years, has produced amazing results for both chromatographers and mass spectroscopists to the point that LC–MS represents one of the most important tools in the characterization of all organic, inorganic and biological compounds. While the evolution of coupling GC and MS has reached its full maturity bringing to a single, relatively simple and inexpensive device, LC–MS interfacing progress, far more challenging than was GC–MS, has far to complete its growth, continuously generating new, high-developed interfaces with radically different designs and better performances. The reason for the complexity of LC–MS is in the fact that the domains of LC and MS are often seen as antagonists: the effort of achieving and maintaining the high vacuum required for the analyzer operation in MS is in opposition with the intrinsic nature of LC, predominantly operating at high pressure. In addition, the low tolerance of mass spectrometers for non-volatile mobile-phase components contrasts with the LC dependence on non-volatile buffers to achieve high-resolution separations. In the effort of overcoming a series of evident difficulties, LC–MS interfaces use different strategies and, as a consequence of that, they are usually expensive and complicated compared to the intrinsic simplicity of GC–MS.

Difficulty in LC–MS is summarized in the following two points:

1. *Solvent restriction.* An HPLC mobile phase (solvent) is a liquid of variable composition, and sample components are led to the detector together with a given solvent composition. Separation and ionization of these components often depend on different and, sometimes, antagonist mobile-phase compositions.
2. *Sample restriction.* Liquid chromatography analyzes all samples that can be dissolved in solvents. The analytes may vary dramatically in weight, polarity and stability and this poses a huge variability in terms of response and system requirements on both interface and mass spectrometer.

Although this is a challenging premise, the demand for new LC–MS methods is increasingly strong in all fields of human development. Countless scientific progresses are made worldwide with the use of LC–MS and J.B. Fenn and K. Tanaka, 2002 Nobel Prize winners, are the living testimonies of the importance of this approach. As a matter of fact, the invention of electrospray ionization brought a new energy to this field, opening a whole new perspective for HPLC and MS.

It might sound strange, but many of the most interesting innovations in LC–MS were possible not only by a brilliant technical revolution but also through a constant reduction, over the years, of the typical HPLC mobile-phase flow rate. The success of first micro- and then nano-HPLC was firmly associated with the advantage of improving signal-to-noise ratio in mass spectral detection. Lab-on-a-chip devices can be seen as the latest frontier of this trend.

In addition, mass spectrometers are arranged in different configurations to produce suitable MS, MS/MS and MS^n spectra of all possible HPLC amenable compounds. Novel interfaces and ionization techniques were developed as well to improve the linkage with the eluate and fight all possible drawbacks. Electron ionization, which pioneered the LC–MS development, is fighting back with new concepts and with the purpose to play a prime role for the analysis of small–medium molecular weight molecules in complex matrices without showing any matrix effect.

In this book, the most recent developments in the field of LC–MS are outlined and carefully described by the people who created or refined them. Particular attention is given to the technical innovations that are bringing new energy in the world of LC–MS and that have gained worldwide attention in the form of certain number of publications. The reader will not miss the fundamental techniques traditionally used in this field such as electrospray (ESI), atmospheric pressure chemical ionization (APCI), and all of them will be covered with plenty of details to satisfy the expert and the neophyte.

As any other technical progress, LC–MS is under a constant growth with the perspective of fascinating evolutions right ahead of us. This book is not just a picture of the current state of the art but it was conceived with the future in mind, and it will provide a complete coverage of LC–MS for a long time.

Achille Cappiello

Achille Cappiello (Editor)
Advances in LC–MS Instrumentation
Journal of Chromatography Library, Vol. 72

Chapter 1

Basic aspects of electrospray ionization

LAURA MOLIN and PIETRO TRALDI

The production of charged droplet is a natural phenomenon, the final products of which was observed by us the first time we looked at a cloudy sky. The first experiments on electrospray were those carried out by the physicist (and Abbé) Jean-Antoine Nollet, who observed in 1750 that the water flowing from an electrified metal vessel shows a tendency to aerosolize when the vessel is placed near the electrical ground. At that time physics, chemistry, physiology and medicine were seen as a unique science and consequently experiments (which nowadays seem to be quite strange) were performed at physiological level. Abbé Nollet observed that "a person, electrified by connection to a high voltage generator (hopefully well insulated from the ground! – authors' note) would not bleed normally if he was to cut himself; blood spray from the wound" [1].

About one century later Lord Kelvin studied the charging between water dripping from two different liquid nozzles, which leads to electrospray phenomena at the nozzles themselves [2].

In the twentieth century a series of systematic studies on electrospray were carried out by Zeleny [3], Wilson and Taylor [4,5] and in the middle of the century electrospray became an effective painting technique, widely applied to vehicles, housewares and various metalic objects [6,7]. Electrospray became of scientific interest in 1968, when Dole and co-workers produced gas-phase, high molecular weight polystyrene ions by electrospraying a benzene/acetone solution of the polymer [8].

The early studies of Yamashita and Fenn [9] brought electrospray in the analytical world and from then on electrospray applications have shown a fantastic growth. This technique, was the ionization method, which all the scientific community was waiting. It is an effective and valid approach for the direct study of analytes present in solution and, consequently, can be easily coupled with LC methods. Furthermore, the behaviour of proteins and peptides (as well as oligonucleotides) in electrospray conditions, resulting in the production of multiply charged ions, makes the ionization method essential in biomedical studies and in proteome investigations. For these reasons the Nobel Prize in 2002 was won by Fenn, with the official sentence "for the development of soft desorption ionization methods for mass spectrometric analyses of biological macromolecules".

This chapter is a concise description of chemical–physical phenomena which form the basis of the electrospray ionization (ESI).

1.1 THE TAYLOR CONE

The instrumental set-up for electrospray ionization experiments is schematized in Fig. 1.1. The solution is injected by a stainless steel capillary (10^{-4} m o.d.). Between this capillary and a counterelectrode, placed a few-tenth millimetres far from it, a voltage in the range of 2–3 kV is applied. In these conditions, the injection of aqueous solvents results in the formation of a solution cone just outside the capillary. The formation of this cone-shaped structure can be justified by the presence of charged species inside the solution which experience the effect of the electrostatic field existing between the capillary and the counterelectrode. What is the origin of this charged species in absence of ionic solute? This experiment emphasizes that even in absence of ionic analyte, protic solvents produces ionic species due to their dissociation. Thus, for example, taking into account that K_w at 20°C is $10^{-14.16}$, the H_3O^+ concentration at 20°C is in the order of 8.3×10^{-8} M. Analogously the $K_a(CH_3OH)=10^{-15.5}$ and $pK_a(CH_3OH)=15.5$. Consequently, the solvents usually employed for electrospray experiments already produces ions in solution, which can be considered responsible for the cone formation. Of course in presence of dopant analytes (*e.g.* acids), as well as traces of inorganic salts, this phenomenon shows a dramatic enhancement.

After the cone production, as extensively studied by Taylor [4], if the applied electrical field is high enough, the formation of droplets from the cone apex is observed, which further migrate, due to their charge, through the atmosphere to the counterelectrode.

It has been proved that the droplets formation is strongly influenced by

- Chemical–physical characteristics of the solvent (viscosity, surface tension, pK_a)
- Concentration of ionic analytes

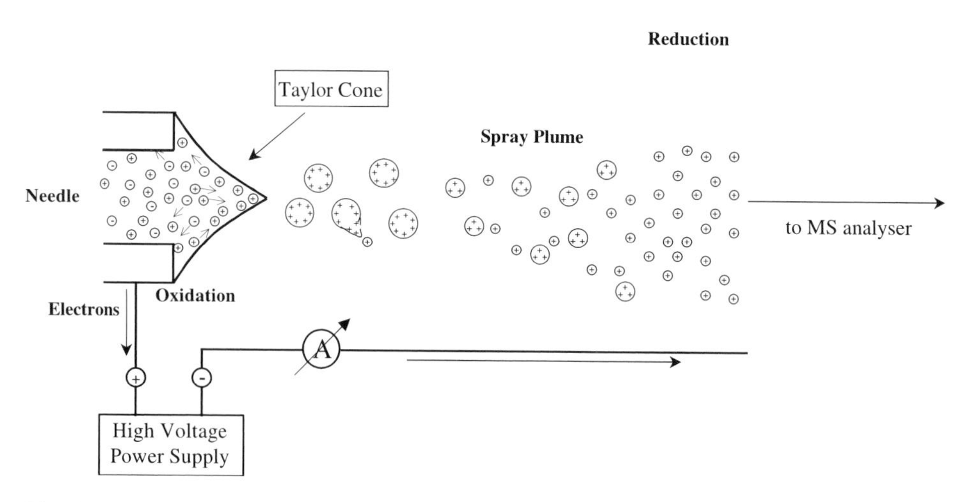

Fig. 1.1. Schematic diagram of an electrospray source showing the production of charged droplets from the Taylor Cone.

- Concentration of inorganic salts
- Voltage applied between capillary and counterelectrode.

Of course, in the case of positive ion analysis the capillary is placed at a positive voltage while the counterelectrode is placed at a negative voltage (this is the case shown in Fig. 1.1). The vice versa is used in the case of negative ion analysis.

The result of this experiment is that the droplets so generated bring a high number of positive (or negative) charges. The formation of the Taylor cone and the subsequent charged droplet generation can be enhanced by the use of coaxial nitrogen gas steam. This is the instrumental set-up usually employed in the commercially available electrospray sources.

Blades *et al.* [10] showed that the electrospray mechanism consists in the early separation of positive from negative electrolyte ions present in the solution. This phenomenon requires necessarily a charge balance with conversion of ions to electrons occurring at the metal–liquid interface of the ESI capillary, in the case of positive ion analysis.

The processes which lead to a major change in composition of ions in the spray solution are those occurring at the metal–liquid capillary interface; and the related oxidation reactions were studied by the use of a Zn capillary tip. This experiment was performed by employing the three different capillary structures shown in Fig. 1.2. The stainless steel capillary normally employed (sometimes covered by a platinum layer but anyway passivated by the presence of a thin layer of chrome oxides) (Fig. 1.2B) was substituted with that

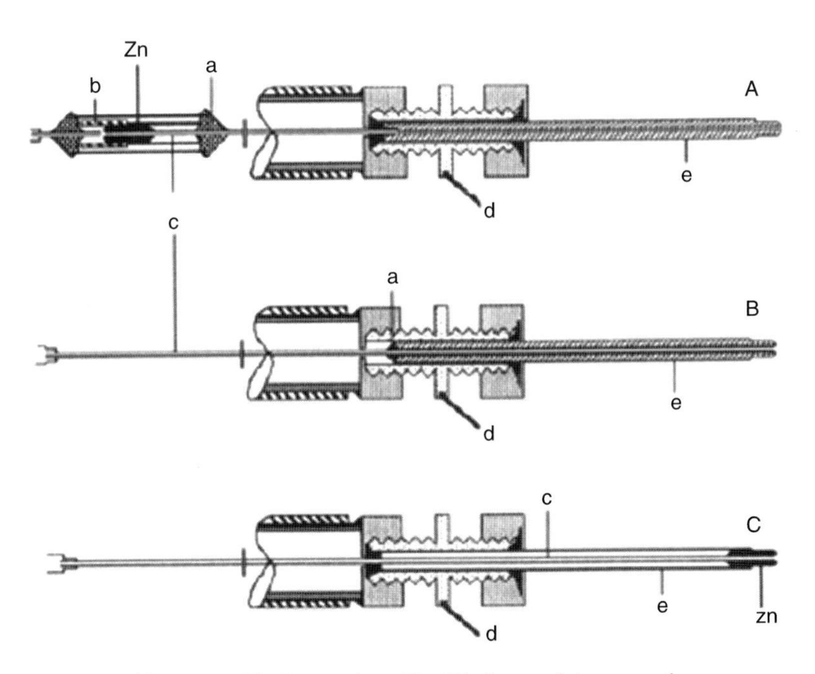

Fig. 1.2. Different capillaries employed by Blades *et al.* to prove the occurrence of oxidation phenomena at the capillary tip. (Reprinted with kind permission from Ref. [10], Copyright 1991 American Chemical Society.)

shown in Fig. 1.2C, in which the tip was made of Zn. Zn was chosen as a suitable material because it has a very low reaction potential.

$$Zn_{(s)} = Zn^{2+} + 2e \quad E^0_{red} = -0.76 \text{ V}$$

In these conditions Zn^{2+} ions were easily detected in the mass spectrum simply by spraying methanol at a flow rate of 20 μL/min. In order to confirm this hypothesis and to exclude other possible origins of the Zn^{2+} production, a further experiment was carried out by placing the Zn capillary before the electrospray capillary line and electrically insulated from it (Fig. 1.2A). In this case Zn^{2+} ions were not detected. These results provide qualitative and quantitative evidence that electrochemical oxidation takes place at the liquid–metal interface of the electrospray capillary tip. In this view the electrolysis cell is constituted by the capillary entrance port and the ion transport does not take place in liquid phase but in the gas phase. In the case of positive ion analysis the electrochemical oxidation reaction would occur at the liquid–metal interface of the capillary tip and the yield of this reaction will depend on the electrical potential as well as on the electrochemical oxidation potentials from the different possible reactions. Kinetics factor can exhibit only minor effects, considering the low current involved.

The effect of oxidation reactions at the capillary tip produces of an excess of positive ions, together with the production of an electron current flowing through the metal (see Fig. 1.1). The excess of positive ions could be the result of two different phenomena, *i.e.* the production of positive ions themselves or the removal of negative ions from the solution.

Reduction reactions could take place at a large counterelectrode shown in Fig. 1.1. They can be considered as either conventional ones, occurring on the solution liquid droplets, or unconventional, occurring directly on gas phase ions. It is reasonable to assume that the former mechanism is the most effective one due to the fact that, as will be described later, the ions reaching the counterelectrode are still solvated by several solvent molecules.

1.2 FATE OF SPRAYED DROPLETS

The electrical current due to the motion of droplets can be easily measured by the amperometer (A) shown in Fig. 1.1. This measurement is surely of interest, because it allows to estimate, from the quantitative point of view, the total number of elementary charges leaving the capillary and which, theoretically, can correspond to gas phase ions. The droplet current I, the droplet radii R and charge q were originally calculated by Pfeifer and Hendricks [11]:

$$I = \left[\left(\frac{4\pi}{\varepsilon}\right)^3 (9\gamma)^2 \varepsilon_0^5\right]^{1/7} (KE)^{3/7} (V_f^{4/7}) \tag{1.1}$$

$$R = \left(\frac{3\varepsilon\gamma^{1/2} V_f}{4\pi\varepsilon_0^{1/2} KE}\right)^{2/7} \tag{1.2}$$

$$q = 0.5\left[8(\varepsilon_0 \gamma R^3)^{1/2}\right] \tag{1.3}$$

where γ is the surface tension of the solvent; K the conductivity of the infused solution; ε the dielectric constant of the solvent; ε_0 the dielectric constant of the vacuum and V_f the flow rate; E is the electric field.

De la Mora and Locertales [12] have found, based on both theoretical calculation and experimental data, the following equations for the same quantities:

$$I = f\left(\frac{\varepsilon}{\varepsilon_0}\right)\left(\gamma K V_f \frac{\varepsilon}{\varepsilon_0}\right)^{1/2} \tag{1.4}$$

$$R \approx \left(\frac{V_f \varepsilon}{K}\right)^{1/3} \tag{1.5}$$

$$q = 0.7\left[8\pi(\varepsilon_0 \gamma R^3)^{1/2}\right] \tag{1.6}$$

where $f(\varepsilon/\varepsilon_o)$ is a function of the $\varepsilon/\varepsilon_o$ ratio calculated by Fernandez de la Mora.

Eqns. 1.1 and 1.4 seem drastically different at the first sight, but they indicate an analogous dependence of I from the two most relevant experimental parameters, *i.e.* the flow rate and the conductivity. Eqns. 1.3 and 1.6 both show a decrease in the dimension of droplets by an increase in the solution conductivity. These relationships are particularly relevant because in solution, when different electrolytes are present, the conductivity K may be obtained as the sum of the conductivities of the different species and is proportional to the ion concentration:

$$K = \sum_i \lambda_{0,m,i} C_i \tag{1.7}$$

where $\lambda_{0,m,i}$ is the molar conductivity of the electrolyte i.

The charged droplets, generated by solution spraying, decrease their radius due to solvent evaporation but their total charge remain constant. In the first step the energy required for the solvent evaporation is from the environmental thermal energy. In a second step this process is enhanced through further heating obtained by the use of a heated capillary or by collisions with heated gas molecules. The maintenance of the total charge during this evaporation phase can be explained by the fact that the ion emission from the solution to the gas phase is an endothermic process. The decrease of the droplet radius can be described by Eqn. 1.8 where $\overline{v}$ is the average thermal speed of solvent molecules in vapour phase:

$$\frac{\mathrm{d}R}{\mathrm{d}t} = \frac{\alpha \overline{v}}{4\rho} \frac{p^0 M}{R_g T} \tag{1.8}$$

where p^0 is the solvent vapour pressure at the droplet temperature; M the solvent molecular weight; ρ the solvent density; R_g the gas constant; T the droplet temperature and α the solvent condensation coefficient.

This relationship gives evidences of all the factors that can influence the droplet dimension and consequently the effectiveness of ESI.

The decrease of the droplet radius with respect to time leads to an increase of the surface charge density. When the radius reaches the Rayleigh stability limit (given by Eqn. 1.9) the electrostatic repulsion is identical to the attraction due to the surface tension. For lower radii the charged droplet is unstable and decomposes through a process generally defined "Coulombic fission" [13]. This fission is not regular (in other words the two parts originated by it do not have necessarily analogous dimensions).

$$q_{R_y} = 8\pi(\varepsilon_0 \gamma R^3)^{1/2} \qquad (1.9)$$

Detailed studies performed by Gomez and Tang [14] allowed to calculate the lifetime and the fragmentations of droplets. An example is shown in Fig. 1.3.

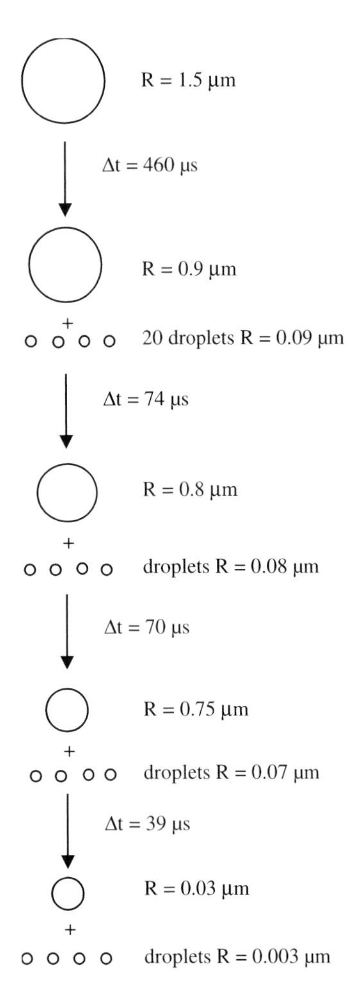

Fig. 1.3. Data obtained by theoretical calculation of drop dimensions and lifetime.

1.3 IONS FORMATION FROM CHARGED DROPLETS

Until now two different mechanisms have been proposed to give a rationale for the formation of ions from small charged droplets. The first of them has been recently discussed by Cole [15] and Kebarle and Peschke [16]. It describes the process as a series of scissions which lead at the end to the production of small droplets bringing one or more charges but only one analyte molecule. When the last, few solvent molecules evaporate, the charges are localized on the analyte substructure, giving rise to the most stable gas phase ion. This model is usually called the "charged residue mechanism" (CRM) (see lower part of Fig. 1.4).

More recently Thomson and Iribarne [17] have proposed a different mechanism in which a direct emission of ions from the droplet is considered. It occurs only after the droplets have reached a critical radius. This process is called "ionic evaporation" (IEM) and is dominant with respect to Coulombic fission of particles with radii $r < 10\,\text{nm}$ (see upper part of Fig. 1.4).

Both CRM and IEM are able to explain many of the behaviours observed in ESI experiment. However a clear distinction between the two mechanisms lies in the way by which an analyte molecule is separated from the other molecules (either of analyte or solvent present in droplets). In the case of IEM this separation takes place when a single analyte molecule, bringing a part of the charge in excess of the droplet, is desorbed in the gas phase, thereby reducing the coulombic repulsion of the droplets. In CRM mechanism this separation occurs through successive scissions, reducing the droplet dimensions until when only one single molecule of analyte is present in them. In general the CRM model is retained valid in the process of gas phase ion formation for high molecular weight molecules.

1.4 SOME FURTHER CONSIDERATIONS

What we discussed above gives an idea of the high complexity of the ESI process: the ion formation depends on many different mechanisms occurring either in solution or during the charged droplets production and ion generation from the droplets themselves.

First of all, the ESI users must take into account that the concentration of the analyte present in the original solution does not correspond to that present in the droplets generating

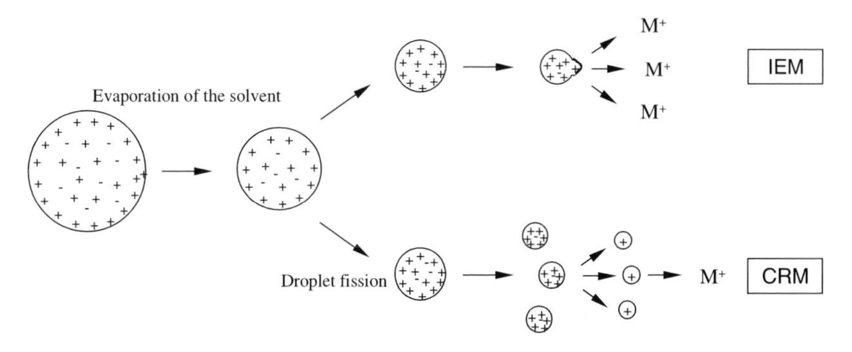

Fig. 1.4. Schematic diagram of ionic evaporation and charged residue mechanisms (IEM and CRM, respectively) leading to the production of gas phase ion from charged droplets.

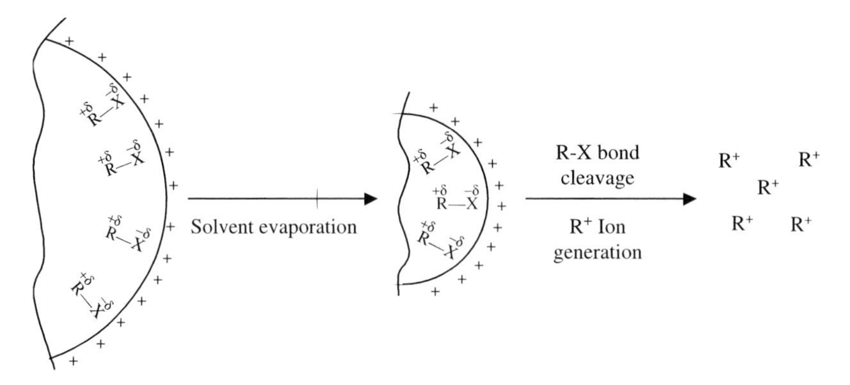

Fig. 1.5. Formation of R^+ species ($[Pt(\eta^3\text{-Allyl})P(C_6H_5)_3]^+$) from $[Pt(\eta^3\text{-Allyl})XP(C_6H_5)_3]$ complexes, due to the formation of ion pair catalysed by the high surface charge density present in the droplet.

the gas phase ions. This point must be carefully considered when the original solution is far from neutrality. In this case the pH will show sensible changes. Gatling and Turecek [18], studying the $Fe^{2+}(bpy)_3$ $Ni^{2+}(bpy)_3$ complex dissociation under electrospray conditions, found an apparent increase of $[H_3O^+]$ in the order of 10^3–10^4 fold with respect to the bulk solution. Furthermore, due to solvent evaporation, the pH value is not homogeneous inside the droplet. A spherical microdroplet is estimated to maintain pH 2.6–3.3 in a 5–27 nm thick surface layer without exceeding the Rayleigh Limit, which implies that complex dissociations occur near the droplet surface of high local acidity.

In the case of polar compounds the surface charge density present in the droplet can activate some decomposition reaction of the analyte. In the study of $[Pt(\eta^3\text{-Allyl})XP(C_6H_5)_3]$ complexes (X=Cl, Br) the formation of any molecular species in ESI condition was not observed [19]. This result is quite surprising, considering that the same compounds lead to molecular species either in fast atom bombardment or in electron ionization conditions, *i.e.* in experimental conditions surely "harder" than ESI. This result has been related to the occurrence of phenomena strictly related to the ESI condition and explained by the high positive charge density present on the droplet surface. It can activate the formation, in the polar molecule under study, of ion pair consisting of X^- and $[M–X]^+$ (X=Cl, Br) (see Fig. 1.5). The latter are the only species detectable in positive ion ESI conditions.

REFERENCES

1 ORNL Review, State of the Laboratory. Vol. 29, no. 1–2 1995. http://www.ornl.gov/info/ornlreview/rev29-12/text/environ.htm.
2 J.N. Smith, *Fundamental Studies of Droplet Evaporation and Discharge Dynamics in Electrospray Ionization*, California Institute of Technology, Thesis (PhD), 2000.
3 J. Zeleny, *Phys. Rev.*, 10 (1917) 1–6.
4 G. Taylor, *Proc. R. Soc. Lond. Ser. A*, 280, (1964) 283.
5 C.T.R. Wilson and G. Taylor, *Proc. Cambridge Philos. Soc.*, 22 (1925) 728.

6 R.L. Hines, *J. Appl. Phys.*, 37 (1966) 2730–2736.
7 R. Tilney and H.W. Peabody, *Brit. J. Appl. Phys.*, 4 (1953) S51–S54.
8 M. Dole, L.L. Mack, R.L. Hines, R.C. Mobley, L.D. Ferguson and M.B. Alice, *J. Chem. Phys.*, 49 (1968) 2240–2249.
9 M. Yamashita and J.B. Fenn, *J. Phys. Chem.*, 88 (1984) 4451–4459.
10 A.T. Blades, M.G. Ikonomou and P. Kebarle, *Anal. Chem.*, 63 (1991) 2109–2114.
11 R.J. Pfeifer and C.D. Hendricks, *AIAA J.*, 6 (1968) 496–502.
12 J.F. de la Mora and I.G. Locertales, *J. Fluid Mech.*, 260 (1994) 155–184.
13 L. Rayleigh, *Philos. Mag.*, 14 (1882) 184–186.
14 A. Gomez and K. Tang, *Phys. Fluids*, 6 (1994) 404–414.
15 R.B. Cole, *J. Mass Spectrom.*, 35 (2000) 763–772.
16 P. Kebarle and M. Peschke, *Anal. Chim. Acta.*, 406 (2000) 11–35.
17 B.A. Thomson and J.V. Iribarne, *J. Chem. Phys.*, 71 (1979) 4451–4463.
18 C.L. Gatling and F. Turecek, *Anal. Chem.*, 66 (1994) 712–718.
19 S. Favaro, L. Pandolfo and P. Traldi, *Rapid Commun. Mass Spectrom.*, 11 (1997) 1859–1866.

Achille Cappiello (Editor)
Advances in LC–MS Instrumentation
Journal of Chromatography Library, Vol. 72

Chapter 2

Atmospheric pressure chemical ionization (APCI): new avenues for an old friend

ANDREA RAFFAELLI

2.1 INTRODUCTION

Atmospheric pressure chemical ionization (APCI) appeared on the market of mass spectrometry coupled to liquid-phase chromatographic around the end of the 1980s [1], even if it dates quite earlier for gas-phase investigations, and immediately induced a large interest among the researchers in the field of liquid chromatography–mass spectrometry (LC–MS) for its sensitivity and ruggedness, compared to the formerly used interfacing/ionization techniques. By the way, its introduction shortly preceded the commercial introduction of electrospray, and probably contributed to the large and rapid diffusion of the latter.

2.1.1 General features of chemical ionization

At the half of the 1960s, Munson and Field presented a new ionization method for mass spectrometry [2]. This new technique took advantage of the ionization, obtained using the "traditional" electron ionization (EI) way, of a large excess of methane that filled the ion source. The ionization of the analyte is carried out by gas-phase ion/molecule reactions. This traduces in a lower extent of energy transferred upon the analyte ion, with a consequently lower extent of fragmentation.

In the absence of the molecular ion in an EI spectrum we are missing the first information needed for a correct interpretation of mass spectral data: the molecular weight. Chemical ionization (CI) mass spectrometry can allow to recover such an important piece of information, facilitating the rest of the interpretation pathway (Fig. 2.1).

Nowadays it is indeed quite common to acquire both EI and CI gas chromatography–mass spectrometry (GC–MS) data when complex mixtures are under investigation: CI provides the information about the molecular weight of the analyte, whereas EI fragmentation data give structural information [3]. Moreover, sometimes the fragmentation pattern of the even-electrons molecular ion obtained usually by chemical ionization–mass spectrometry

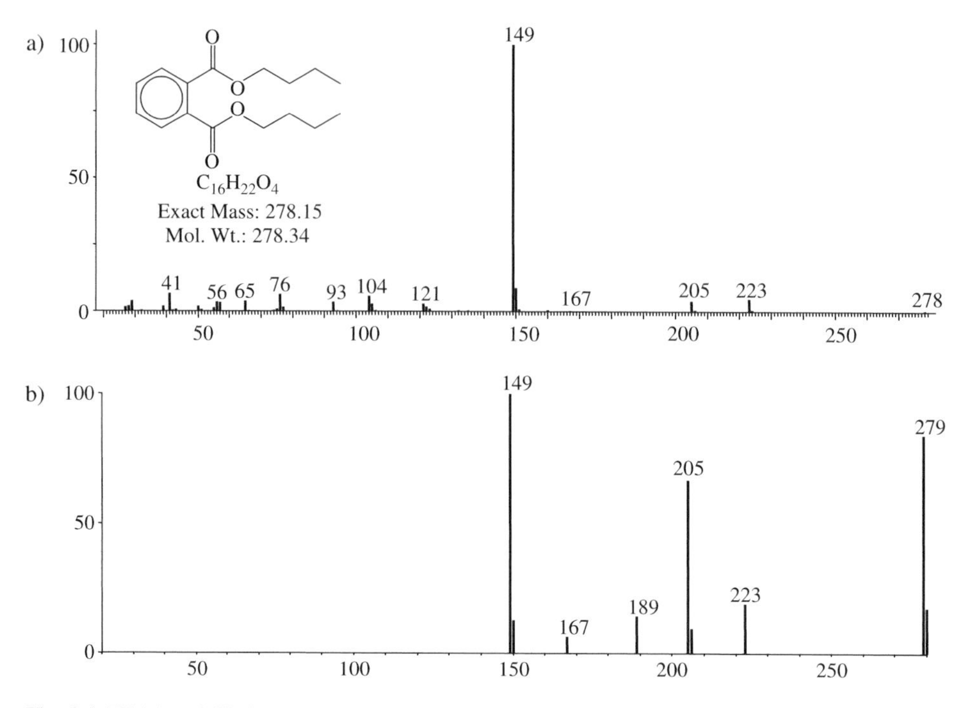

Fig. 2.1. EI (a) and CI (b) mass spectra of dibutyl phthalate. The molecular ion (should be *m/z* 278) is not present in the EI spectrum.

(CI–MS) is slightly different, and complementary, with respect to the odd-electrons molecular ion typical for electron ionization–mass spectrometry (EI–MS), this feature adding further information about the target analyte.

Another important new point related to CI deals with the relatively high efficiency for negative ions formation with respect to EI. The possibility to acquire spectra in both ion polarities further provides complementary data useful for structural elucidation.

2.1.2 General features of atmospheric pressure chemical ionization

The ionization mechanism of CI involves, as indicated above, gas-phase ion/molecule reactions. In particular, a reactant ion transfers its charge to an analyte molecule with formation of an analyte ion, usually by acid–base reactions. This action requires efficient collisions among the reactant ions and the neutral analyte molecules present in the source. If the process takes place at atmospheric pressure, the number and efficiency of such collisions is extremely higher than under the vacuum present inside a "traditional" CI source (10^{-2}–10^{-3} Torr). In principle, hence, a CI source operating at atmospheric pressure should be more efficient than a traditional vacuum one. As it will be described later, the ionization mechanism of APCI is somehow more complex with respect to the traditional CI, but APCI sensitivity will be higher and, mainly, it can easily be interfaced with LC. APCI is widely used nowadays in different fields, especially in pharmaceutical industry, that takes advantage of its sensitivity, selectivity and ruggedness. Looking at the literature, we can say that the papers presenting APCI applications are roughly 1/10 of those dealing with electrospray ionization (ESI) applications, but very likely this does not reflect the real ratio of usage, owing to the secretness of many pharmaceutical applications. The co-presence of ESI and APCI on the same instrument is quite common, as the main section of the ion source is practically the same, just the probe needing to be changed. Moreover some MS companies recently presented combined ESI/APCI sources, where the instrument can use both the ionization techniques in the same time.

2.1.3 Historical background of APCI

APCI is older than ESI or ionspray ionization (ISI), but it was used only for gas analysis up to the mid 1970s. A lot of literature reports from the late 1960s to the late 1970s report about the use of alpha particles radiation [4,5], a ^{63}Ni β-particle emitter [6] and needle-to-plane corona discharge [7,8] ion sources, but this last was later used more widely. In 1969, Carroll proposed the use of these techniques for the coupling to ion mobility spectrometry [9]. The use with standard mass analyzers working under high vacuum conditions came in 1973 by Horning [10], who shortly afterwards illustrated the combination of APCI–MS with LC for the first time [11]. The coupling with high performance liquid chromatography (HPLC) created immediately a large interest in the scientific community, as it was, under several aspects, the ideal HPLC–MS interface: "The best interface is 'no interface at all'". Indeed APCI was the closest approach to this situation, offering also the advantage of flow rates up to 1 or even 2 ml/min, so that a direct coupling to 4.6 mm I.D. analytical HPLC columns was possible without any splitting.

During these studies the acronym "API" (for atmospheric pressure ionization) was introduced, and rapidly became widely used. Nowadays the term "APCI" is preferred since it suggests the use of CI techniques at very high pressures (up to atmospheric or even slightly above) rather than other high-pressure ion sources such as plasma discharge and so on. Moreover, at present the term "API" denotes an ion source working at atmospheric pressure, which can accommodate different interfaces, such as ESI/ISI, APCI, atmospheric pressure photoionization (APPI), etc., whereas the term "APCI" denotes just one of the interfaces used in an API source.

In the same period Sciex Inc. described a hypersensitive API analytical system for trace analysis in air [12], later being implemented on the TAGA (TAGA is the acronym for **T**race **A**tmosphere **G**as **A**nalyzer) mass spectrometer series. The first TAGA system, the TAGA 3000, used a single quadrupole equipped with an APCI source and was designed to be sufficiently rugged to be mounted in a van. This was the first commercial, truly mobile mass spectrometer which could provide real-time air analysis at trace levels, even when in motion [1]. In 1981, SCIEX introduced the first commercial tandem quadrupole system, the TAGA 6000, also equipped with an APCI source, which dramatically improved the specificity of the system. On that instrument, Henion and coworkers introduced for the first time the use of APCI with tandem mass spectrometry–liquid chomatography (LC–MS–MS) [13]. Subsequent improvements to that interface led to the SCIEX heated nebulizer LC/MS. Inlet, the first commercial APCI LC–MS interface, presented in 1988, which provided the capability for detecting a wide variety of polar, semi-volatile compounds

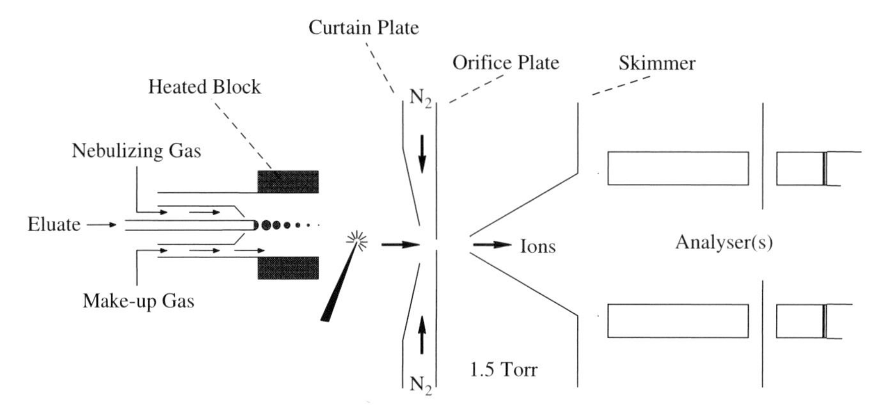

Fig. 2.2. Schematic representation of an API source with a heated nebulizer interface for APCI.

and quasi-molecular ions with minimal thermal degradation [1]. Later, the diffusion of electrospray [14] and ionspray [15] initiated their "gold rush" to become the most widely used LC–MS interfaces.

The features and investigations related to APCI developments were described by Carroll and co-workers in a comprehensive review published in 1981 [16].

2.2 PRINCIPLES AND APPARATUS

Fig. 2.2 shows a schematic representation of an API source with a heated nebulizer interface for APCI.

The sample solution flows through a capillary inserted in a coaxial pneumatic nebulizer. A make up gas (usually air) is added to assist the ionization process. The mixture of air and nebulized solution passes through a heated zone that assist the solvent evaporation. Despite the high temperature of the heater (400–600°C), minimal degradation of the sample occurs, since the heat provided is used just for the vaporization of the solvent and the sample temperature usually do not exceed 80–100°C. A corona discharge needle kept at high voltage (5–6 kV) is positioned between the exit of the heated nebulizer and the source interface plate.

2.2.1 Corona discharge ionization mechanism

The needle generates a discharge current of *ca.* 2–3 μA, which ionizes air producing primary ions (mainly $N_2^{+\cdot}$, $O_2^{+\cdot}$, $H_2O^{+\cdot}$ and $NO^{+\cdot}$ in positive mode, $O_2^{-\cdot}$, $O^{-\cdot}$, $NO_2^{-\cdot}$, $NO_3^{-\cdot}$, $O_3^{-\cdot}$ and $CO_3^{-\cdot}$). The primary ions react very rapidly (within 10^{-6} sec) transferring their charge to solvent molecules, in a reaction controlled by the recombination energy of the primary ions themselves, to produce the effective CI reactant ions. These are characterized

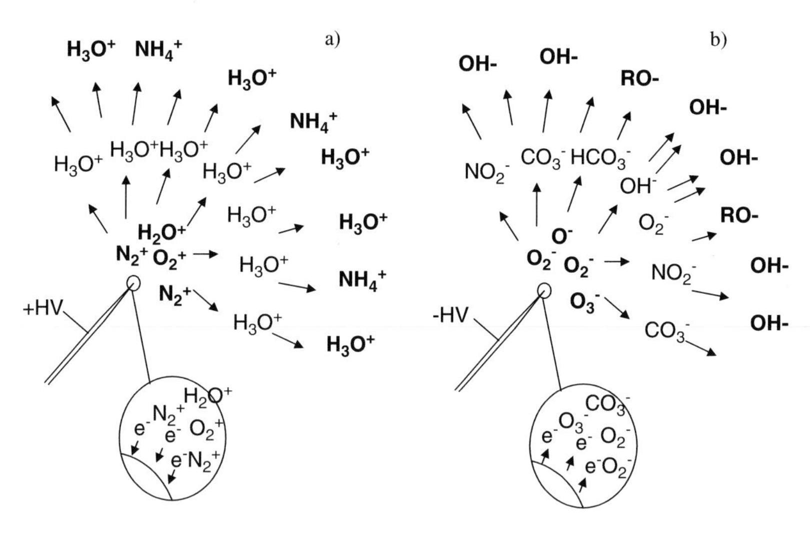

Fig. 2.3. Corona discharge ionization cascade in positive (a) and negative (b) ion modes.

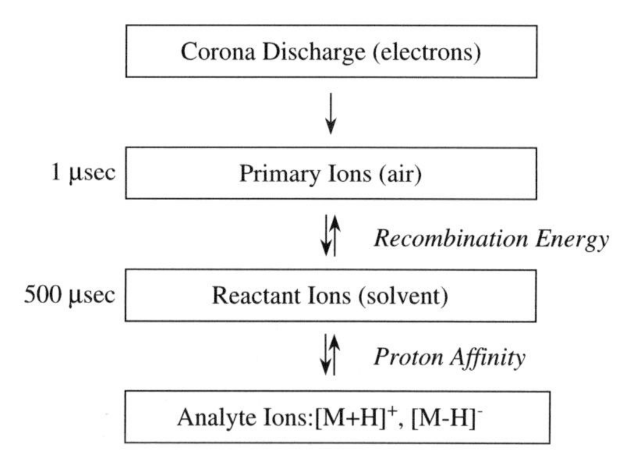

Scheme 2.1. Diagram of the ionization processes in APCI.

by a longer lifetime (about 0.5×10^{-3} sec) and react with analyte molecules to produce analyte quasi-molecular ions by charge or proton transfer reactions, according to the proton affinity of the analyte itself (Fig. 2.3).

The total reaction time in the source corresponds in practice with the final proton transfer (about 0.5×10^{-3} sec) as the time of the preceding solvent ionization can be disregarded. The whole ionization cascade is represented in the Scheme 2.1.

Overall we can observe $[M+H]^+$ ions in positive mode and $[M-H]^-$ ions in negative mode. Adduct ions are not readily formed in APCI. Even polyoxygenated molecules that normally produce mainly ammonium and sodium adducts under ISI conditions, such as

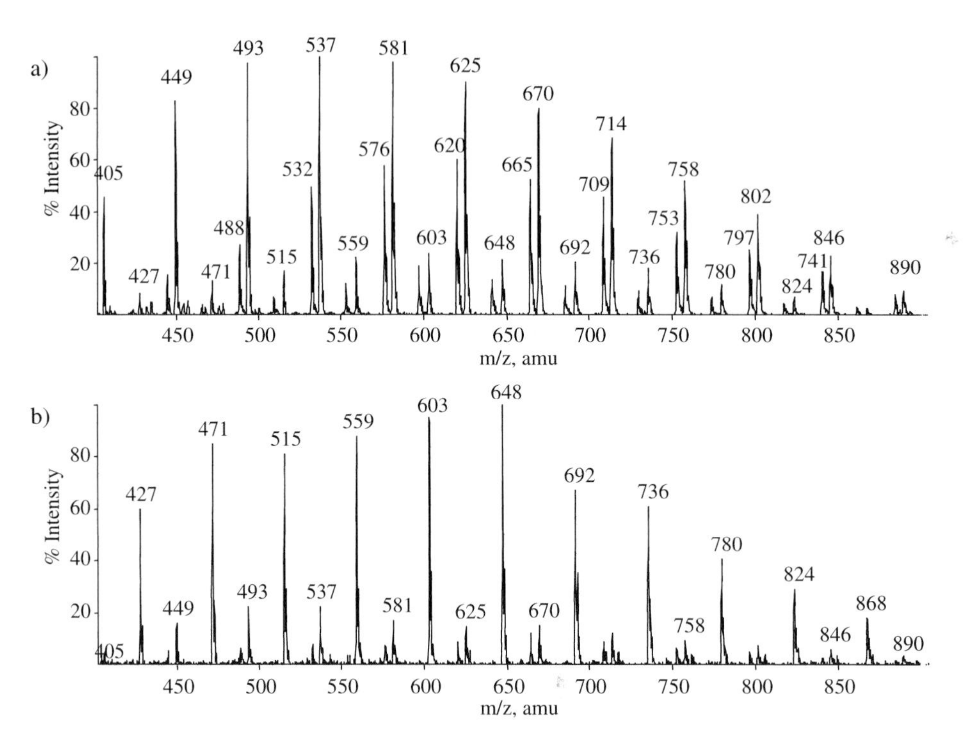

Fig. 2.4. ESI (a) and APCI (b) mass spectra of a sample of a polyethoxylated fatty alcohol.

non-ionic surfactants, form predominant protonated molecules when analyzed by APCI–MS. An example is shown in the Fig. 2.4.

Polyethoxylated non-ionic surfactants are oligomeric mixtures giving rise to family of ions equally spaced by 44 mass units (the $-CH_2-CH_2-O-$ subunit). Under ionspray conditions (Fig. 2.4a) we can observe ammonium [M+18] and sodium [M+23] adducts as the predominant ions (see for instance the ions at *m/z* 620 and 625, related to a 602 Da oligomer), whereas the protonated molecules give quite weak signals, if any. The same sample, analyzed by APCI–MS (Fig. 2.4b), shows the protonated molecules as the most abundant ions (see for instance the ion at *m/z* 603 for the same oligomer), whereas the ammonium and sodium adducts are below 20%. On the basis of the above explained ionization mechanism, we can observe that the residue adduct ions observed in an APCI spectrum originate from some parallel ion evaporation process.

2.2.2 Parallel processes

The non-thermal plasma involved in the APCI mechanism is rather complex, and some secondary processes can take place. Some of these parallel reactions could be detrimental to a good efficiency in the APCI ionization process. The above mentioned ion evaporation process can occur as the conditions are quite close to those typical to the ion evaporation

$$XH^+ + B \longrightarrow X + BH^+$$

$$BH^+ + M \longrightarrow MH^+ \quad ???$$

Scheme 2.2. Competition processes in APCI.

ionization with an induction electrode [17,18]. Indeed, in APCI we produce a pneumatic nebulization of the eluate, and the discharge needle can act as an induction electrode.

This is not the only parallel process possible in an APCI source, and surely it is not the nastiest. We have to take into consideration that we operate at atmospheric pressure, so that all the components of air are present and contribute to the plasma. As mentioned before, nitrogen and oxygen take an active role in the ionization cascade, but other minor components can produce undesired species inhibiting the main process. Moreover we introduce our HPLC eluate: the solvent, including its impurities and additives, enters to be part of the plasma as well. Again, the solvent plays its role in the APCI mechanism, but sometimes we could observe different phenomena attributable to the impurities and, more important, to the additives. According to the above described reaction mechanism, the recombination energy of the species involved in the formation of the primary ions and the proton affinity of the species interested to the last proton transfer strongly influence the whole process. This means that the presence of high recombination energy or proton affinity components can alterate the APCI pathway. Low recombination energy species can be profitably used in particular cases, as described later. We need to pay attention, on the contrary, to high gas-phase basicity (or acidity for negative ions measurements) as they can originate a deep ion suppression. For instance, it is quite common to use strong bases, such as triethylamine or triethanolamine, as modifiers for HPLC mobile phases. They can receive the protons from the reactant ions to form their protonated forms (Scheme 2.2). The successive proton transfer to the anlyte can occur only if this is more basic than triethylamine or triethanolamine.

If this is not true, we cannot observe the protonated molecular ion! If the gas-phase basicity of B and M are comparable, then we can often observe adducts such as $[M{\cdot}B{\cdot}H]^+$. In the same way, strongly acidic additives in negative ion mode can lead to the formation of adducts with anions originating from them, such as chloride, formate, acetate, trifluoroacetate and so on. We must take into consideration the fact that, in a chemical ionization cascade, the last reactant ion (in positive ion mode) is that coming from the strongest base present in the mixture. Analogously, in negative ion mode, the final reactant species comes out from the strongest acid present. These parallel processes must be considered when we set up the HPLC conditions, in particular the composition of the mobile phase.

2.2.3 APCI main features and fields of applications

As stated in the above considerations, APCI is a soft ionization technique that operates in the gas-phase, even if it is a very rugged and reliable LC–MS interface. The ions, once formed, are driven into the mass analyzer through a small diameter orifice (100–300 μm) or a thin capillary by the electric field generated between the discharge needle, the interface plate, the orifice plate and the rest of the mass spectrometer. A curtain of dry nitrogen normally prevents neutral particles from entering the vacuum region. At the same time this curtain gas assists the declustering process to obtain free ions from their solvated forms.

APCI allows a very robust and versatile LC–MS coupling. All together, the main advantages of APCI mass spectrometry can be summarized as follows:

- CI-like spectra, with molecular weight information.
- Suitable for volatile or semi-volatile analytes.
- Relatively easy to install and "friendly" for the user.
- Very rugged by an analytical point of view.
- Flow rates up to 1–2 ml/min can be used without any problem.
- Very good sensitivity.
- It is enough "universal".
- Good chromatographic fidelity.

There are of course, some disadvantages, as usual. These disadvantages are not determinants and can be summarized as follows:

- No fragmentation, and hence no structural information.
- Some thermal degradation can occur.
- Inorganic buffers can create problems if too concentrated (above 5–10 mM).

On the basis of the advantages and disadvantages listed above, we can identify the ideal mass spectrometric system for the implementation of APCI. That is a tandem mass spectrometer such as a triple quadrupole, an ion trap or one of the recent hybrid systems, so that the disadvantage of the lack of fragmentation can be overcome using MS–MS. MS–MS can assist also in improving the sensitivity and selectivity when we have to determine trace components present in complex matrices. The use of selected reaction monitoring (SRM), indeed, allows to reach very good detection limits for quantitation with a good specificity as well, much better than in selected ion monitoring (SIM) available on single MS instrumentation, such as single quadrupole or time of flight mass spectrometers.

As far as the compounds amenable to APCI is concerned, we can summarize here the main features of this ionization technique:

- APCI is suitable for not so polar compounds, which may be charged in solution, but typically are not strong acids or bases.
- pH does not have a strong influence on the response.
- The normal flow rate used in APCI is 0.5–2 ml/min. A typical operating flow rate can be 1 ml/min. The most recent APCI sources can operate in a satisfactory manner also at 0.2 ml/min or even at lower flow rates.
- Usually under APCI conditions the parameter to set is the discharge current, typically limited to 2–3 μA. This induces a needle voltage comparable with the values normally set for ISI.
- APCI is not able to generate multiply charged ions in any case.

On the basis of the above considerations, compounds ranging from low to mid polarity, volatile or semivolatile are suitable for APCI–MS analysis. Sometimes APCI can provide a better sensitivity with respect to ESI even with molecules amenable to ESI, for instance if a phosphate buffer is used. Indeed ESI is quite sensitive to inorganic salts present in solution, whereas APCI can tolerate buffers more readily.

APCI is able to provide an effective and powerful detection also with some classes of organic compounds that, in principle, do not look amenable to its ionization mechanism. Compounds characterized by a very low gas-phase basicity but a relatively high electronegativity, such as polychlorinated biphenyls (PCBs) or polycyclic aromatic hydrocarbons (PAHs), can be readily ionized under APCI conditions by the use of a simple trick. We need only to add traces of benzene (at ppm levels) to the nebulizing or auxiliary gas [19,20]. Primary ions transfer their charge to benzene due to its lower recombination energy. Benzene ions, consequently, act as reactant ions and transfer their charge to the analytes, as their recombination energy is lower with respect of benzene itself. We could nominate this trick as "dopant-assisted" APCI. This approach has been used, for instance, for the analysis of PAHs [21] and fullerenes [19,20,22]. Recently it was taken again into account for the analysis of hydrophobic compounds of significant environmental relevance which are not detectable with the ordinary APCI techniques [23]. Among them, good sensitivity has been found for highly chlorinated biphenyl derivatives such as dichlorodiphenyltrichloroethane (DDT), dichlorodiphenyldichloroethane (DDD) and dichlorodiphenyldichloroethene (DDE); cyclopentadienes such as Aldrin and its epoxy derivatives Dieldrin and Endrin; and dibenzofurans and dibenzoparadioxins such as 2,3,7,8-tetrachlorodibenzofuran (2,3,7,8-TCDF) and 2,3,7,8-tetrachlorodibenzoparadioxin (2,3,7,8-TCDD).

APCI HPLC–MS is a very powerful and rugged technique for routine analysis in different fields:

- Pharmaceutical chemistry and pharmacology (structural characterization of drugs and their impurities, metabolic and pharmacokinetic investigations, etc.).
- Clinical chemistry and biochemistry (quantitation of endogenous and xenobiotic compounds having diagnostic significance, reference and definitive clinical chemistry analytical methods, etc.).
- Biotechnologies ("on-line" monitoring of fermentations, characterization of the compounds obtained, etc.).
- Food chemistry and agriculture (food analysis, devoted to nutritional compounds and residual components determination, characterization and determination of additives, etc.).
- Environmental control (identification, characterization and quantitation of polar water pollutants, determination of pesticides, surfactants and their metabolites or degradation products, etc.).
- Organic and organometallic chemistry (characterization of synthesis products or intermediates, purity and composition check of starting materials, etc.).
- and so on.

Its capability to provide only one single ion from most analytes (protonated molecule in positive ion mode, deprotonated molecule in negative mode) makes it ideal for quantitation methods based on selected ion monitoring (SIM). Also, the high MS–MS sensitivity allows to take advantage of the large specificity typical of SRM.

2.3 APPLICATIONS

APCI LC–MS is widely used in pharmaceutical industries for performing low-matrix effect biodeterminations of drugs and metabolites in biological fluids. It is more "universal" than ESI for the analysis of not so polar compounds and can provide better sensitivity and trouble-free analytical determination when non-volatile buffers are used in the chromatographic separation. This short account on the APCI aspects and features presents here a few examples of applications in environmental control and in food chemistry.

2.3.1 Determination of thiourea in wastewater [24]

Thiourea is one of the most toxic substances for nitrifying bacteria. Its toxicity limit, that is the concentration that inhibits more than 50% of their activity, is 76 ng/ml [25]. Normally it is not present in wastewater, but it can become a serious problem for industrial wastewater treatment plants, in particular when the wastewater comes from tannery factories. Thiourea or some possible precursors can be used in some stages of the tanning process. Therefore, a fast and reliable method for its determination is strongly needed. The approach to the determination of this class of pollutants, very polar, and water soluble, needs to face four main problems:

1. The samples of industrial wastewater to be analyzed contain a large number of different substances, some of which present at a relatively high concentration.
2. Thiourea is very soluble in water and practically insoluble in organic solvents such as ethyl ether and methylene chloride so that its extraction from the waste samples is almost impossible. Solid-phase extraction fails as well, with a recovery below 20–30% in the best cases.
3. Thiourea is relatively unvolatile and its GC analysis is not trivial, at least using common apparatus and columns.
4. Preliminary HPLC measurements caused the irreversible degradation of the HPLC column after a few injections of real samples, due to the build up of some components of the matrix onto the stationary phase.

These points clearly indicate that the samples must be analyzed directly, without any extraction, and that no chromatographic separation technique can be used. Preliminary ESI measurements on thiourea water solutions revealed a strong tendency to form sodium adducts, whose collision induced dissociation fragmentation in difficult and unreliable, and hence not useful for a SRM method. APCI tandem mass spectrometry using SRM can provide a brilliant solution to this problem, owing to the high sensitivity and selectivity related to MS–MS methods. The analytical procedure was developed using a Perkin Elmer Sciex API III *plus* triple quadrupole mass spectrometer. This instrument allowed to achieve detection limits down to 1 ng/ml, with a really minimal sample pre-treatment. Fig. 2.5 shows the APCI–MS and MS–MS spectra of thiourea.

The possibility to use the two fragmentations 77→60 and 77→43 for SRM on a triple quadrupole mass spectrometer allows to quantitate thiourea directly by flow injection analysis (FIA) in the waste sample with a high sensitivity and specificity. The quantitation limit is around 1 ng/ml in the matrix. Calibration is linear up to 5 μg/ml (that is 3.5 decades) with good statistical parameters (linear through zero with 1/x weighting regression, $y = 400\times$, $R = 0.99993$, $n = 5$, C.V. 8% at 10 ng/ml, 2% at 1 μg/ml).

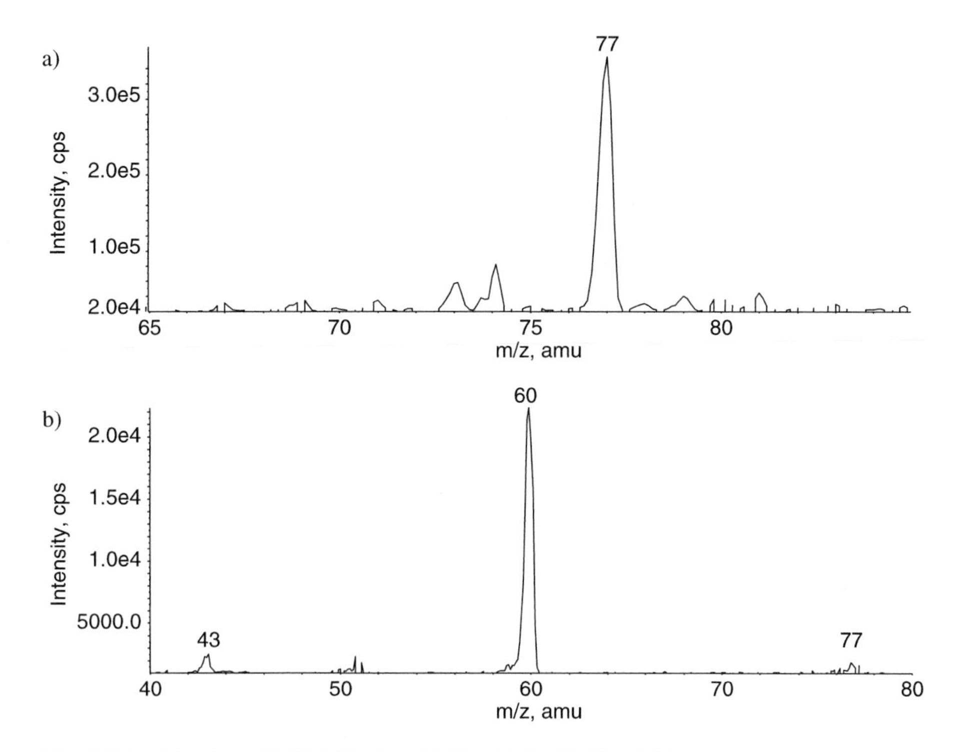

Fig. 2.5. Positive ions APCI–MS (a) and MS–MS (b, 30 eV) of thiourea.

The reliability of this analytical procedure was tested by the analysis of "spiked" samples and it resulted very good. It is also worthy to note that we never found thiourea levels above 50 ng/ml when the plant was working properly, and we never found thiourea levels below 100 ng/ml when something was wrong and there was an ammonia release from the plant. This means that the experimental amounts found were in a very good agreement with the toxicity limit known of 76 ng/ml.

The specificity of the method is also very high. Two fragmentations are monitored and this minimizes the possibility of any interference. It is worthy to note that some samples (all coming from the same tanning factory) showed a partial interference: there was a compound in the matrix giving the 77→43 fragmentation, but not the 77→60 fragmentation (the most intense).

2.3.2 Determination of compounds of nutritional interest in olive oil [3,26]

Natural antioxidants present in several vegetable species and in their derivatives play an active role in the protection of human organism against the action of free radicals and can protect them with respect to the development of cardiovascular diseases and cancer. For example, polyphenols present in olive oil and wine constitute the basis of the healthy "mediterranean diet".

APCI–MS can be very helpful in the quantitation of these components, even without any chromatographic separation, using the same kind of approach above described for the analysis of thiourea in wastewater. FIA–SRM–MS–MS, ESI and APCI–MS provide a rapid and reliable method for the direct determination of oleuropeine in the phenolic extract from olive oil [27] and of resveratrol and other nutritional polyphenols directly in wine samples [28].

APCI–MS can be very useful also for a more extensive characterization of olive oil, by a nutritional point of view [3,29]. Olive oil contains a lot of important nutritional components, responsible for its anti-oxidant action. In particular, tocopherols (vitamin E), carotinoids (i.e. β-carotene), retinol (vitamin A), and phylloquinone (vitamin K) are present in a significantively high amount. These "good" nutritional components are unfortunately flanched by some "bad" ones, such as cholesterol. All these compounds are lipophylic, and can be found in the unsaponifiable matter when the olive oil is treated with KOH in methanol. APCI ionization can provide a method for the rapid, high throughput screening of these parameters, as ESI does not appear a good choice, due to its lower sensitivity for some of the components (such as β-carotene), as well as to its higher sensitivity to suppression and matrix effects.

The APCI–MS and MS–MS (product ions scan) of α, γ (or β, they are isobars), and δ-tocopherols, retinol, phylloquinone, β-carotene and cholesterol are summarized in the Table 2.1. Retinol and cholesterol do not show any protonated molecular ion, but a strong

TABLE 2.1

APCI MS AND MS–MS SPECTRAL DATA AND SRM TRANSITIONS OF SOME COMPONENTS OF NUTRITIONAL INTEREST PRESENT IN OLIVE OIL

Compound	MS	MS–MS* *m/z* (%)	SRM
α-Tocopherol	431 ($[M+H]^+$)	431 (20), 343 (4), 205 (5), 165 (100), aliph. ser. 50–130 (5–20)	431–165
δ-Tocopherol	417 ($[M+H]^+$)	717 (20), 374 (3), 191 (5), 151 (100), aliph. ser. 50–120 (5–20)	417–151
γ-Tocopherol	403 ($[M+H]^+$)	403 (20), 267 (5), 177 (7), 137 (100), aliph. ser. 50–120 (5–20)	403–137
Retinol	269 ($[M–H_2O+H]^+$)	269 (10), 239 (10), aliph. ser. 40–230 (5–55), 93 (100)	269–93
Phylloquinone	451 ($[M+H]^+$)	451 (50), 393 (7), 297 (8), aliph. ser. 170–290 (5–30), 187 (100)	451–187
β-Carotene	537 ($[M+H]^+$)	537 (30), 413 (10), 321 (40), aliph. ser. 50–300 (5–60), 177 (100)	537–177
Cholesterol	369 ($[M–H_2O+H]^+$)	269 (45), 287 (15), 259 (20), aliph. ser. 50–250 (5–90), 161 (100)	369–161

* The table lists the most significant fragment ions obtained at a collision energy of 20 eV (argon, 230 molecules/cm^2 CGT).
Aliph. ser. means a distribution of ions spaced at 14 ± 2 Da. The base peak is indicated separately when it is part of the distribution.

$[M-H_2O+H]^+$, probably due to thermal dehydration in the APCI source. On the basis of the spectra, it is possible to set-up a rapid FIA analysis for the simultaneous determination of these seven analytes taking advantage of the SRM technique. The transitions used are summarized in the Table 2.1. All the transitions are completely independent (different precursor ions, different product ions), so that the simultaneous quantitation can be easily performed by FIA. The analysis of suitable standard mixtures allows us to obtain a good linearity and a good dynamic range (10–5000 ng/ml). Fig. 2.6 shows the total ion current (TIC) the extracted ion current (XIC) traces related to a 100 ng/ml standard solution. This method provides speed (tocopherols are normally determined by HPLC, with long analysis times) and versatility (the other components are not commonly determined on olive oil) measuring seven important nutritional parameters of olive oil in just 1 min. Six olive oil samples, four extra-virgin oil samples from Tuscany, obtained under controlled extraction conditions, and two commercial samples were analyzed using this method. The results are shown in Fig. 2.7. The figure shows the content of the three tocopherols and of β-carotene. Retinol, phylloquinone and cholesterol were below the detection limit in all cases. The content of α-tocopherol is much higher with respect to other tocopherols and β-carotene, which are present at comparable levels. The quantity is expressed as μg/ml in the unsaponifiable matter: the conversion in mg/kg of original oil was not possible because of the lack of a suitable matrix for recovery tests. Rectified olive oil, the normal matrix used in this cases, still contains some residue of the analytes, in particular α-tocopherol and β-carotene, which are not removed or degraded during the rectification process. An accurate evaluation requires, hence, the use of isotopic dilution methods using labelled standards. The preparation of the deuterated tocopherols has been described in our laboratory [30–32], and work is in progress to set up the synthesis of labelled retinol, phylloquinone

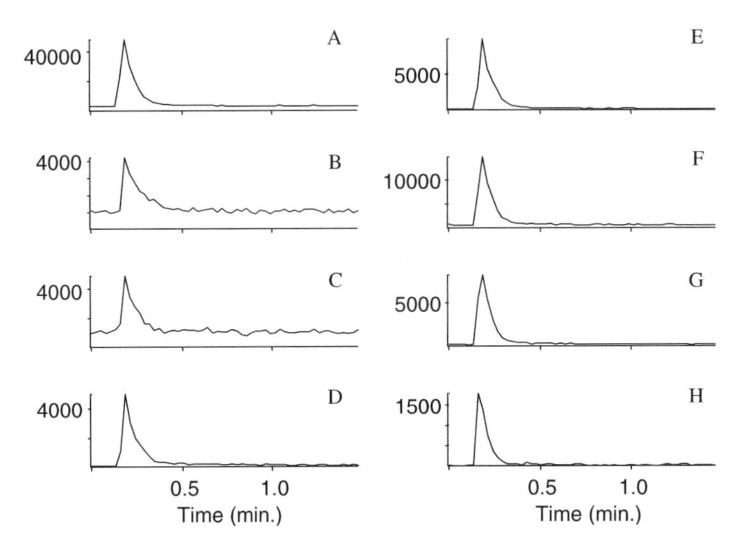

Fig. 2.6. TIC (a) and XIC chromatograms related to the FIA–SRM–APCI–MS–MS of a standard mixture of retinol (b), cholesterol (c), δ-tocopherol (d), γ-tocopherol (e), α-tocopherol (f), phylloquinone (g) and β-carotene (h).

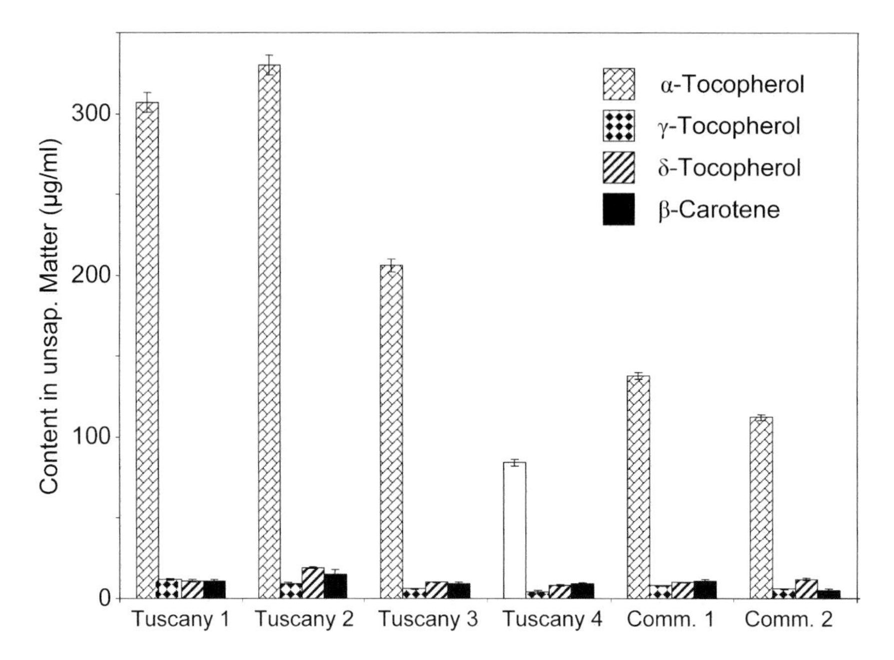

Fig. 2.7. Graphic representation of the tocopherols and β-carotene content in six olive oil samples determined by FIA–SRM–APCI–MS–MS.

and β-carotene. Deuterated cholesterol is available commercially. This application, even if not yet complete, is a good demonstration of the versatility of MS for the rapid determination of trace components in complex matrices, In this case we can get information on seven compounds of nutritional interest in olive oil in just 1 min injection.

REFERENCES

1 Perkin Elmer Sciex, *The API Book*, Perkin Elmer Sciex Instruments, Toronto, Canada, 1992 and references therein.
2 M.S.B. Munson and F.H. Field, *J. Am. Chem. Soc.*, 88 (1966) 2621–2630.
3 See for example: A. Raffaelli and A. Saba, in R.M. Caprioli (Ed.), *The Encyclopedia of Mass Spectrometry Vol. 3: Biological Applications Part B: Carbohydrates, Nucleic Acids and other Biological Compounds*, Elsevier, Oxford, 2006, pp. 327–340.
4 P. Kebarle and A.M. Hogg, *J. Chem. Phys.*, 42 (1965) 668–674.
5 A.M. Hogg, R.N. Haynes and P. Kebarle, *J. Am. Chem. Soc.*, 88 (1966) 28–31.
6 T.W. Martin, R.E. Rummel and C.E. Melton, *Science (Washington, DC, United States)* 138 (1962) 77–83.
7 M.M. Shahin, *J. Chem. Phys.*, 45 (1965) 2600–2605.
8 M.M. Shahin, *J. Chem. Phys.*, 47 (1967) 4392–4398.
9 D.I. Carroll, R.F. Wernlund and M.J. Cohen, Franklin GNO Corp., U.S. Patent 3.639.757, February 1, 1972.

10 E.C. Horning, M.G. Horning, D.I. Carroll, I. Dzidic and R.N. Stillwell, *Anal. Chem.,* 45 (1973) 936–943.
11 D.I. Carroll, I. Dzidic, R.N. Stillwell, M.G. Horning and E.C. Horning, *Anal. Chem.,* 46 (1974) 706–710.
12 J.B. French and N.M. Reid, *Dynamic Mass Spectrometry, Vol. 6*, Heyden and Son, London, 1980, pp 220–233.
13 J.D. Henion, B.A. Thomson and P.H. Dawson, *Anal. Chem.,* 54 (1982) 451–456.
14 C.M. Whitehouse, R.N. Dreyer, M. Yamashita and J.B. Fenn, *Anal. Chem.,* 57 (1985) 675–679.
15 A.P. Bruins, T.R. Covey and J.D. Henion, *Anal. Chem.,* 59 (1987) 2642–2646.
16 D.I. Carroll, I. Dzidic, E.C. Horning and R.N. Stillwell, *Appl. Spectr. Rev.,* 17 (1981) 337–406.
17 B.A. Thomson, J.V. Iribarne and P.J. Dziedzic, *Anal. Chem.,* 54 (1982) 2219–2224.
18 B.A. Thomson and J.V. Iribarne, *J. Chem. Phys.,* 71 (1979) 4451–4463.
19 J.F. Anacleto, H. Perrault, R.K. Boyd, S. Pleasance, M.A. Quilliam, P.G. Sim, J.B. Howard, Y. Makarovsky and A.L. Lafleur, *Rapid Commun. Mass Spectrom.,* 6 (1992) 214–220.
20 J.F. Anacleto, R.K. Boyd, S. Pleasance, M.A. Quilliam, J.B. Howard, A.L. Lafleur and Y. Makarovsky, *Can. J. Chem.,* 70 (1992) 2558–2568.
21 J.F. Anacleto, L. Ramaley, F.M. Benoit, R.K. Boyd and M.A. Quilliam, *Anal. Chem.,* 67 (1995) 4145–4154.
22 J.F. Anacleto and M.A. Quilliam, *Anal. Chem.,* 65 (1993) 2236–2242.
23 C. Perazzolli, I. Mancini and G. Guella, *Rapid. Commun. Mass Spectrom.,* 19 (2005) 461–469.
24 A. Raffaelli, S. Pucci, R. Lazzaroni and P. Salvadori, *Rapid Commun. Mass Spectrom.,* 11 (1997) 259–264.
25 W.W. Eckenfelder, Jr., *Industrial Water Pollution Control*, McGraw Hill, New York, NY, 1966, Chapter 6.
26 E. Perri, F. Mazzotti, A. Raffaelli and G. Sindona, *J. Mass. Spectrom.,* 35 (2000) 1360–1361.
27 E. Perri, A. Raffaelli and G. Sindona, *J. Agric. Food Chem.,* 47 (1999) 4156–4160.
28 A. Raffaelli, A. Cuzzola, S. Pucci and P. Salvadori, *49th ASMS Conference on Mass Spectrometry and Allied Topics*, Chicago, IL, USA, May 27–31, 2001, ThPN 348.
29 A. Raffaelli, Unpublished results, presented at *MASSA 2001 – IONISSEA NELLO SPAZIO, An International Symposium on Mass Spectrometry*, Marina di Campo, Elba Island (LI), Italy, June 24–27, 2001, KN-5, Abstracts, p. 35.
30 F. Mazzini, E. Alpi, P. Salvadori and T. Netscher, *Eur. J. Org. Chem.,* (2003) 2840–2844.
31 F. Mazzini, A. Mandoli, P. Salvadori, T. Netscher and T. Rosenau, *Eur. J. Org. Chem.,* (2004) 4864–4869.
32 F. Mazzini, T. Netscher and P. Salvadori, *Tetrahedron,* 61 (2005) 813–817.

Achille Cappiello (Editor)
Advances in LC–MS Instrumentation
Journal of Chromatography Library, Vol. 72

Chapter 3

Electron ionization in LC–MS: a technical overview of the Direct EI interface

ACHILLE CAPPIELLO and PIERANGELA PALMA

3.1 INTRODUCTION

High-performance liquid chromatography (HPLC) and electron ionization mass spectrometry (EIMS) are two analytical techniques that, in principle, seem to be incompatible. However, because these two approaches share a great number of applications in the analysis of suitable molecules, typically <1000 Da, much effort has been devoted by the scientific community to attempt to develop a reliable, easy-to-use, and flawless interface. As a matter of fact, the first attempts in combining HPLC and MS, including the Moving Belt interface developed more than 25 years ago, included EI because no other ionization technique was readily available for this purpose at that time. The first successful and commercially available device to combine EI and HPLC was developed by Willoughby and Browner in 1984 [1]. It was based on the conversion of the liquid effluent into a gas phase and transformed into a beam of solute particles, after the evaporation of the spray droplets and the elimination of the solvent vapors through a multi-stage momentum separator.

Surprisingly, the initial success of EI in coupling liquid chromatography and mass spectrometry, offered by the particle beam interface, was not followed by any notable progress or, at least, refinement of the original design. In fact, the development of new soft, liquid-based ionization techniques generated a family of atmospheric pressure ionization–based interfaces (API) that polarized most researchers' attention and caused the rapid disappearance of the previous EI-based concepts, although a considerable number of small-medium molecular weight analytes can provide very good EI spectra.

The simple operating principle of EI can induce the ionization of any molecule present in the gas phase, regardless of its chemical background, as long as it can withstand the typical EI source conditions (high vacuum and temperature). This attribute represents a point of strength when compared to API-based interfaces. In EI, in fact, the absence of ion–molecule or ion–ion reactions minimizes the influence of the mobile phase in the ionization process and brings to an end all those weaknesses that are typical of electrospray ionization (ESI) such as signal suppression, influence of the mobile phase composition, need for desalting, or post-column solvent modifications. Moreover, the typical EI spectrum is highly informative, and its high reproducibility allows an easy comparison with thousands of spectra from commercially available sources (such as NIST or Wiley). EI detection can benefit from recently developed sophisticated algorithms such as, for instance, the one developed by the National Institute of Standards and Technology (NIST) and called AMDIS (Automated MS Deconvolution and Identification System), which extracts the analyte's mass spectrum in complex chromatographic mixtures with several overlapping peaks [2].

The fact that EI- and HPLC-operating conditions remain distant cannot, however, be ignored, and it might explain some of the frustration observed in the everyday use of a typical LC–EIMS interface of the past: poor sensitivity, signal instability, reduced linearity, and disappointing optimization. The growth of new instruments that rely on ESI or atmospheric pressure chemical ionization (APCI) solved most of those drawbacks and widened extraordinarily the field of possible applications; however, they also introduced new analytical problems such as signal suppression or adduct-ion formation that are often found in certain applications.

Although thousands of papers have been published on new and improved API interfaces, only a few authors have contributed to the attempt of efficiently generating EI spectra from a liquid effluent. Kientz *et al.* [3] were able to generate EI spectra based on an eluent-jet formation by means of inductive heating of the micro-LC effluent, and momentum separation in a jet separator; Dijkstra *et al.* [4] improved that technique with particular emphasis on its use under chemical ionization CI–MS conditions; Amirav and Granot [5] developed a new apparatus to obtain high-quality, library-searchable EI spectra based on supersonic molecular beam mass spectrometry (SMB–MS) approach. This valuable attempt allows conventional mobile-phase flow rates and enhances molecular ion response for an improved molecular weight determination.

Our research group has devoted a considerable effort to update and improve the design and performance of the particle beam interface [6]. Significantly improved sensitivity for any mobile-phase composition was achieved by drastically reducing the mobile-phase flow rate. The microparticle beam (PB) interface and its evolution, called capillary EI (CapEI, commercialized by Waters, Milford, MA, USA in 1999), use micro-scale flow rates (1 – 5 μL/min), and were successfully employed in the analysis of several classes of compounds and with different modifiers added to the mobile phase [7,16]. It is noteworthy to point out that this approach showed, for the first time, a significantly improved tolerance for nonvolatile buffers in mass spectrometry, as demonstrated in two applications [17,18].

To further reduce the adverse effects of mobile-phase vapors and system complexity, we designed a new interface, in which the eluate is directly introduced into the EI source [19–22] at a nanoscale flow rate (<500 nL/min). The name of the new interface is Direct-EI to emphasize the direct coupling to a mass spectrometer, with no need for any intermediate apparatus. The commercial availability of good and reliable nanoscale HPLC columns facilitated this straightforward approach. Our goal was to bring an LC–MS interface to the level of simplicity and performance typical of a gas chromatography GC–MS, using the same mass detector and the same software with only a minimum of instrument adjustment. Direct-EI is a simple device through which the HPLC liquid effluent directly enters the EI source. The lack of any particular transport mechanism involved in the interfacing process reduces sample losses, thus significantly improving the detection limits to the low picogram level for most substances. Neither interface clogging nor CI interferences are usually observed. Similar to Cap-EI, Direct-EI is compatible with nonvolatile buffers in the mobile phase, opening the door to a wide range of challenging applications.

Typical chromatographic parameters such as various mobile-phase compositions and flow rate were investigated with different mass spectral conditions, such as source temperature or focusing electric potentials. The results were compared and evaluated in terms of limits of detection, mass spectral quality, response linearity, reproducibility, and nonvolatile buffer tolerance. Five test compounds were used for each experiment to highlight any possible difference. Finally, a brief list of documented applications is reported.

3.2 THE INTERFACE

The interfacing mechanism operates inside a common EI source, similar to that found in any GC–MS system. If a GC–MS is to host the interface, the liquid phase can be admitted from the GC capillary column port via a capillary connecting tubing, sealed to prevent vacuum loss. The mechanism is based on the formation of an aerosol in high-vacuum conditions, followed by a quick droplet desolvation and final vaporization of the solute prior to the ionization. The completion of the process is achieved within the ion-source volume. In Fig. 3.1, a scheme of the interface is shown. The core of the interface is represented by the micronebulizer, which consists of a cone-shaped tip that can be slightly bent sideways to accommodate different source designs, and with an orifice of approximately 5 – 10 μm. The nebulizer tip protrudes into the ion source so that the spray expansion is completely contained inside the ion-source volume. It is important to note that the eluate must emerge from the nebulizer as liquid phase, and that any premature in-tube solvent

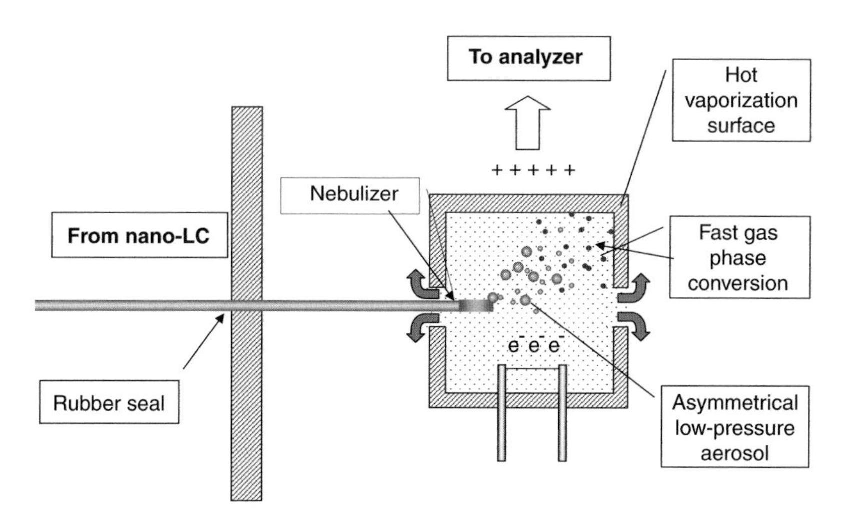

Fig. 3.1. Scheme of the Direct-EI interface.

evaporation must be prevented by thermally insulating the nebulizer and the connecting tubing from the source heat. Premature mobile-phase evaporation impairs the nebulizer's ability to generate an aerosol and the consequent solute precipitation might obstruct the small tip orifice impeding the normal liquid passage. Once released by the nebulizer, the liquid is converted into small spray droplets that are directed toward a specific target surface in the ion source. The high temperature of the ion source, between 200°C and 300°C, has a double function: to compensate for the latent heat of vaporization during the droplet desolvation, and to convert the solute into the gas phase upon contact with the hot target ion-source surface. The nebulizer is connected to the analytical nano-column by a 30-μm-i.d. fused silica capillary tubing. An additional opening in the ion source might speed up the removal of solvent vapors and avoid any ion–molecule reaction, thus preventing the formation of the typical chemical ionization spectrum with $(M + H)^+$ ions. It is noteworthy to point out that the impact of a sub microliter-per-minute liquid flow generates approximately less than a milliliter-per-minute of vapor inside the ion source, clearly within the pumping capacity of any GC–MS system. Neither the electron path nor the electric fields are influenced by the limited interface intrusion into the ion source volume. If all components of this simple system are correctly placed and sized, then each solute separated by the nano-column is smoothly converted into the gas phase, the peak profile is nicely reproduced, and high-quality mass spectra are generated. Inefficient thermal insulation of the interface leads to high background noise and sputtered chromatographic peaks. The lowest flow rate required by the interface to generate a fine and homogeneous spray is approximately 100 nL/min, whereas the highest flow rate compatible with the interface is compound-specific; however, in general, it depends on capability of the ion source design to remove the solvent vapors.

The major advantage of this technical solution is the extreme compactness and simplicity without any need for additional complex devices between the liquid chromatograph and the mass spectrometer. The lack of any particular transport mechanism involved in the interfacing process reduces sample losses, enhancing sensitivity and extending the range of possible applications. In other words, the interfacing process has a very low influence on the sample transition, and the only limitation is represented by the chemical–physical properties of the analytes.

The interface was tested so far using a Hewlett-Packard 5989A, single-stage quadrupole mass spectrometer, but experiments with an Agilent 5475 MSD are on the way. The operating pressure was approximately 7×10^{-5} torr measured in the manifold of the ion source at the highest operating flow rate (1 μL/min) and 1.5×10^{-5} torr at 300 nL/min, using a 50:50 (v:v) mixture of water and acetonitrile as the mobile phase. Mass spectrometer tuning and calibration were performed automatically using perfluorotributylamine (PFTBA) as a reference compound and monitoring the ions at *m/z* 69, 219, and 502. No mobile phase was allowed into the ion source during calibration. The dwell times during SIM and scan times for full spectrum acquisition were adjusted to obtain 0.5 cycles/s and a mean of 10–15 acquisition samples for each HPLC peak (0.8 – 1 scan/s). Higher scan frequency can be set for quantitative results. As reported above, the source temperature is, of course, an important issue in this interfacing mechanism, and its effect was extensively investigated. In fact, once the solute particles land on the hot metal surface, the heat is distributed throughout them before their complete gas-phase conversion. If not sufficiently rapid, then this crucial process may cause thermal decomposition of the most thermally sensitive analytes. We should always remember that the major limitation of this approach is represented by the solute gas-phase conversion.

The very small particles generated by the Direct-EI interface increase the surface-to-mass ratio to speed up vaporization, reduce decomposition, and increase the number of possible suitable analytes. The combination of a good nebulization and appropriate temperature constitutes the recipe for a correct analysis, and must be evaluated case by case.

Liquid chromatography was performed on a Kontron Instruments 420 single-pump, conventional HPLC system (Kontron Instruments, Milano, Italy) that was upgraded with a multiposition valve nanogradient (MP-valve) generator. This new device [23,24] is based on the use of a 14-port switching valve, to generate smooth gradients at nano- and microflow rates. The valve is equipped with six loops, and each one contained a selected mixture of eluents: the first one is filled with the weakest eluent (100% water), and the last one is filled with the strongest one (100% acetonitrile). The loops are loaded manually, and their volume is sufficient for 40–50 chromatographic runs. An electric switch allows selection of a given loop at a time, thus generating a specific solvent gradient through a sequence of controlled segments of precise mobile-phase compositions. The combination between an electric actuation that can be controlled by a CPU and an adequate number of loops permits not only to exactly reproduce the programmed slope, but also to achieve different gradient shapes (i.e., linear, convex, and concave) for different separation needs, even at the lowest flow rate. The typical stepwise gradient profile is smoothed out by a mixing chamber of a given volume that is positioned right before the injector. The new device is designed to be easily accommodated on any preexisting conventional HPLC system, converting it into an efficient nano- or micro-HPLC. Nanoliter-per-minute flow rates can be obtained with a laboratory-made splitter placed between the pumping system and the MP-valve. The splitter is only responsible for scaling down the flow rate isocratically and feeding the MP-valve for the gradient formation. For a large-volume sample injection, relative to the extremely low mobile-phase flow rates used in this work, a Valco injector equipped with a 60- or 500-nL internal loop was employed. Laboratory-made and commercial 75-μm-i.d. columns with C18, 3-μm particles were used for the chromatographic separations. An Agilent 5975 inert MSD coupled with a 1100 nano-HPLC system is under testing.

3.3 INTERFACE PERFORMANCE

In principle, the LC–MS interface is a subtle device that only bridges the gap between the two instruments without interfering with either one. It should provide an absolute freedom in the choice of the best chromatographic conditions, and at the same time it shouldn't spoil the detection potential of the mass spectrometer or induce artifacts in the expected mass-spectrometric signal. None of the LC–MS interfaces present in the market is even close to this ideal concept, though they are efficiently used in thousands of well-established applications. As a matter of fact, GC–MS put the principle into practice, though for a relatively limited number of substances when compared to LC–MS. Direct-EI, being very close to the GC–MS approach, copes very well with the ideal concept until its use is prevented by a high molecular weight or by thermal instability. Its efficiency and reliability is demonstrated by five test compounds that belong to several different compound classes:

1. Azinphos ethyl
2. Ibuprofen

3. Caffeine
4. Bisphenol A
5. Testosterone

employed in a series of key experiments:

- Limits of detection;
- Reproducibility;
- Signal intensity versus mobile-phase composition, mobile-phase flow rate, ion-source temperature;
- Mass spectrum quality versus mobile-phase flow-rate;
- Monitoring of crucial instrumental parameters before and after the introduction of a non-volatile buffer.

To prove that Direct-EI is a valid competitor for the solution of different analytical tasks, several real-world applications are also shown.

3.3.1 Limits of detection

Unfortunately, EI is a low-efficiency ionization technique because $<$1/10000 of the gas-phase sample molecules is ionized, therefore impressive detection limits cannot be expected. However, in GC–MS, a careful control over the operative conditions allows the attainment of a limit of 100 fg or lower for selected analytes. Picogram-level detection is normally expected in SIM mode for most substances while two orders of magnitude higher amounts are required to record an interpretable mass spectrum. Direct-EI performs considerably well in this context, and outclasses all its predecessors; a sensitivity in the low picogram level for many substances is the rule. However, soft-ionization techniques such as ESI are, in some cases, far more efficient in generating fewer fragment ions, maintaining molecular ion information, and reducing signal dispersion. The cost of this exceptional sensitivity is paid in terms of structural information so that a second analyzer to generate MS/MS spectra is routine. A typical EI spectrum, in general, has extensive structural information, and a single-stage mass spectrometer might be sufficient for analyte characterization or identification.

In this context, detection limits (Table 3.1) were calculated for all five test compounds in the full-scan, extracted-ion, and SIM modes. In full-scan mode, a full spectrum was acquired, and the detection limit was calculated as the minimum amount of substance that generated an interpretable, library-matchable spectrum. In the other two modes of operation, the detection limit was obtained for a signal-to-noise ratio of 5:1. Those data were collected in the flow injection analysis (FIA) mode, using a mobile phase of water and acetonitrile (50:50, v:v) at 200 nL/min and an injection loop of 60 nL. Each limit is the mean value of five replicates. Slightly better results can be achieved with a column and a gradient program because of the focusing effect and better signal-to-noise ratio. Single- and multiple-ion detection limits were also calculated for each selected analyte. As expected, SIM showed the best results for all compounds. It is noteworthy to point out that the typical, less efficient HPLC separation, when compared to capillary GC, produces wider peaks for a generally much worse signal-to-noise ratio. This accounts for at least

TABLE 3.1

LIMITS OF DETECTION

Compound	LODs (ng)						
	SCAN	Extracted ion			SIM		
		3 ions	2 ions	1 ion	3 ions (*m/z*)	2 ions (*m/z*)	1 ion (*m/z*)
Azinphos ethyl	6	3	2	2	0.024 (105; 132; 160)	0.020 (132; 160)	0.017 (132)
Ibuprofen	0.150	0.600	0.150	0.150	0.060 (161; 163; 206)	0.024 (161; 163)	0.024 (161)
Caffeine	1.500	–	0.375	0.250	–	0.024 (109; 194)	0.012 (194)
Bisphenol A	0.300	0.300	0.300	0.120	0.017 (119; 213; 228)	0.012 (119; 213)	0.010 (213)
Testosterone	12	12	12	3	0.300 (124; 246; 288)	0.300 (124; 288)	0.170 (124)

a tenfold decrease of the detection limit for the same compound passing from GC–MS to LC–MS.

3.3.2 Reproducibility

Direct-EI is greatly influenced by the ion-source geometry, temperature, and vacuum conditions, but once settled into a specific configuration, it shows a very stable behavior that allows a smooth transition from liquid to gas phase, generating an even, predictable ion profile. Quantitative analyses take a great advantage from those attributes, and good signal reproducibility is expected. We performed 30 consecutive injections of 6 ng for each analyte in SIM, recording the relative peak areas. The flow rate of 500 nL/min was decreased only for testosterone because above 250 nL/min, some chemical ionization signal was observed. As reported in Table 3.2, the relative standard deviation that ranges between 5% and 10% demonstrated a very stable signal response and confirmed its role as a reliable quantitative platform.

3.3.3 Signal intensity versus mobile-phase composition

As stated before, the ideal interface should tolerate different chromatographic conditions and leave plenty of choice for a successful sample separation. Different from API, Direct-EI physically separates the highly volatile solvent vapors from the higher molecular weight solute and produces an independent ionization of solvents and solutes in the gas

TABLE 3.2

REPRODUCIBILITY OF THE PEAK AREAS

Compound (*m/z*)	Peak area RSD (%)
Azinphos ethyl (132)	5.72
Ibuprofen (161)	5.41
Caffein (194)	10.23
Bisphenol A (213)	7.32
Testosterone (124)	6.66

phase. Most solvent vapors are removed quickly and efficiently by the vacuum before reaching the electron path zone, whereas the solute is more exposed to the electrons impact. The remaining solvent ions do not interfere with the solute signal, and their effect is limited to creating low-mass background noise. In this context, the response under SIM conditions is not influenced by the mobile-phase composition, and different solvents can be safely used. It is obvious that an increasing solvent pressure in the ion source does not go unnoticed, eventually promotes chemical ionization, and might spoil the overall mass spectrum quality. However, that effect is limited to the highest flow rates, and is not observed under SIM conditions where single ions are monitored. The independence from different chromatographic conditions was evaluated by simulating gradient conditions and using a mixture of water with acetonitrile or methanol as organic modifiers. Figs. 3.2A and B show the signal response recorded for each test compound versus different concentrations of water–acetonitrile and water–methanol, respectively. The amount injected was 6 ng in all cases. The same *m/z* values reported in Table 3.2 were monitored in the SIM mode, and the mean value was calculated from five replicates. Because of the higher viscosity offered by the water–methanol combination, the flow rate was reduced from 500 to 300 nL/min. An additional flow rate reduction (250 nL/min) was necessary to suppress the chemical ionization shown by testosterone. The flat response in all conditions and for all compounds was an important result, and demonstrated the excellent compliance of the interface to the different chromatographic conditions that are common for most of the applications in reversed-phase chromatography.

3.3.4 Signal intensity versus mobile-phase flow rate

Because the mobile-phase in Direct-EI does not play any specific role in promoting solute ionization, one would not expect any significant variation in signal response for different flow rates. To investigate that aspect of the interface, we measured the intensity of the response for all five compounds, monitored a single ion for each analyte as previously reported, and varied the mobile-phase flow rate from 1000 to 200 nL/min. A fixed composition of acetonitrile–water 1:1 (v:v) was used, five replicates were collected to calculate the mean value, and 6 ng of substance were injected via a 60-nL loop. The results were

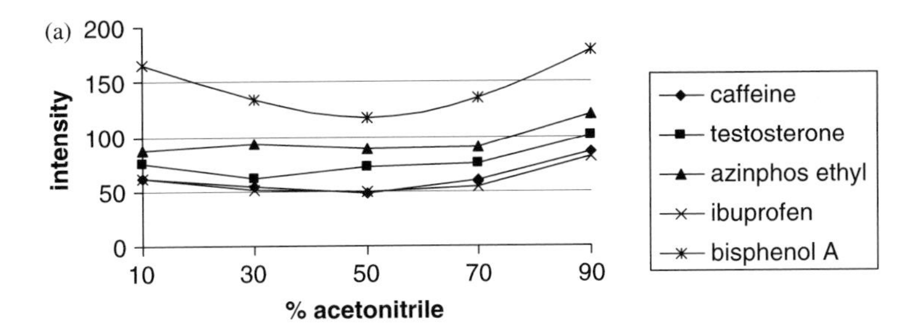

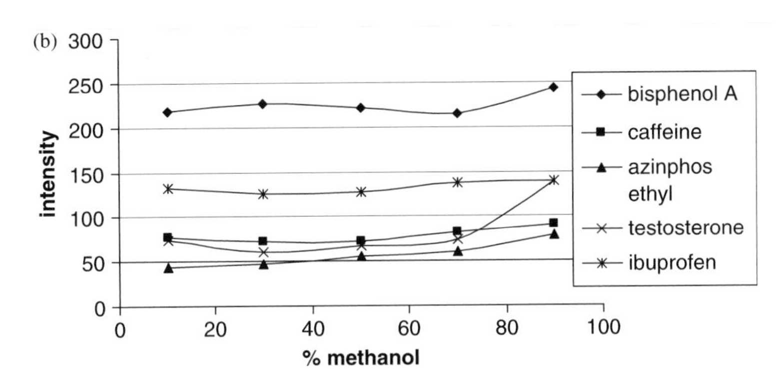

Fig. 3.2. Signal intensity versus mobile-phase composition. The organic modifiers are (a) acetonitrile and (b) methanol.

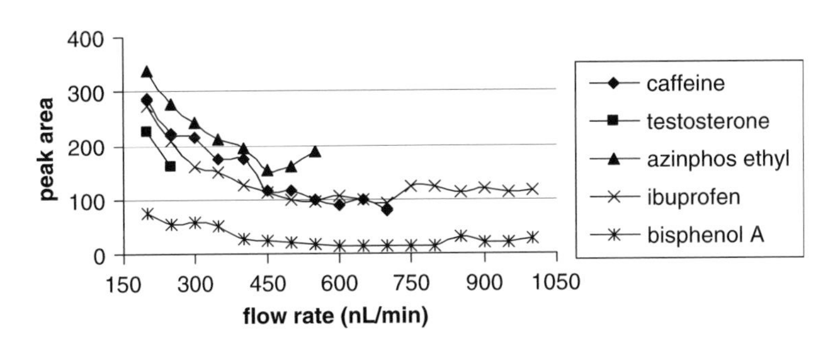

Fig. 3.3. Signal intensity versus mobile-phase flow rate.

plotted as shown in Fig. 3.3. Lack of experimental data for several compounds at the highest flow rates revealed the appearance of chemical ionization with evident $(M + H)^+$ ions; those results were not considered. Testosterone was found particularly prone to CI; in fact, for that compound the highest tolerable flow rate was 250 nL/min. This value does not imply an insuperable limit, especially in SIM when only specific ions are monitored, but rather a warning sign that the overall spectral quality is deteriorating. However, some CI interference does not compromise the overall mass spectral quality and ability of an electronic library to recognize a specific compound. Experimentally, a lower flow rate is clearly beneficial in terms of sensitivity, with a threefold increase for most compounds at 200 nL/min. A possible explanation can be found not only in an improved signal-to-noise ratio, but also in the higher ionization efficiency achieved when there is less competition in the gas phase for the available electrons.

3.3.5 Signal intensity versus ion source temperature

Temperature is the driving force of electron ionization, and has a double function in the Direct-EI interface: to compensate for the latent heat of mobile-phase evaporation, and to promote the gas-phase conversion of the solute. Too much or too little of this crucial parameter can heavily affect the overall quality of the mass spectral results. During the interface development, we noticed that when heat was applied prematurely on the transfer capillary before the nebulizer, the solvent vapors formed sooner than the spray, leading to a sputtering of the liquid phase and, as a consequence, an uneven chromatographic profile. Because the nebulizer protrudes in the heated zone of the ion source, its connection to the column must be thermally insulated. In this way, the liquid phase is preserved up to the very end of its journey inside the capillary tubing, and a stable spray is produced. However, heat prevents the freezing of the solvent at the nebulizer orifice, and compensates for the latent heat of changing state. On the other side of the ion source, where the solute particles are collected on the hot source surface, heat is applied to convert them rapidly into the gas phase. In that case, a quick heat distribution is mandatory to avoid sample decomposition. The optimal temperature was selected by introducing 6 ng of each compound at 200 nL/min with a mobile phase of water and acetonitrile (50:50, v:v), and by monitoring a specific ion signal for each compound (see Table 3.2 for *m/z* values). Five experiments were performed for each compound, and mean values are plotted in Fig. 3.4. As shown in the figure, the optimal temperature was 250°C for all substances. Caffeine is the most volatile compound, and it was less affected by this parameter, showing a flat response.

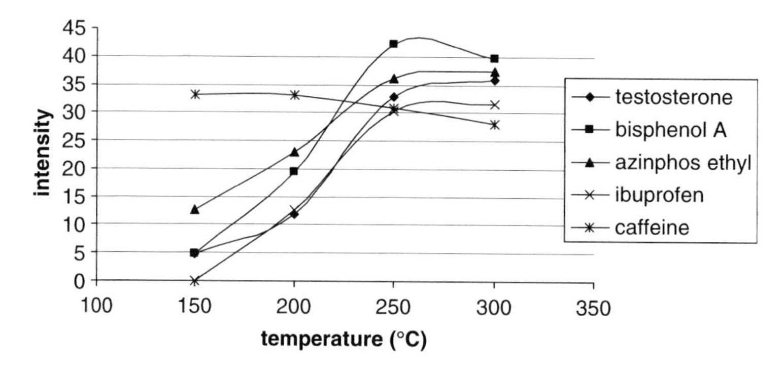

Fig. 3.4. Signal intensity versus ion-source temperature.

3.3.6 Mass spectrum quality versus mobile phase flow rate

The ability to record electron ionization spectra is the benchmark of the Direct-EI interface. Soft ionization techniques require MS/MS for characterization and/or identification. To accomplish the job, Direct-EI generates high-quality, library-matchable, readily interpretable electron ionization mass spectra for prompt compound characterization and confirmation. The reduced flow rate of this interface has proven a clear advantage in many occasions in terms of sensitivity and signal stability. The quality of the experimental mass spectrum, at different flow rates, was measured as the degree of overlap (%) obtained when compared to the reference spectrum that is recorded in the Wiley electronic library. Only "pure" EI spectra were considered, so that CI data were discarded. The results are plotted in Fig. 3.5. Six nanograms of each compound were injected, using a 1:1 (v:v) mixture of water and acetonitrile. Decreasing the mobile-phase flow rate is clearly advantageous for the overall mass spectral quality and especially for caffeine, azinphos ethyl, and bisphenol A that give a fairly high response over the widest flow rate range. In Fig. 3.6, the average mass spectrum of bisphenol A is acquired close to its limit of detection (1 ng) and at the lowest flow rate (200 nL/min). The recorded spectrum is very similar to that of the library with a matching quality of 94%; however, in our case the mass

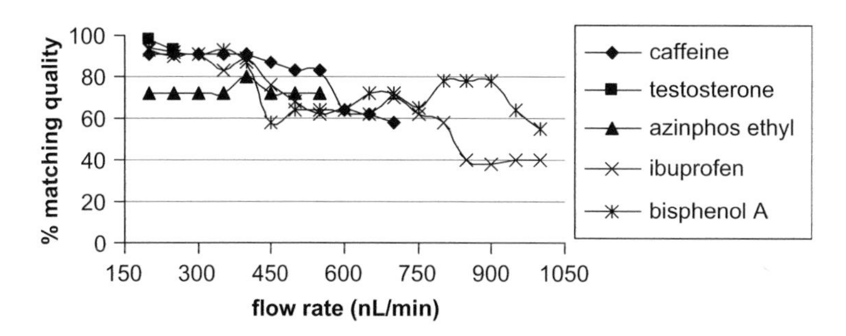

Fig. 3.5. Mass spectrum quality versus mobile-phase flow rate.

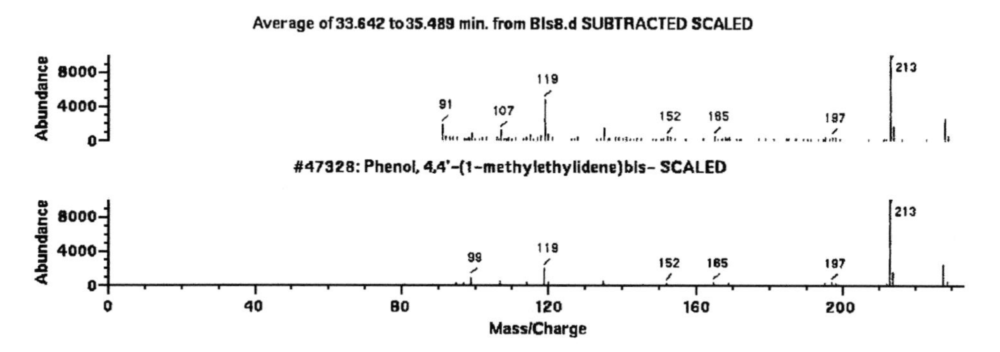

Fig. 3.6. Comparison between recorded (top) and Wiley library (bottom) mass spectra (1 ng injected) of bisphenol A.

range was intentionally limited to above *m/z* 90 to reject most of the chemical noise and enhance signal response.

3.3.7 Introduction of a nonvolatile buffer

A direct consequence of the reduced flow rate is the possibility to introduce a steady stream of a nonvolatile buffer as a mobile-phase modifier. A micro-particle-beam interface before, and Cap-EI later, demonstrated that the combination of a reduced flow rate and electron ionization enhanced the ruggedness and flexibility to tolerate chemically "aggressive" mobile phases [17,18]. Direct-EI, with its nanoscale approach, expands this unique behavior. One of the most challenging applications in LC–MS is to enhance the tolerance toward the nonvolatile buffers that are added to the mobile phase to improve chromatographic separation in several applications of biological and environmental interest. As a matter of fact, the presence of nonvolatile, ionic species (phosphate and sulfate buffers) in the ESI spray is deleterious. Nonvolatile species cause salt depositions on the metal surfaces, and could completely block ion transmission. In addition, if the anion and cation pair too strongly with the analyte, then the analyte ions might be prevented from carrying the excess charge on the droplet surface; as a result, the ESI response may be very low. For those reasons, when ESI is interfaced with HPLC or capillary electrophoresis (CE), volatile buffers composed of weak acids and bases must replace nonvolatile modifiers. Also, strong acids such as trifluoroacetic (TFA), heptafluorobutyric, and hydrochloric acid, which are used as ion-pairing agents in HPLC, tend to mask the analyte signal in ESI-MS. Electron ionization, used in Direct-EI, is not influenced by the presence of preformed ions in the form of salts or strong acids and bases so that the analyte can be ionized independently. Because of the extremely reduced liquid intake, Direct-EI slowly displaces a negligible salt deposition within the ion chamber as the only sign of a salt presence. After many days of continuous operation, the salt deposition is barely visible in the ion source, and can be removed readily by routine maintenance procedures before any changes in the performance is observed. Using this method, a 10-mM concentration of phosphate buffer, used to optimize the chromatographic separation, was admitted into the system, and several parameters were continuously monitored for a reliable follow-up of the instrument during a period of 32 h of total acquisition time (4 working days). Entrance lens, AMU gain, repeller, multiplier potentials, and caffeine and perfluorotributylamine ion signals were recorded at the beginning and at the end of each day. No parameter (Fig. 3.6) was appreciably influenced by the long-time buffer intake. An estimated total of 0.26 mg of phosphate salt introduced into the system during this study was barely visible as a fine, even dispersion of white powder inside the ion source.

3.4 APPLICATIONS

A hint of the possible fields of application in which the interface can operate is given briefly in this section. We chose those applications where Direct-EI can excel in terms of simplicity and reliability or offer a straightforward alternative to other interfacing techniques, namely, ESI and APCI. A large number of small-medium molecular-weight molecules can fall into this category with the only condition being the capacity to withstand a gas-phase ionization.

The chromatographic plots relative to the analysis of a selected number of pesticides, nitro-PAHs, hormones, and endocrine-disrupting compounds (EDCs) are shown in Figs. 3.7–11. Laboratory-made C18 (75 μm i.d. × 15 cm, 3-μm particle size) and LC Packings nanocolumns (C18, 75-μm i.d. × 15-cm, 3-μm particle size, loop size 60 nL), were used in all applications at a mobile-phase flow rate of 250 nL/min except for the EDCs where 200 nL/min was used. In the latter application, the injection volume was 600 nL, thus operating in the large-volume injection mode. A linear gradient from 100% water to 100% CH_3CN was obtained in 40 min for the separation of a 20-ng/μL mixture of pesticides and for the EDCs, with a concentration that ranged from 0.084 to 4.2 ng/μL. A 30-min gradient was required for the separation of a 25-ng/μL mixture of nitro-PAHs and for the hormones. The concentration of the mixture of hormones was 25 ng/μL for all except medroxyprogesterone acetate, which had a concentration of 150 ng/μL. The gradient was achieved by filling the six loops with a specific solvent composition, namely, loop 1: 100% water–0% CH_3CN; loop 2: 80% water–20% CH_3CN; loop 3: 60% water–40% CH_3CN; loop 4: 40% water–60% CH_3CN; loop 5: 20% water–80% CH_3CN; and loop 6: 0% water–100% CH_3CN. It is noteworthy that the analysis of 29 EDCs was obtained in a single experiment, and each analyte was characterized by using a single quadrupole. In contrast, because of the different acidic-basic behaviors, ESI would require a redistribution of the selected compounds between

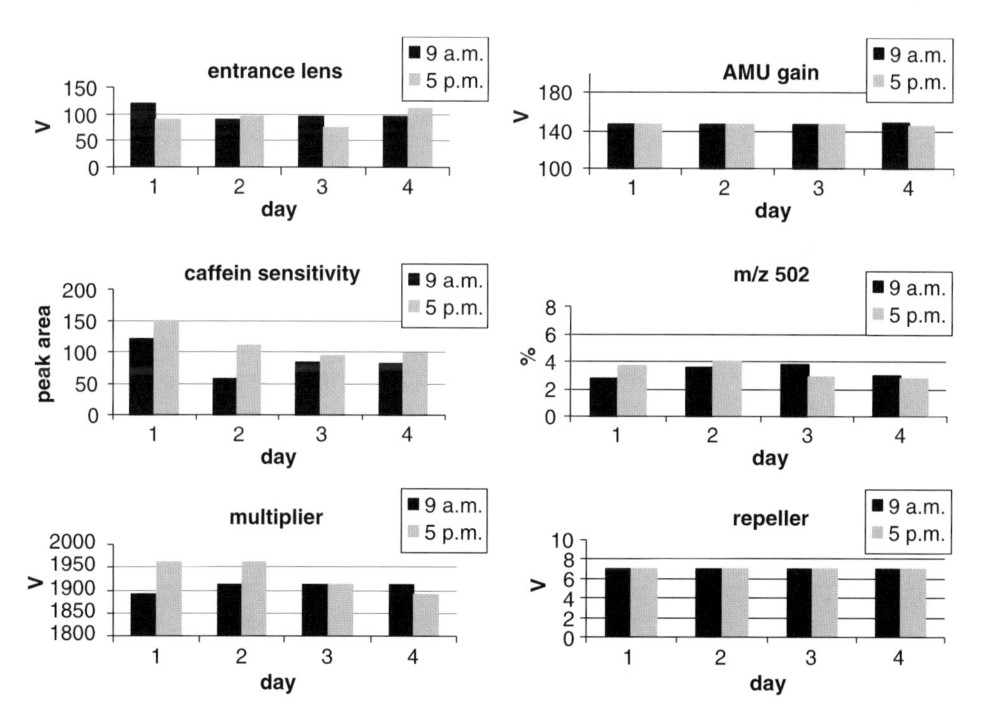

Fig. 3.7. Crucial instrumental parameters during phosphate-buffer introduction (10 mM KH_2PO_4; 0.26 mg total amount introduced). Mobile-phase flow rate: 200 nL/min.

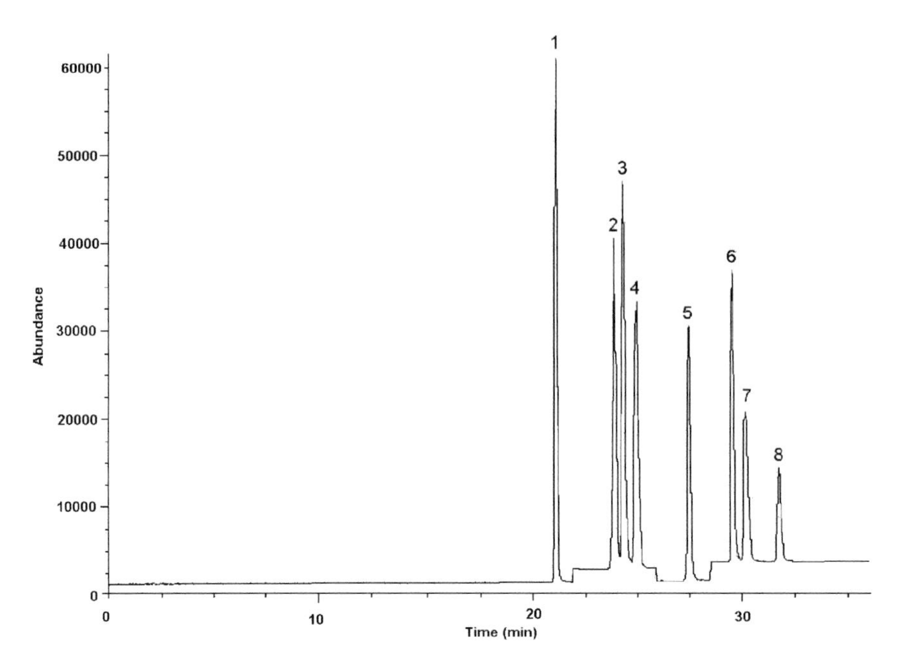

Fig. 3.8. Reconstructed ion chromatogram of a 20-ng/μL mixture of pesticides. LC Packings C18, 75 μm i.d. × 15 cm, 3 μm particle size nanocolumn; flow rate 250 nL/μL; injection volume 60 nL. Linear gradient from 100% H_2O to 100% CH_3CN in 40 min. (1) monuron; (2) chlortoluron; (3) fluometuron; (4) isoproturon; (5) propham; (6) linuron; (7) chlorbromuron; (8) chlorpropham.

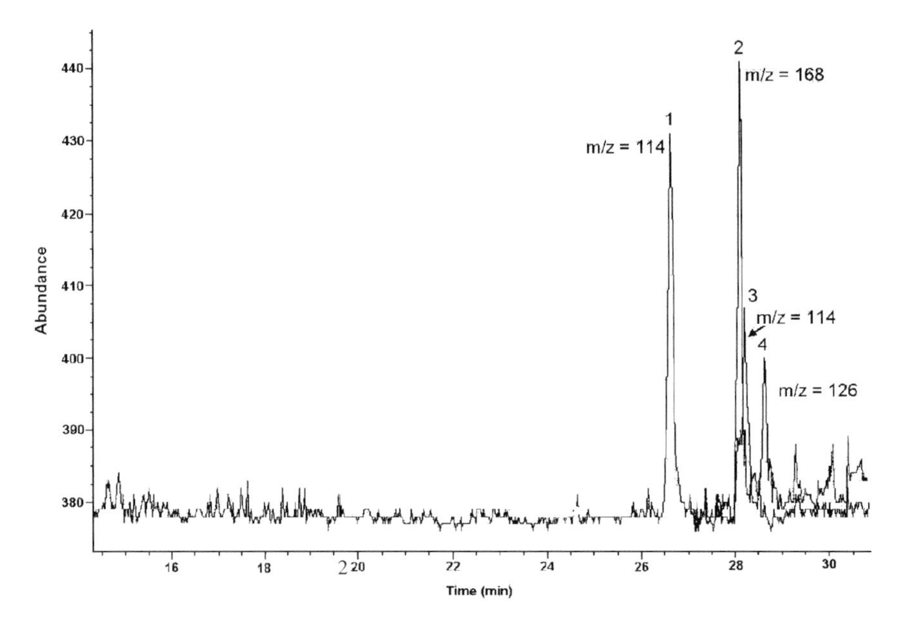

Fig. 3.9. SIM chromatogram of a 25-ng/μL mixture of nitro-PAH. LC Packings C18, 75 μm i.d. × 15 cm, 3 μm particle size nanocolumn; flow rate 250 nL/μL; injection volume 60 nL. Linear gradient from 100% H_2O to 100% CH_3CN in 30 min. (1) 1,8-dinitronaphthalene; (2) 2,2′-dinitrobiphenyl; (3) 1,5-dinitronaphthalene; (4) 1,3-dinitronaphthalene.

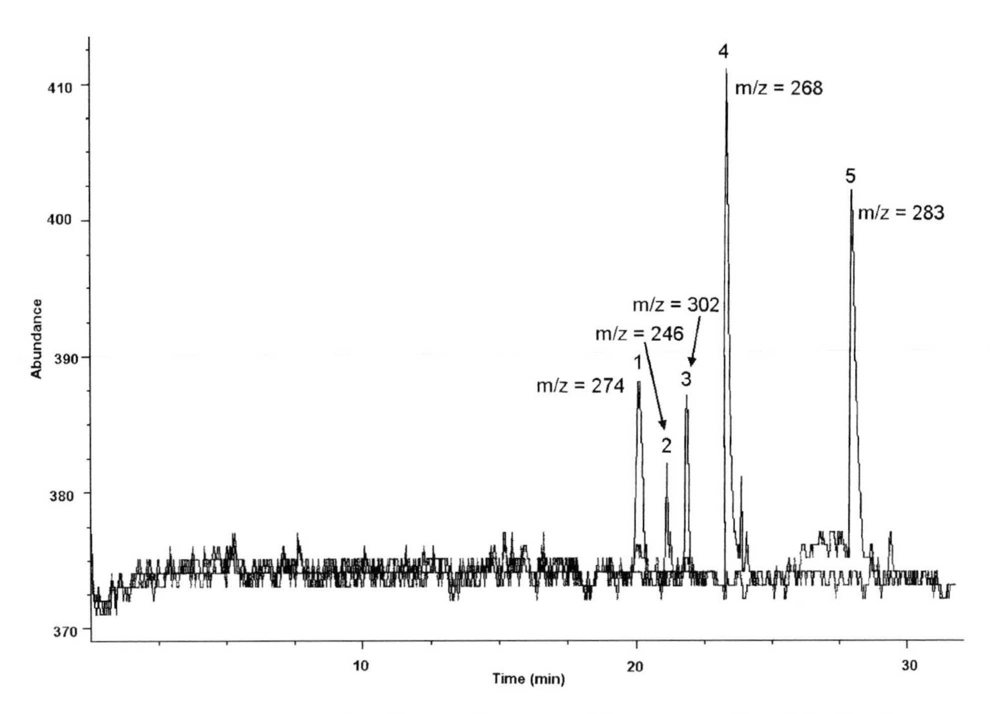

Fig. 3.10. SIM chromatogram of a mixture of hormones. The concentration of the first four compounds is 25 ng/μL, and that of medroxyprogesterone acetate is 150 ng/μL. LC Packings C18, 75 μm i.d. × 15 cm, 3 μm particle size nanocolumn; flow rate 250 nL/μL; injection volume 60 nL. Linear gradient from 100% H_2O to 100% CH_3CN in 30 min. (1) 19-nortestosterone; (2) testosterone; (3) 17-methyltestosterone; (4) diethylstilbestrol; (5) medroxyprogesterone acetate.

positive- and negative-ion detection mode with two different chromatographic runs, each one with an appropriate mobile phase, not considering the added complexity of the MS/MS detection for a complete analyte characterization.

Another interesting application is the analysis of fragrances and other perfumery compounds that are not completely amenable to GC and give some problem of uneven response when detected in ESI. Fig. 3.12 shows a scan profile of 36 substances at a concentration of 1 mg/mL, almost completely separated by the column. For all of them it was possible to record interpretable mass spectra as well as for three interferences present in the original mixture. The analysis was performed in gradient conditions starting from 0% CH_3CN in H_20 to 100% CH_3CN in 32 min at a flow rate of 200 nL/min. Mobile phase was added with 0.1% formic acid.

Another aspect in which Direct-EI excels over ESI is when very complex matrices are analyzed. These kinds of samples enclose various interferences that can alter signal response of the analyte. Under ESI conditions, where the ionization takes place in a liquid phase, this "matrix effect" can induce either signal suppression or enhancement with poor or unreliable quantitative results. The gas-phase ionization of Direct-EI is a strictly individual process where the analyte is not influenced by other co-eluted substances present in

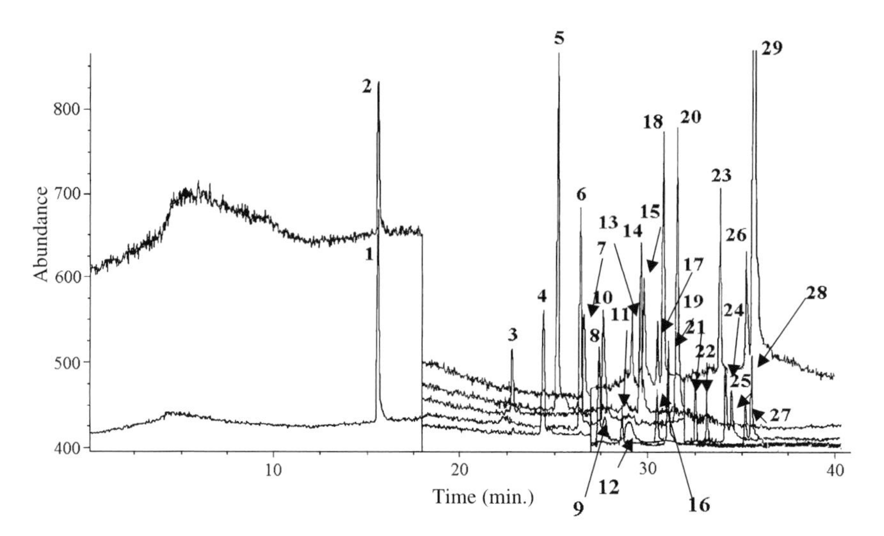

Fig. 3.11. SIM chromatogram of a mixture of endocrine-disrupting compounds with a concentration range of 0.084 to 4.2 ng/μL. LC Packings C18, 75 μm i.d. × 15 cm, 3-μm particle size nanocolumn; flow rate 200 nL/μL; injection volume 600 nL. Linear gradient from 100% H_2O to 100% CH_3CN in 40 min. (1) methomyl; (2) caffeine; (3) aldicarb; (4) cyanazine; (5) dimethyphthalate; (6) atrazine; (7) carbaryl; (8) bisphenol A; (9) 17-β-estradiol; (10) testosterone; (11) propazine; (12) 17-α-ethynilestradiol; (13) 4-*tert*-butylphenol; (14) 4-*sec*-butylphenol; (15) diethylstilbestrol; (16) terbutrin; (17) 4-*tert*-amylphenol; (18) 2-*tert*-butylphenol; (19) 4-cumylphenol; (20) naphthalene; (21) lindane; (22) mestranol; (23) 4-*tert*-octylphenol; (24) anthracene; (25) phenanthrene; (26) 4-*n*-nonylphenol; (27) fluoranthene; (28) pyrene; (29) 4-*n*-octylphenol.

a complex matrix. As a consequence of this, ion generation is not influenced by other solutes dissolved in the mobile phase and the typical ion patterns are expected.

3.5 CONCLUSION

Direct-EI interface fills effectively the gap that exists between GC–MS and API-based LC–MS. This interface generates high-quality, library-searchable, electron ionization spectra of molecules that are too large or too thermally unstable for GC–MS, but that can produce a typical EI spectrum. Direct-EI is a robust and stable platform where a "direct", simple approach combines with high operational flexibility and a large extent of information. It offers a very good overall performance, scarcely influenced by compound structure and mobile-phase composition. In addition, the reduced flow-rate allows the introduction of nonvolatile modifiers for added chromatographic choice and freedom. Its simple and straightforward mechanism permits an easy conversion, in theory, of any GC–MS system on the market, using the same MS software but widely expanding the overall analysis capability. It shows its limit when the molecular weight increases or when EI cannot compete in terms of super-low detection limits for specific analytes. When compared to API techniques, direct-EI shines for signal stability and superior confidence

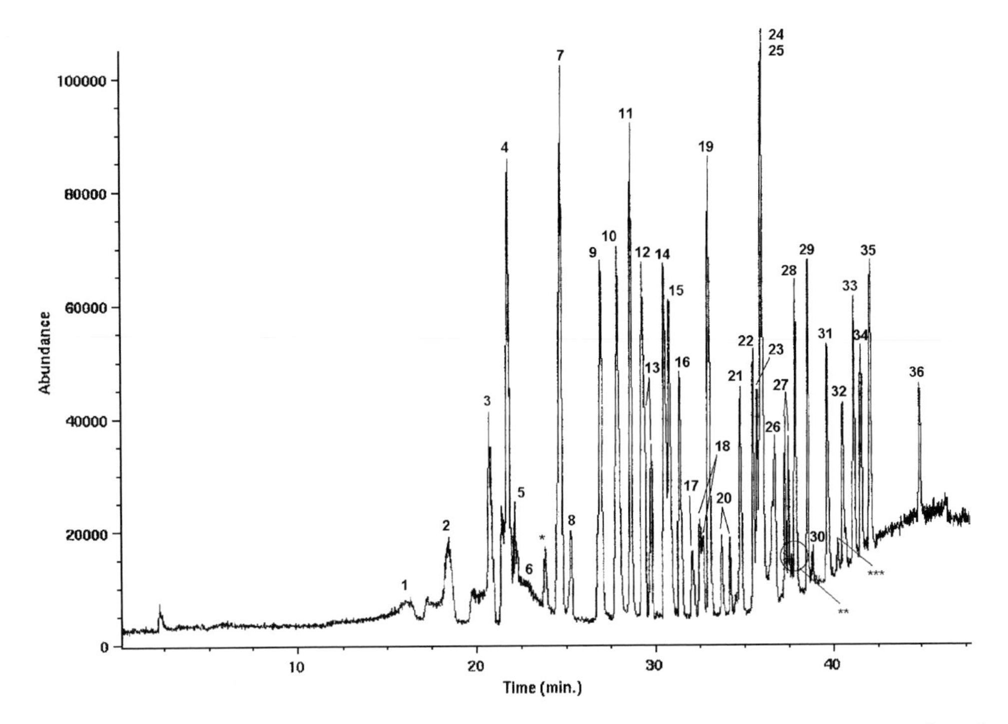

Fig. 3.12. Full-scan chromatogram of a mixture of 36 perfumery compounds at a concentration of 1 mg/mL. The gradient used for the separation of the standard mixture was 0% CH_3CN in H_2O to 100% CH_3CN in 32 min at a flow rate of 200 nL/min. Mobile phase was added with 0.1% formic acid. (1) ethyl acetoacetate; (2) benzyl alcohol; (3) fructone; (4) phenethylol; (5) pipol; (6) firlone; (7) cinnamic alcohol; (8) florol; (9) methyl anthranilate; (10) indol; (11) benzyl acetate; (12) viridine; (13) stemone; (14) camphor; (15) jasmine; (16) linalol; (17) pelargodicnal; (18) myroxyde; (19) paradisone; (20) geranyl nitrile; (21) corps UA8174; (22) corps UA123; (23) pipol isobutyrate; (24) butylcinnamic aldehyde; (25) firsantol; (26) lilial; (27) iralia; (28) nirvanol; (29) iffocimene; (30) violettyne; (31) cetalox; (32) helvetolide; (33) ambrettolide; (34) muscone; (35) cedramber; (36) isopropyl myristate. The peaks indicated by *, **, and *** are impurities that were not identified, but their mass spectra were recorded.

in quantitative determinations. The absence of signal suppression-enhancement phenomena, due to matrix effects or co-eluted interferences, increases reliability and simplifies method development.

REFERENCES

1 R.C. Willoughby and R.F. Browner, *Anal. Chem.*, 56 (1984) 2625–2632.

2 P. Ausloos, C.L. Clifton, S.G. Lias, A.I. Mikaya, S.E. Stein, D.V. Tchekhovskoi, O.D. Sparkman, V. Zaikin and D. Zhu, *J. Am. Soc. Mass Spectrom.*, 10 (1999) 287–299.

3 C.E. Kientz, A.G. Huist, A.L. De Jong and E.R.J. Wils, *Anal. Chem.*, 68 (1996) 675–681.

4 R. Dijkstra, B.L.M. Van Baar, C.E. Kientz, W.M.A. Niessen and U.A. Th. Brinkman, *Rapid Commun. Mass Spectrom.*, 12 (1998) 5–10.
5 A. Amirav and O. Granot, *J. Am. Soc. Mass Spectrom.*, 11 (2000) 587–591.
6 A. Cappiello and F. Bruner, *Anal. Chem.*, 65 (1993) 1281–1287.
7 A. Cappiello, G. Famiglini and F. Bruner, *Anal. Chem.*, 66 (1994) 1416–1423.
8 A. Cappiello and G. Famiglini, *Anal. Chem.*, 67 (1995) 412–419.
9 A. Cappiello, G. Famiglini, F. Mangani and B. Tirillini, *J. Am. Soc. Mass Spectrom.*, 6 (1995) 132–139.
10 A. Cappiello, G. Famiglini and B. Tirillini, *Chromatographia*, 40 (1995) 411–416.
11 A. Cappiello, G. Famiglini, P. Palma, A. Berloni and F. Bruner, *Environ. Sci. Technol.*, 29 (1995) 2295–2300.
12 A. Cappiello, *Mass Spectrom. Rev.*, 15 (1996) 283–296.
13 A. Cappiello, G. Famiglini, A. Lombardozzi, A. Massari and G.G. Vadalà, *J. Am. Soc. Mass Spectrom.*, 7 (1996) 753–758.
14 A. Cappiello and G. Famiglini, *J. Am. Soc. Mass Spectrom.*, 9 (1998) 993–1001.
15 A. Cappiello, G. Famiglini, F. Mangani, M. Careri, P. Lombardi and C. Mucchino, *J. Chromatogr.*, 855 (1999) 515–527.
16 A. Cappiello, M. Balogh, G. Famiglini, F. Mangani and P. Palma, *Anal. Chem.*, 72 (2000) 3841–3846.
17 A. Cappiello, G. Famiglini, L. Rossi and M. Magnani, *Anal. Chem.*, 69 (1997) 5136–5141.
18 A. Cappiello, G. Famiglini, F. Mangani, S. Angelino and M.C. Gennaro, *Environ. Sci. Technol.*, 33 (1999) 3905–3910.
19 A. Cappiello, G. Famiglini, F. Mangani and P. Palma, *Mass. Spectrom. Rev.*, 20 (2001) 88–104.
20 A. Cappiello, G. Famiglini, F. Mangani and P. Palma, *J. Am. Soc. Mass Spectrom.*, 13 (2001) 265–273.
21 A. Cappiello, G. Famiglini, P. Palma, and F. Mangani, *Anal. Chem.*, 74 (2002) 3547–3554.
22 A. Cappiello, G. Famiglini and P. Palma, *Anal. Chem.*, 75 (2003) 497A–503A.
23 A. Cappiello, G. Famiglini, C. Fiorucci, F. Mangani, P. Palma and A. Siviero, *Anal. Chem.*, 75 (2003) 1173–1179.
24 A. Cappiello, G. Famiglini, F. Mangani, P. Palma and Siviero, *Anal. Chim. Acta.* 493 (2003) 125–136.

Achille Cappiello (Editor)
Advances in LC–MS Instrumentation
Journal of Chromatography Library, Vol. 72

Chapter 4

Electron ionization LC–MS with supersonic molecular beams

ORI GRANOT and AVIV AMIRAV

4.1 INTRODUCTION

Liquid chromatography–mass spectrometry (LC–MS) experienced significant growth in recent years and has become an established, widely used technique. Electrospray ionization (ESI) [1–3] is by far the most widely used LC–MS ionization method due to its superior sensitivity, robustness and extended sample molecular weight range, enabled by the formation of multiply charged ions. ESI is supplemented with atmospheric pressure chemical ionization (APCI) [4,5] which, in some cases, shows better performance with relatively small and less polar compounds. Recently, atmospheric pressure photo ionization (APPI) [6–8] was introduced, which further helps in the analysis of certain non-polar compounds that are weakly or non-ionized by ESI or APCI. However, ESI, APCI and APPI still suffer from limitations in the ionization of non-polar compounds, and are also characterized by non-uniform compound-specific response. More important, all of these atmospheric pressure ionization (API) methods are soft techniques based on ion reaction chemistries that usually produce only the protonated or deprotonated molecular ion (with or without adducts), hence requiring MS–MS or high resolution accurate MS for further analyte characterization. As a result, they are not well suited for the analysis of true unknown compounds as routinely performed in gas chromatography–mass spectrometry (GC–MS) with its automated 70 eV electron ionization (EI) library based sample identification.

Particle beam LC–MS (LC–PB–MS) [9–16], with its electron ionization, is thus far the only LC–MS method that enables library searchable EI mass spectra. This type of identification can be achieved automatically by non-experts, can be legally defensible and it contains structural information that helps in the identification of unknown compounds [17]. Thus, it has been argued that despite its lower sensitivity and smaller range of compounds, LC–PB–MS is still a vital and useful technology [16]. However, recent PB–MS research is sparse, performed mostly by Cappiello and co-workers [18–23] by the combination of capillary and nano LC with EI–MS.

While standard 70 eV EI is a powerful ionization method for unknown sample identification, it is not ideal. About 30% of the NIST library compounds either have a weak (below 1% relative abundance) or no molecular ion. This problem is further exacerbated for LC related compounds that are larger and more thermally labile than standard GC–MS compounds. Furthermore, they are usually less volatile and hence require higher EI ion-source temperatures with the consequence of further intra ion-source degradation and weaker molecular ion production. Without a molecular ion, EI based sample identification is not as trustworthy. Furthermore, for compounds that are not included in the EI libraries the absence of the molecular ion severely hampers the usability of EI mass spectra. Thus, the "ideal" ionization method should provide the informative, library searchable EI fragments combined with an enhanced molecular ion (relative to standard thermal EI), whose observation as the highest mass spectral peak should be trusted [24,25].

In a preliminary communication [26] and recent publication [27], we described a new method and instrumentation for the combination of LC–MS and EI through the use of supersonic molecular beams (SMB) as a medium for EI of vibrationally cold sample compounds (hence named Cold EI) [28–33]. The new approach of LC–(Cold-EI)–MS with SMB [26,27,34] is based on spray formation at high pressure, followed by full thermal sample particle vaporization prior to sample expansion as isolated molecules from a supersonic nozzle. This first step is similar to sample vaporization in APCI, however, instead of the next APCI step of Corona discharge for inducing chemical ionization, the sample according to our method expands from the supersonic nozzle. Then, the supersonic free jet is collimated and forms a supersonic molecular beam that contains vibrationally cold sample molecules. These molecules proceed axially along a fly-through brink type EI ion-source for obtaining Cold EI mass spectra with an enhanced molecular ion. In the preliminary report and paper [26,27], the problem of vaporization of intact thermally labile compounds was addressed through achieving fast sample vaporization followed by supersonic expansion cooling. The issue of liquid solvent load on the vacuum pumps was addressed by differential pumping. Cluster formation was practically eliminated by using a relatively large diameter and separately temperature-controlled nozzle. Vaporized solvent molecules served as the SMB carrier gas without adding another seeding gas.

In this paper we review our LC–EI–MS method with emphasis on the unique features and improved MS information content provided by Cold EI, and we describe recently obtained results.

4.2 EXPERIMENTAL – LIQUID SAMPLING EI–MS WITH SMB

The LC–MS with supersonic molecular beam (LC–SMB–MS) apparatus, which is schematically shown in Fig. 4.1, is based on our modified home-made GC–MS with supersonic molecular beam system that was previously described [29–32]. Samples were introduced using a model 1100 HPLC (Agilent Technologies, Waldbroon Germany). In most cases samples were injected with a HPLC injector (model 9725i, Rheodyne, Rohnert Park CA, USA) using a 5 μl injection loop, directly into a liquid transfer line without a separation column. A 600 μl home-made injection loop was also used for the semi-continuous feeding of liquid for the study of Cold EI mass spectra and its dependence on various

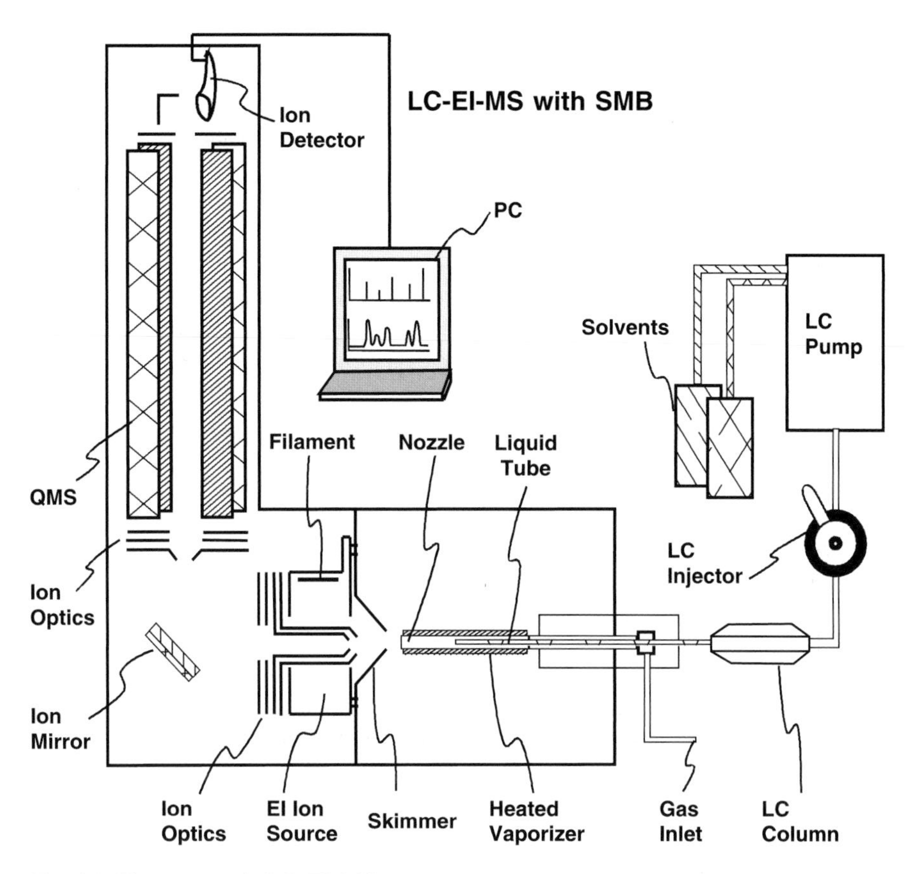

Fig. 4.1. The supersonic LC–EI–MS apparatus.

experimental parameters. The liquid transfer line was typically made of fused silica with 110 μm inner diameter (ID) and 170 μm outer diameter (OD) provided by PolyMicro Technologies (Phoenix Arizona, USA). Methanol (HPLC grade, J. T. Baker Philippsburg NJ, USA) was used as the solvent of choice at flow rates in the 50–250 μl/min range. Acetonitrile (HPLC grade, J. T. Baker Philipsburg NJ, USA), water (Chromatography grade, Merck Darmstadt, Germany), heptane, benzene, cyclohexane (all Spectroscopy grade, Merck, Darmstadt, Germany) and mixtures of methanol and water or acetonitrile were also used. Helium or nitrogen gas was provided in some cases through a 0.53 mm ID standard GC transfer line capillary at low flow rates in the 0–40 ml/min range. The reason for restricting the added gas flow rate was that our system is based on full direct discharge of the vaporized HPLC liquid flow into the first nozzle vacuum chamber and thus each one ml/min added gas flow reduced the upper acceptable liquid flow rate by about one μl/min.

The sample and its solvent were vaporized inside the soft thermal vaporization nozzle (STVN) chamber that is described below. The gaseous mixture of solvent, sample and helium or nitrogen (if added) expanded through the supersonic nozzle into the first vacuum chamber that was pumped by a turbo molecular pump having 250 l/s pumping speed

(Navigator 301, Varian Inc. Torino, Italy). The SMB was collimated by a skimmer with 0.8 mm diameter. This pumping capability and skimmer diameter are identical to what we have in our supersonic GC–MS system [33,35]. After skimming, the collimated SMB entered a second vacuum chamber pumped by two Balzers 63 mm diffusion pumps with 135 l/s pumping speed each and total net pumping speed of 210 l/s (after considering the gate valve's gas conductivity). The two diffusion pumps were backed by the nozzle chamber turbo molecular pump which was backed by a single 200 l/min Edwards RV12 rotary pump (Edwards, Crawley, UK).

The SMB sample compounds were ionized in our home-made dual cage design "fly-through" EI ion-source [36], and the ions were deflected 90° through an ion mirror (deflector) and analysed by an Extrel 2000 amu mass range quadrupole mass analyzer. The mass analysed ions were detected by a channeltron ion detector (DeTech, Palmer MA, USA), and the data were processed for identification and quantification using the Merlin software of Extrel and NIST 98 mass spectral library.

The "heart and soul" of our system is the STVN chamber. The STVN accepts the liquid flow from the LC or flow injection liquid transfer line tube, and converts it into supersonic molecular beams of vibrationally cold undissociated sample compounds that are amenable for EI and mass analysis.

In Fig. 4.2 the STVN used in most of our experiments is schematically pictured. The liquid sample solution enters from the liquid transfer line tubing that is connected with a union to a heated liquid transfer line for obtaining thermally assisted spray as in thermospray [37,38] and thermally assisted particle beam [39]. The thermally assisted spray is

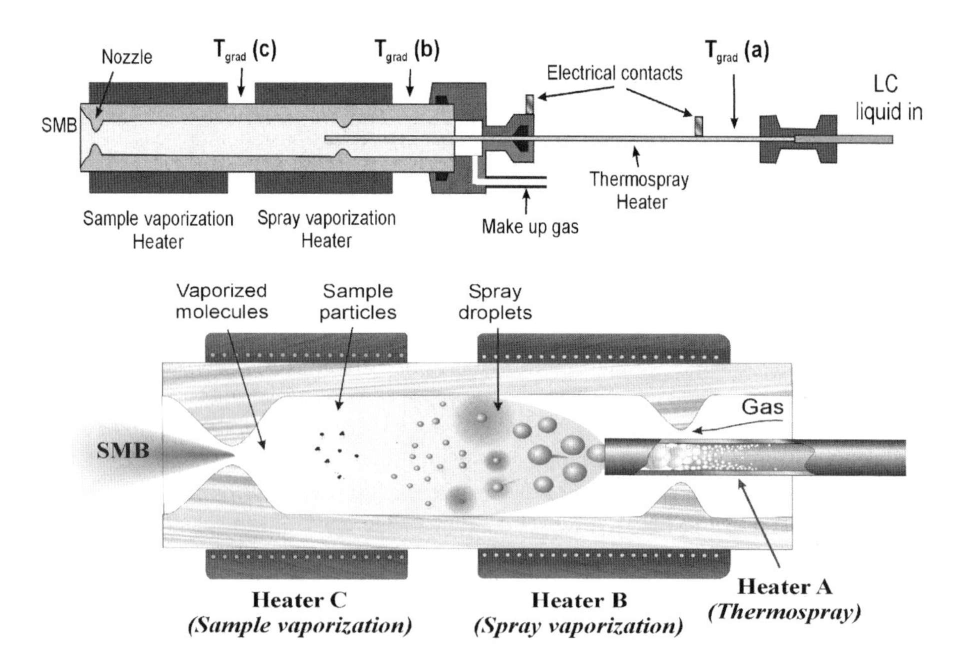

Fig. 4.2. The soft thermal vaporization nozzle (STVN).

formed via direct current heating of a stainless steel tube of 0.53 mm ID and 0.77 mm OD (Restek, Bellefonte PA, USA) containing fused silica capillary tubing inside the metal oven tubing. In this device, about 5 W power was enough to vaporize 20–60% of the solvent (while inside the capillary tubing), which was sufficient for inducing a stable spray.

The spray vaporization chamber was made from fused silica tubing with 1.4 mm ID and 3 mm OD, and it was terminated with a home-made nozzle (using glass blowing tube size reduction) of 0.3–0.35 mm diameter. The STVN chamber itself was separately heated by two heaters made from Kanthal AF wire (Kanthal Hallstahmmar Sweden) (0.42 mm diameter) coiled around alumina ceramic tubing that matched the outer diameter of the STVN fused silica chamber. The rear heater was used for spray droplet vaporization followed by particle vaporization while the front heater was used for the final soft particle vaporization into intact unionized sample molecules and to establish the optimal nozzle temperature for cluster-free vibrational cooling. The rear heater power was typically 20 W, which attained external temperatures of about 500–800°C. The rear heater length was 30 mm and the total STVN heated length with its two heaters was 50 mm. According to Covey and co-workers [40] the rear heater in APCI should preferably be heated up to 900°C, above the onset of the Leidenfrost effect [41], so that the spray droplets would self-vaporize during their approach towards impacting the oven surface. The vaporization rate should be sufficient to form self-repulsion of the spray droplets without adsorption to the walls that may lead to peak tailing and sample degradation. We similarly found that the use of a hotter rear heater is beneficial, but in our case, the use of two separate heaters was essential for the proper separate control of the supersonic nozzle temperature which is critical in order to eliminate post expansion clusters and/or induce effective sample vibrational cooling. We observed that the internal STVN walls (as well as the entire heater) were significantly cooled by the spray which could lead to STVN-induced peak tailing, depending on the specific design, and this tailing was noticeably lower with 1.4 mm ID STVN tubing than with 0.9 mm ID tubing. Another unexpected observation was that the spray droplets and sample particles experienced backward scattering, thus the helium gas was used (up to 40 ml/min) mostly to block this backward scattering and not to help the nebulization process. We attribute this backward scattering phenomenon to an increased pressure at the heated zone due to the vaporization of the liquid droplets. The nozzle was never clogged in our studies and our major reliability (robustness) problem was in the occasional clogging of the solvent delivery tubing at its outer edge that was placed a few mm inside the rear STVN heated zone plus in the thermospray heated zone. A major reason for such clogging was a sudden stop of the LC flow for various reasons (such as replacement of a solvent or a change of LC flow parameters etc.). Consequently, we used a separate thermocouple located at the thermospray heated tube for controlling a circuit breaker in case the LC liquid flow stopped and as a result the thermospray heater temperature exceeded a certain predetermined temperature.

The nozzle diameter is an important parameter that is linked to the useful liquid flow range. It must be sufficiently large to eliminate post expansion cluster formation [42], yet small enough to enable effective vibrational cooling. Cluster formation of the sample with the solvent molecules depends on three body collisions, hence on P^3D, while the vibrational cooling depends on PD, where P is the pressure behind the nozzle and D is the nozzle diameter. Thus, for a given liquid solution flow rate, the doubling of the nozzle diameter leads to the reduction of the vaporized solvent pressure by a factor of 4, hence to the reduction of cluster formation by a factor of 32. Meanwhile, such doubling of the nozzle diameter had only a minor factor of two penalty on the efficiency of vibrational cooling, that was far superior to that of pure helium in view of methanol being a heavier molecule without velocity slip [43,44]. As a result, our STVN possesses three separately heated thermal zones and three temperature gradients between them and these temperature gradients must be appropriately included and considered in the STVN design. We note that the temperature gradients between the heated zones and liquid transfer line was also important in preventing premature solvent vaporization that could cause sample condensation on relatively cool surfaces and eventually to clogging of the transfer tube. Similarly a fourth radial temperature gradient between the heated tube and inner solvent delivery tube was also important to suppress the clogging of the solvent delivery tubing.

We estimate that the thermally assisted spray was formed in the STVN chamber at about 0.1 atm with a methanol flow rate of 120 μl/min. Under these conditions, appropriate vibrational cooling was achieved combined with practical elimination of clusters of sample compounds with solvent molecules.

Initially, we assumed that achieving the fastest possible sample vaporization would lead to the softest sample vaporization due to minimizing the time during which the sample can degrade [26]. However, in view of our recent results with experiments aimed at improving the analysis of thermally labile compounds in GC–MS, we came to the opposite conclusion that it is better to employ slower vaporization at cooler conditions since temperature has a greater effect than time on the promotion of thermal degradation [45]. As a result, we now feel that the use of larger diameter STVN chamber is preferable and certainly it is more robust in terms of requiring less frequent internal cleaning.

Our LC–EI–MS approach can be perceived as a synthesis of a few current and past approaches, but in fact, it is a combination of a few of these approaches plus its own unique ingredients. The initial step of spray formation relates to thermospray [37,38] or to the ThermaBeam Particle Beam system of Willoughby and co-workers [39], but unlike thermospray our spray is only thermally assisted, formed at non-vacuum, relatively high pressure and without the step of thermospray ionization. Unlike the ThermaBeam [39] and all other Particle Beam systems, our approach is characterized by complete sample particle vaporization prior to its supersonic expansion. Our approach seems more akin to APCI in that it involves full sample vaporization at relatively high pressures, but unlike typical APCI, our spray is fully thermally assisted without using any gas for pneumatic nebulization. Furthermore, no high pressure CI is involved as our Cold EI is a full in-vacuum ionization method. On the other hand, to a first approximation our ability to handle thermally labile compounds is similar to that of APCI since both methods share similar high pressure full thermal sample vaporization. However, in contrast to APCI, our Cold EI can ionize all the sample compounds that were vaporized, regardless of their polarity, and no ion molecule reaction (collision free molecular beam) can interfere in the obtained mass spectra. While our approach uses EI as the ionization method, our EI is Cold EI (in contrast to the particle beam methods) with the ionization of sample compounds while they are vibrationally cold in the SMB, without any scattering from interior surfaces of the EI ion-source. Consequently, our EI ion-source is a unique fly-through dual cage EI ion-source aimed at obtaining Cold EI mass spectra without vacuum background [36] and with enhanced molecular ion and mass spectral information.

4.3 RESULTS – COLD EI MASS SPECTRA

In Fig. 4.3, the Cold EI mass spectrum of corticosterone in methanol solution is shown in the upper trace, and is compared with the standard NIST 98 EI library mass spectrum shown in the lower trace. Note the similarity of the library mass spectrum to that obtained

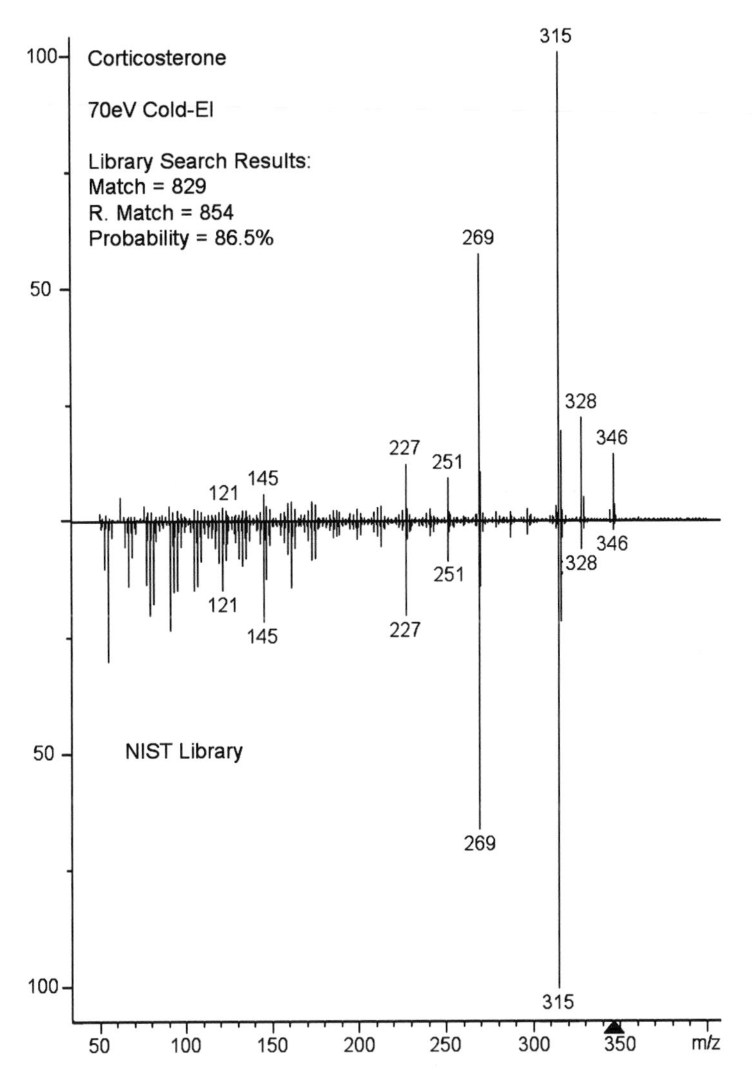

Fig. 4.3. A comparison of Cold EI mass spectrum of corticosterone obtained with the supersonic LC–EI–MS system and its fitted NIST library mass spectrum, including the NIST library matching factors and probability of identification. Note the enhanced molecular ion ($m/z = 346$) exhibited. Corticosterone was flow injected using 200 μl/min methanol solvent flow rate with 1 ng/μl corticosterone sample concentration.

with the SMB apparatus. All the major high mass ions of *m/z* 227, 251, 269 and 315 are with practically identical relative intensity and thus good library search results are enabled with NIST library matching factor of 829, reversed matching factor of 854 and 86.5% confidence level (probability) in the corticosterone identification. In addition, the molecular ion at $m/z = 346$ is now clearly observed while it is practically missing in the library (very small in the shown mass spectrum and absent in the other three replicate mass spectra). The relative abundance of the high-mass ion at $m/z = 328$ is also enhanced. On the other hand, low mass fragments are suppressed and the overall effect of vibrational cooling on the appearance of the Cold EI mass spectrum is of shifting intensity from low mass fragments to high mass fragments and in particular to the molecular ion. Overall, the obtained NIST matching factors are high but usually are not as high as obtained with standard EI. However, the probability factors and confidence level in the identification is actually higher with Cold EI as found in the analysis of many pesticides [24]. These higher probabilities and superior confidence level in sample identification emerge from three main reasons: (a) The ion-source temperature is typically higher in standard EI, such as with particle beam LC–MS, than the ion-source temperature used to obtain the NIST mass spectra, since in real analyses the ion-source temperature must be raised to prevent peak tailing of the least volatile sample compound and maintain ion-source cleanliness. Thus, at increased ion-source temperatures the degree of molecular ion fragmentation is increased due to greater intra-ion vibrational energy content. As a result, while in Cold EI the molecular ion is enhanced, in standard EI it is suppressed in comparison with its relative abundance in the NIST library, resulting in reduced identification probabilities with standard EI compared with Cold EI; (b) thermally labile compounds such as corticosterone can degrade on the hot ion-source surface in particle beam LC–MS or GC–MS; (c) the availability of the molecular ion is of critical importance for correct sample identification with high confidence level. For example, in the NIST hit list other candidates appear after corticosterone, such as corticosteroneacetate (molecular weight of 388 amu), which is listed as number two with 10.8% probability, since its standard EI mass spectrum is almost identical to that of corticosterone. The reason for this is that the MS of both compounds exhibit no molecular ion while having exactly the same fragments (and structure) after the molecular ion fragmentation of $m/z = 315$ and even a small $m/z = 328$. This observation is typical in a series of homologous compounds and well known for large aliphatic hydrocarbons that all show mostly $m/z = 43$, 57, 71 and 85 fragments. Consequently, having both a clear molecular ion peak at $m/z = 346$ and lack of a molecular ion peak at $m/z = 388$ unambiguously indicates that the sample compound is corticosterone and not corticosteroneacetate. Similarly, several other candidates in the NIST hit list are eliminated resulting in almost 100% probability that the sample is corticosterone or one of its isomers (1.2% probability). Thus, Fig. 4.3 demonstrates the usefulness of the SMB-liquid sampling-MS approach, in both the analysis of a thermally labile compound as well as in showing the benefit of a mass spectrum which shows a combination of enhanced molecular ion and standard fragments for correct sample identification.

While EI is very powerful in automated sample identification via the use of 70 eV EI libraries, LC–MS with ESI finds growing use of accurate mass measurements such as with time of flight mass analyzers for the provision of elemental formula tables with declining order of deviation from the measured mass. While this method of accurate mass measurements requires expensive MS instrumentation, it provides additional information that can help in sample identification.

Our approach of LC–EI–MS with SMB is unique in the provision of compatibility with an alternative method for obtaining elemental formulas that is based on accurate measurement of isotope abundances [46]. It is well known and established that the relative abundances of the various isotopomers (molecular ion peaks with different isotopes) can provide accurate elemental formulas information [47–49]. While such isotope abundance analysis (IAA) method is documented in the literature for many years, it is currently basically ignored and unused due to experimental difficulties in the accurate measurement of isotope abundances. The reasons for this are that over 30% of the samples do not show a molecular ion in EI while with the rest of the compounds self chemical ionization and vacuum background can severely interfere with the accuracy of isotope abundance measurements. However, with our fly-through ion-source, vacuum background is significantly reduced [27], the molecular ion is significantly enhanced and the collision-free SMB conditions eliminate self CI ion–molecule reactions. Thus, only cluster chemical ionization can distort the experimentally obtained isotope abundance.

While cluster CI can help in the confirmation of molecular weight identity [25], it can be a detrimental aspect to isotope abundance analysis that could serve for a further independent elucidation of the sample empirical formula. The ability to obtain Cold EI mass spectra with clean and informative isotope abundance patterns, free from any cluster CI effects is demonstrated in Fig. 4.4. A few molecules were investigated and in Fig. 4.4 the relative (to the lowest mass molecular ion) isotope abundances of the molecular ion isotopomers of phenothiazine and CAS# 2082-79-3 compound (molecular weight of $m/z = 530$) are given, and compared with the calculated isotope abundances using the NIST isotope calculator. Very good fits are observed between the experimental and calculated results, demonstrating the practical elimination of residual cluster CI effects as well as vacuum background interference. While cluster CI can affect the isotope abundance peak height distribution, it can be suppressed by using a relatively low solvent flow rate (120 μl methanol in Fig. 4.4) or higher nozzle temperature. In addition, one can use solvents such as acetonitrile, or others that do not have labile hydrogen atoms (non-alcohols) and with these solvents cluster CI cannot produce protonated molecular ions, thus its adverse effect on isotope abundance analysis is eliminated. We note that the observation of such "clean" isotope abundance patterns of the "true" (non-protonated) molecular ion is unique to Cold EI.

A novel IAA method and software was recently developed at our laboratory [46]. It can automatically convert experimental mass spectral data into elemental formulas information and/or significantly improve the confidence level in library identification [46]. Our novel IAA method and software automatically confirms or rejects NIST library identification. In case of confirmation, sample identification is unambiguous in view of its confirmation by an independent set of data and method. In case of rejection, IAA independently provides a list of elemental formulas with declining order of matching to the experimental data, in similarity to costly accurate mass measurements. However, unlike with accurate mass our LC–EI–MS can benefit from the combination of two methods for sample identification that provides ultimate confidence level in the identification of a broad range of compounds, with low cost single quadrupole MS instrumentation.

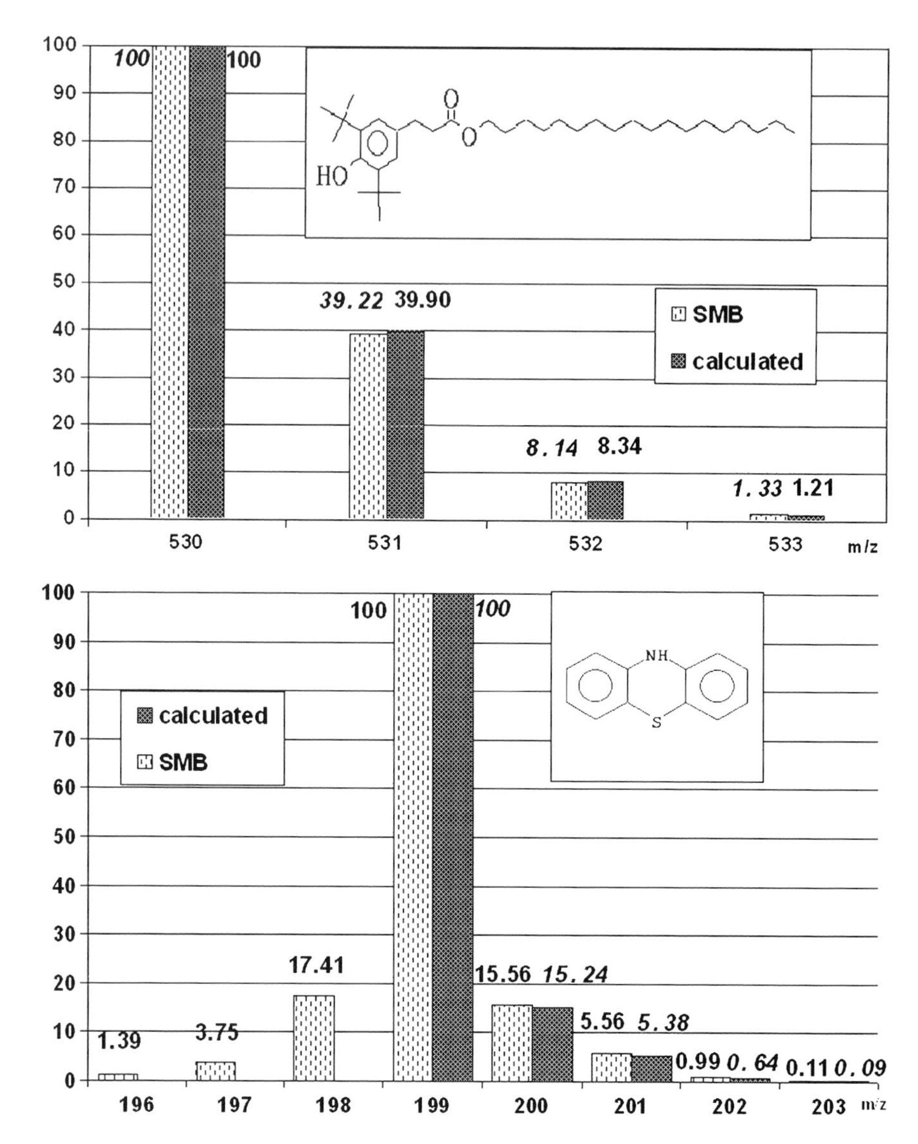

Fig. 4.4. Relative isotope abundances obtained with Cold EI for Phenothiazine and $C_{35}H_{63}O_3$ (CAS# 2082-79-3) and its favourable comparison with the calculated values using the NIST isotope calculator.

In Fig. 4.5 the Cold EI mass spectra of the indicated six drugs, carbamate pesticides and non-polar compounds (reserpine MW = 608, scopolamine MW = 303, aldicarb MW = 190, methomyl MW = 162, pyrene MW = 202 and CAS# 2082-79-3 MW = 530) are shown with rich and informative fragmentation patterns, as expected from EI. Good matching to the NIST library mass spectra plus enhanced molecular ions were obtained for these compounds.

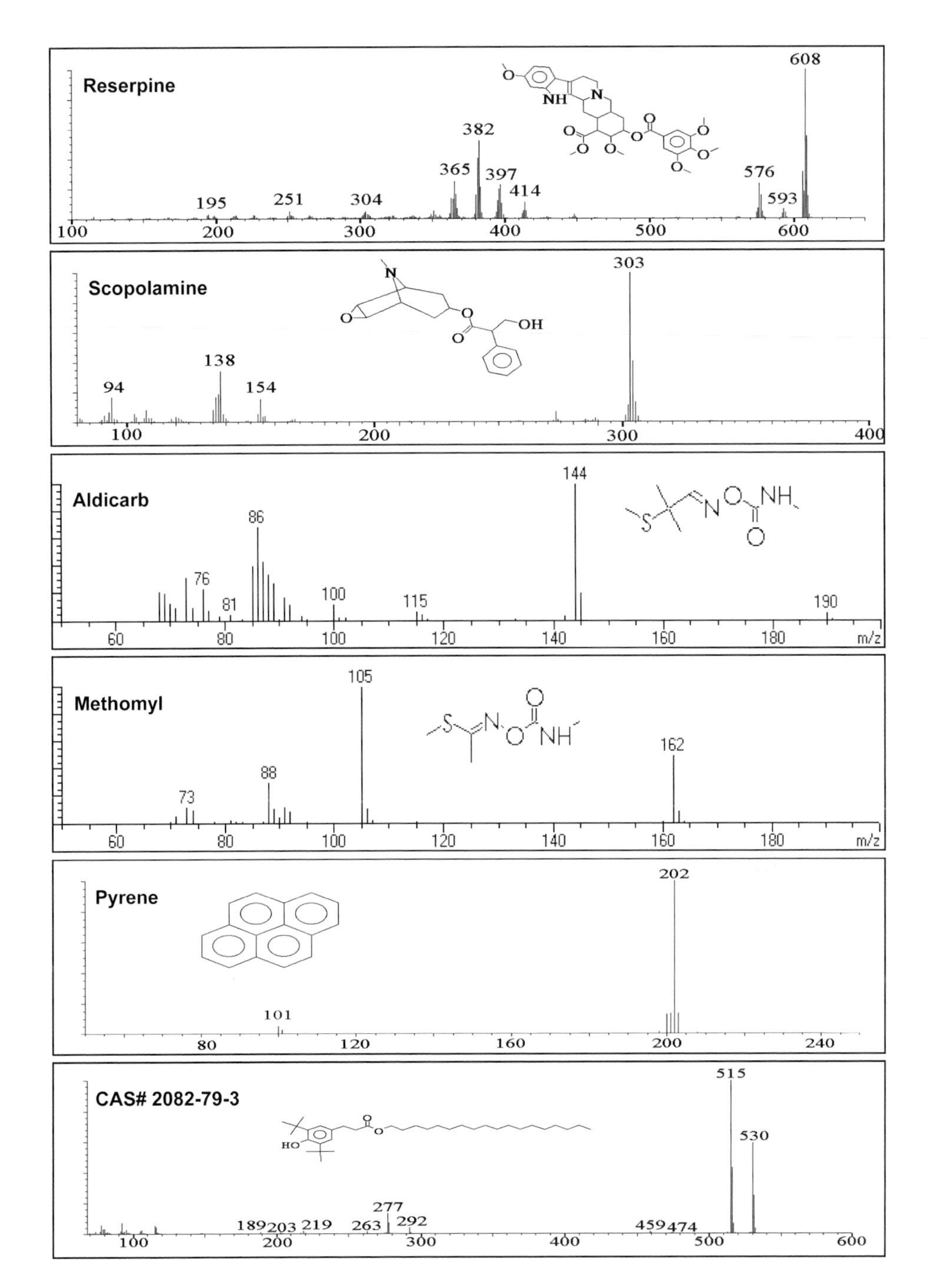

Fig. 4.5. Cold EI mass spectra of the indicated drugs, carbamate pesticides and non-polar compounds. Enhanced molecular ion peaks are observed together with the standard library searchable EI fragments. Samples were flow injected using 120 μl/min methanol solvent flow rate and 100 ng/μl sample concentrations.

A relatively important attribute of EI as an ionization method is that unlike APCI or ESI, its ionization efficiency is uniform and independent of the sample compound polarity. The Cold EI mass spectra of pyrene and CAS# 2082-79-3 are shown in Fig. 4.5 and additional examples including benzene and hexadecane were shown in reference [27]. While pyrene, for example, is easy to analyse by GC–MS, its LC–MS analysis with ESI or APCI is very hard/practically impossible. We found that in our system, pyrene can be easily ionized and analysed at low ppb concentrations, and this similarly applies to all other non-polar volatile and semi-volatile compounds. The CAS# 2082-79-3 compound was given to us as a challenging example of a compound that is hard to analyse by ESI or APCI [50], yet for us it was as expected just another easy compound.

The EI feature of approximately compound independent uniform ionization efficiency is very important for LC–MS since it provides an immediate quantitative estimate of unknown compounds without their identification, which is essential with ESI or APCI for the performance of sample calibration. Thus, having such semi-quantitative response enables the exclusive identification of compounds that are above a certain concentration level while ignoring all other compounds that are at concentrations below a certain predetermined level.

4.4 RESULTS – SELECTED APPLICATIONS

Triacetonetriperoxide (TATP) is an explosive that was recently widely used by terrorist groups such as Palestinians and others since it is easy to prepare from easily available materials such as acetone. Since it does not contain the NO_2 group it is hard to detect by a few standard explosive analyzers. TATP is a thermally labile explosive that tends to degrade in its GC–MS analysis and when it elutes its standard EI–MS is void of molecular ion and it is characterized only by low mass fragments [51]. TATP cannot be ionized by ESI but it can be analysed by LC–MS with APCI when ammonium ions are added to enable the formation of its molecular ion with ammonium adduct [52]. Relatively high detection limit was reported for TATP with APCI of 0.8 ng in single ion monitoring (SIM) [52]. We investigated the analysis of high explosives including TATP by GC–MS with supersonic molecular beams [45] and TATP showed two isomers (conformers) with dominant Cold EI molecular ion for both. In Fig. 4.6, the LC–EI–MS analysis of TATP is demonstrated and the two TATP conformers are properly separated. An informative Cold EI mass spectrum was obtained for both conformers (shown for the dominant conformer in the insert of Fig. 4.6), which is characterized by a dominant $m/z = 222$ molecular ion. The Cold EI–MS of the two conformers is different in the intensity ratio of the molecular ion to the low mass fragment ions. The details of the TATP separation are given in the caption of Fig. 4.6.

Polystyrene is commercially available as various mixtures of size selected oligomers with a given average molecular weight. As hydrocarbons polystyrenes are not easy to analyse by ESI or APCI, Jones *et al.* successfully analysed polystyrene oligomers by particle beam LC–MS [39]. However, while they showed molecular ions all the way to $(\text{polystyrene})_{18}$ with $m/z = 1928–1938$, the relative abundance of the molecular ion of $(\text{polystyrene})_{13}$ at $m/z = 1411$ was already below 0.1%.

In Fig. 4.7 we show our LC–EI–MS analysis of a mixture of polystyrenes with average molecular weight of 760 amu. The main observation is that with our Cold EI a dominant

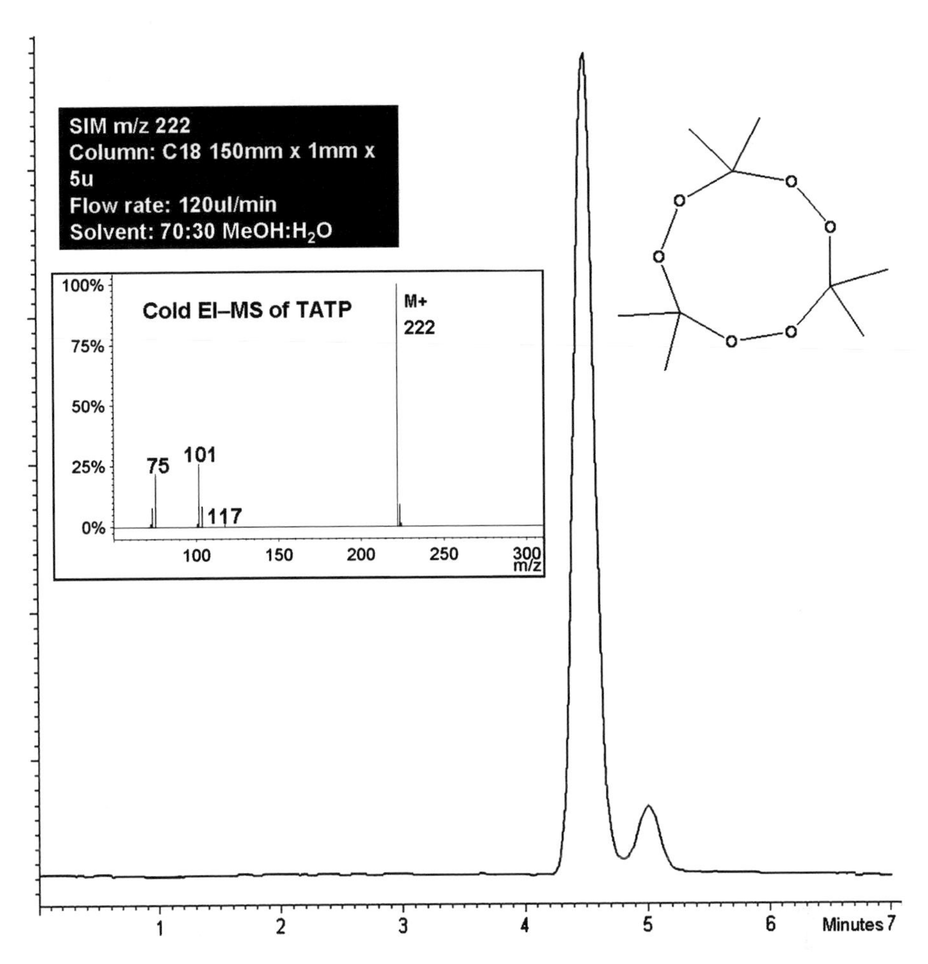

Fig. 4.6. LC–EI–MS analysis of TATP. SIM was used at $m/z = 222$ of the molecular ion. Column used was 150 mm long, 1 mm ID with 5μ C18 particles. MeOH:H_2O (70:30) solvent was used at 120 μl/min flow rate. 9 ppm sample in methanol was injected from a 5 μl loop. The mass spectrum was taken from another full scan injection.

molecular ion is provided to all these oligomers as shown in the two inserts of MS. Furthermore, the ionization efficiency of Cold EI is approximately oligomer size independent. While the ionization cross section is approximately linearly increased with the molecular weight and similarly the efficiency of our jet separation is slightly increased with the sample size, the ion to electron conversion yield of the ion detector and particularly the quadrupole mass analyzer transmission is reduced with mass. In the overall we obtain an approximately size independent response up to about 1200 amu, above which we have a response reduction with mass. However, these response factors can be calibrated. In Fig. 4.8 we show a comparison of LC chromatogram obtained with a UV detector and flow injection analysis obtained with our Cold EI–MS. A close similarity is observed between the various

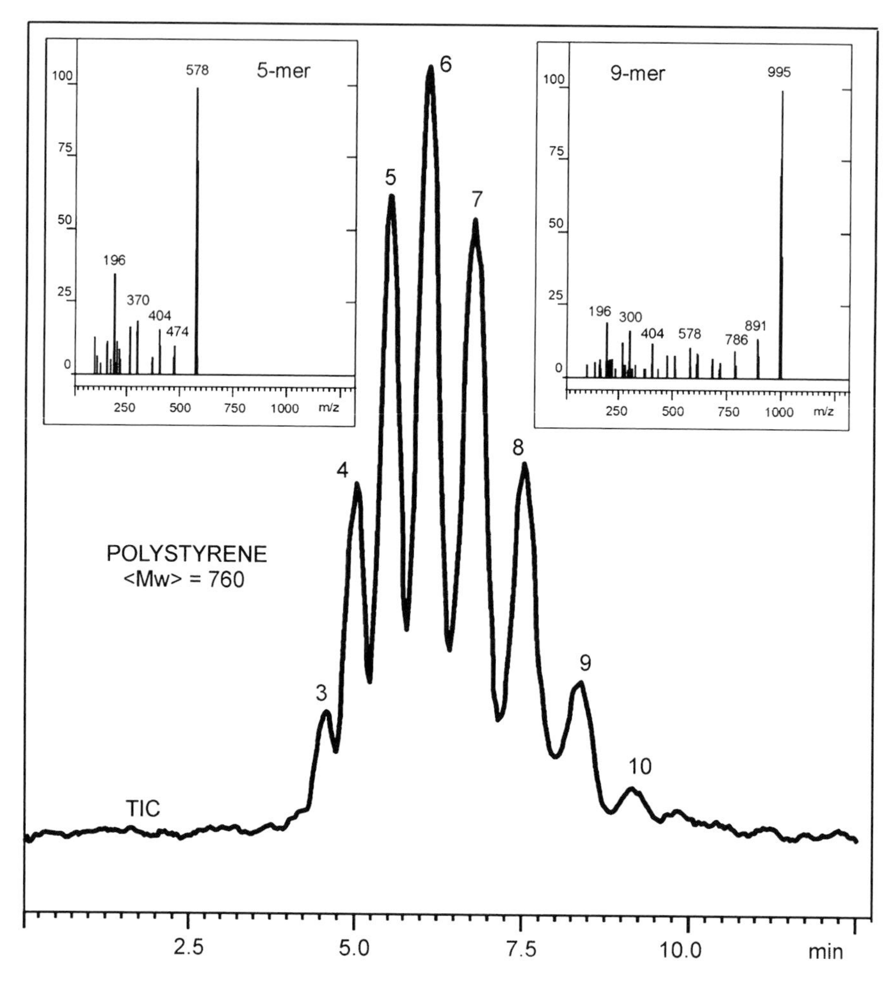

Fig. 4.7. LC–EI–MS analysis of polystyrene oligomers mixture with average molecular weight of 760 amu. The column used was 150 mm long, 2 mm ID with 5μ C18 particles. Solvent was THT:MeOH mixture, initially 20:80 with 10 min gradient to 60:40 and 5 min constant mixture later on. Flow rate was 120 μl/min, and 1000 ppm total mixture concentration sample was injected from a 5 μl loop.

molecular weight peaks and LC chromatographic peaks and thus, a one LC–UV chromatogram can serve for the calibration of the response of our flow injection Cold EI–MS system and after that polystyrene can be quickly analysed by its distribution of molecular weights in the mass spectra as demonstrated in Fig. 4.8. Consequently, we consider our uniform EI response as an important feature which in some cases, as for polystyrenes, can serve for a fast flow injection analysis while using the separation power of the MS without the need for additional separation of the LC in order to save analysis time. Fig. 4.8 demonstrate how flow-injection Cold EI–MS nicely reproduces the LC separation, demonstrating our ability to analyse this mixture with the much faster method of flow injection analysis.

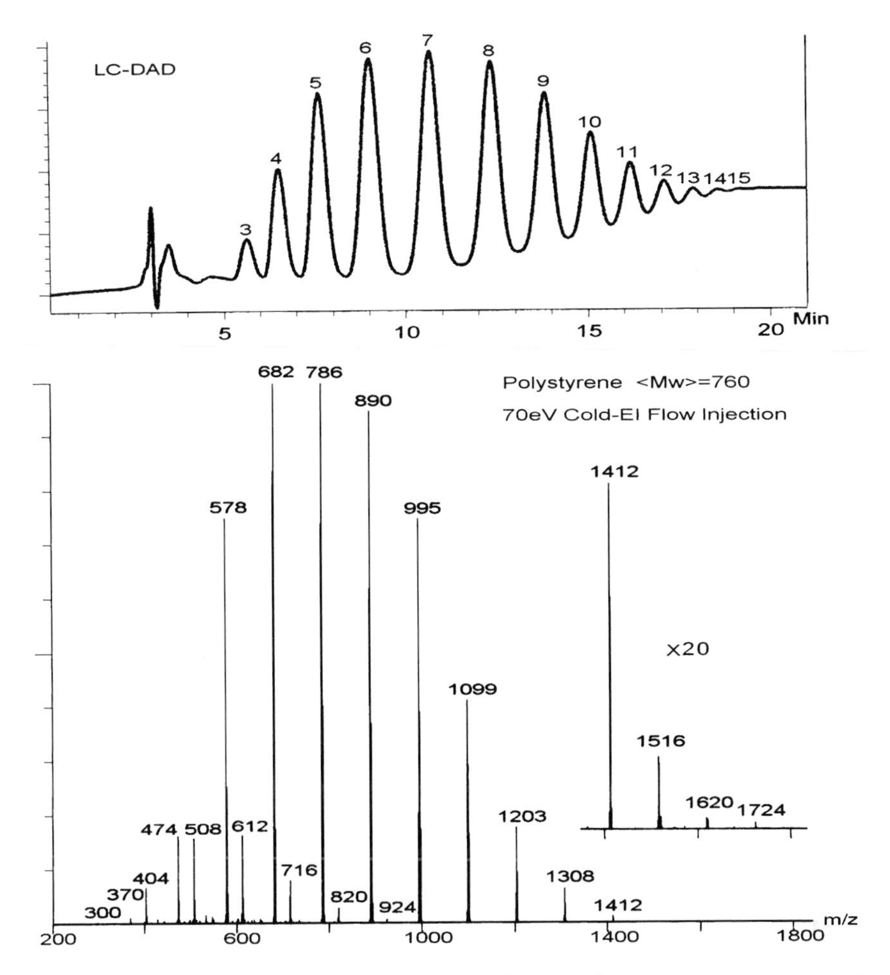

Fig. 4.8. Flow injection analysis of polystyrene oligomer mixture with average molecular weight of 760 amu and its comparison with LC analysis using diode array detector at 320 nm. 120 μl/min methanol flow rate was used. 1000 ppm sample was injected from a 600 μl loop.

In Fig. 4.8 we demonstrate polystyrene oligomers molecular weight (MW) peaks up to and over 1700 amu ($m/z = 1724$ of (polystyrene)$_{16}$). More recently we clearly observed also the $m/z = 1828$ of (polystyrene)$_{17}$. To the best of our knowledge this is the highest MW compound ever analysed by EI–MS that showed a molecular ion. For example, the largest compound in the NIST library has MW of 1674 but all the NIST library compounds with MW > 1500 do not have a molecular ion, thus their ionization as the intact sample compound cannot be trusted. Jones *et al.* showed below 1% molecular ion for m/z-1411 (polystyrene)$_{13}$ in their particle beam EI.

We note that this result of significantly enhanced molecular ion in Cold EI of large hydrocarbons is to be expected since the heat capacity of these large molecules is very large and thus, once they are cooled, the electron induced vibrational excitation is dissipated among many vibrational degrees of freedom and after intra-ion vibrational energy redistribution its average vibrational energy per chemical bond is low and cannot lead to molecular ion fragmentation. A similar observation was made with GC–MS with SMB, in which dominant molecular ions and even a doubly charged ions were observed for long chain linear hydrocarbons in the C_{46}–C_{84} range [33]. In our Cold EI with LC–MS the vibrational cooling with methanol is much more effective than with helium in GC–MS with SMB in view of reduced velocity slip effects [43,44] and van der Waals cluster induced vibrational cooling [53]. Thus, the observation that polystyrene ionization by Cold EI produces dominant molecular ions is to be expected.

4.5 DISCUSSION AND POTENTIAL ADVANTAGES

A new method and instrumentation were described for obtaining high quality, informative, library searchable EI mass spectra for a relatively broad range of samples in liquids. The presented method provides a few major advantages:

1. A library mass spectral search is enabled, unlike with APCI and ESI. This is an important advantageous feature, shared with Particle beam LC–MS and "Direct EI" that enables fast, automated, legally defensible and reliable molecular identification.
2. Extended mass spectral information is revealed including enhanced molecular ion, accurate isotope abundance pattern and extended isomer and structural mass spectral information, which is superior to that provided by particle beam LC–MS and any other standard thermal EI methods.
3. A broad range of compounds are amenable for analysis with sample vaporization softness similar to APCI but without restrictions on vaporized sample polarity. Thus, Cold EI can serve for the analysis of the full range of non-polar molecules as well as polar compounds in the same sample.
4. Mass spectral deconvolution software could further enable fast LC–MS analysis combined with automated sample identification under coelution conditions [27,54–56].

In addition to the above main advantages, our LC–EI–MS approach is characterized by a broad linear dynamic range (LDR) of over four orders of magnitude [27], in contrast to particle beam. Our current detection limit is about 1–2 pg depending on the compound used [27]. Our current signal level with octafluoronaphthalene (OFN) is ~100 ions per picogram in SIM mode on the molecular ion ($m/z = 272$) while the noise is mostly mass independent background noise at a rate of 50 counts/s. Active research is currently being made on improving this detection limit and on the combination of our approach with a new Varian 1200 LC–MS system.

We feel that our Cold EI can serve as a useful LC–MS method and excel especially in the analysis of unknown compounds or for the universal analysis of a broad range of target compounds such as pesticides (and other chemical contaminants) in agricultural products. Although it can be designed as an add-on unit to existing LC–ESI–MS systems, such combination is not easy in view of ESI being an atmospheric pressure ionization method while EI is inherently an in-vacuum ionization method. Our vision is that Cold EI with its unique benefits will serve as a dedicated low cost LC–EI–MS system, with high performance for the analysis of small molecules. We feel that it can be characterized by relatively low cost in view of the following facts: (a) EI does not require costly MS–MS for sample identification, thus it can operate with a single quadrupole; (b) If ESI is not included, the mass range of the MS can be restricted to 1200 amu; (c) no nitrogen generator is required as the system gas requirement is relatively very low or none; (d) the same MS and vacuum system can also serve as a supersonic GC–MS [33,35] with the replacement of only one flange.

It is worthwhile to briefly compare our Cold-EI with SMB approach to the Direct EI approach of Cappiello and co-workers [19–23]. In direct EI, the LC and MS interface is as simple as it gets, and the MS of a GC–MS can be used with only minor added EI ion-source and interface modifications. However, the LC column flow rate is restricted to below 1 μl/min which necessitates the use of packed capillary nano LC columns with its resulting lower sample loading ability (and potential loss of concentration sensitivity). We feel that the major advantage of our Cold EI approach is the higher quality mass spectra obtained which adequately compensates for the added instrumental complexity.

Cold EI requires further investigation for its establishment as a useful technique and in order to explore further applications and expose additional unique features continuous research towards these goals is underway.

ACKNOWLEDGEMENTS

This research was supported by a grant from the Israel Science Foundation founded by the Israel Academy of Sciences and Humanities, a Research Grant Award No. US-3500-03 from BARD, the United States–Israel Binational Agricultural Research and Development Fund and by the James Franck Center for Laser Matter Interaction Research.

REFERENCES

1 M. Yamashita and J.B. Fenn, *J. Phys. Chem.*, 88 (1984) 4451–4459.
2 C.M. Whitehouse, R.N. Dreyer, M. Yamashita and J.B. Fenn, *Anal. Chem.*, 57 (1985) 675–679.
3 R.B. Cole (Ed.), *Electrospray Ionization Mass Spectrometry*, John Wiley & Sons, New York, 1997.
4 D.I. Carroll, I. Dzidic, E.C. Horning and R.N. Stillwell, *Appl. Spectrosc. Rev.*, 17 (1981) 337–406.
5 D.I. Carroll, I. Dzidic, R.N. Stillwell, M.G. Horning and E.C. Horning, *Anal. Chem.*, 46 (1974) 706–710.
6 D.B. Robb, T.R. Covey and A.P. Bruins, *Anal. Chem.*, 72 (2000) 3653–3659.
7 J.A. Syage and M.D. Evans, *Spectroscopy*, 16 (2001) 14–21.
8 A. Raffaelli and A. Saba, *Mass. Spectrom. Rev.*, 22 (2003) 318–331.
9 R.C. Willoughby and R.F. Browner, *Anal. Chem.*, 56 (1984) 2626–2631.
10 P.C. Winkler, D.D. Perkins, W.K. Williams and R.F. Browner, *Anal. Chem.*, 60 (1988) 489–493.
11 R.D. Voyksner, C.S. Smith and P.C. Knox, *Biomed. Environ. Mass. Spectrom.*, 19 (1990) 523–534.

12 T.D. Behymer, T.A. Bellar and W.L. Budde, *Anal. Chem.*, 62 (1990) 1686–1690.
13 A.P. Tinke, R.A.M. van der Hoeven, W.M.A. Niessen, U.R. Tjaden and J. van der Greef, *J. Chromatogr.*, 554 (1991)119–124.
14 A. Cappiello and F. Bruner, *Anal. Chem.*, 65 (1993) 1281–1287.
15 A. Cappiello, G. Famiglini and F. Bruner, *Anal. Chem.*, 66 (1994) 3970–3976.
16 A. Cappiello and G. Famiglini, *J. Am. Soc. Mass. Spectrom.*, 9 (1998) 993–1001.
17 F.W. McLafferty, *Interpretation of Mass Spectra, University Science Books*, Mill Valley, CA. 1980.
18 A. Cappiello, M. Balogh, G. Famiglini, F. Mangani and P. Palma, *Anal. Chem.*, 72 (2000) 3841–3848.
19 A. Cappiello, G. Famiglini, P. Palma and F. Mangani, *Anal. Chem.*, 74 (2002) 3547–3554.
20 A. Cappiello, G. Famiglini, F. Mangani and P. Palma, *J. Am. Soc. Mass. Spectrom.*, 13 (2002) 265–273.
21 A. Cappiello, G. Famiglini, F. Mangani, P. Palma and A. Siviero, *Anal. Chim. Acta.*, 493 (2003) 125–136.
22 A. Cappiello, G. Famiglini and P. Palma, *Anal. Chem.*, 75 (2003) 1173–1179.
23 A. Cappiello, G. Famiglini, P. Palma and A. Siviero, *Mass. Spectrom. Rev.*, 24 (2005) 978–989.
24 M. Kochman, A. Gordin, P. Goldshlag, S.J. Lehotay and A. Amirav, *J. Chromatogr. A*, 974 (2002) 185–212.
25 A.B. Fialkov and A. Amirav, *Rapid. Com. Mass. Spectrom.*, 17 (2003) 1326–1338.
26 A. Amirav and O. Granot, *J. Am. Soc. Mass. Spectrom.*, 11 (2000) 587–591.
27 O. Granot and A. Amirav, *Int. J. Mass. Spectrom.*, 244 (2005) 15–28.
28 A. Amirav, *Org. Mass. Spectrom.*, 26 (1991) 1–17.
29 A. Amirav and A. Danon, *Int. J. Mass. Spectrom. & Ion. Proc.*, 97 (1990) 107–113.
30 S. Dagan and A. Amirav, *J. Am. Soc. Mass. Spectrom.*, 6 (1995) 120–131.
31 S. Dagan and A. Amirav, *J. Am. Soc. Mass. Spectrom.*, 7 (1996) 737–752.
32 A. Amirav, S. Dagan, T. Shahar, N. Tzanani and S.B. Wainhaus, "Fast GC–MS with Supersonic Molecular Beams." A Review Chapter in the Book *Advances in Mass Spectrometry*, Volume 14, Elsevier, Amsterdam, 1998, pp. 529–562.
33 A.B. Fialkov, U. Steiner, L. Jones and A. Amirav, *Int. J. Mass. Spectrom.*, 251 (2006) 47–58.
34 A. Amirav, Israel patent application number 127217, submitted November 1998 (approved). USA, Japan and European Patent Applications, November 1999.
35 A. Amirav, A. Gordin and N. Tzanani, *Rapid. Com. Mass. Spectrom.*, 15 (2001) 811–820.
36 A. Amirav, A.B. Fialkov and A. Gordin, *Rev. Sci. Instrum.*, 73 (2002) 2872–2876.
37 C.R. Blakley and M.L. Vestal, *Anal. Chem.*, 55 (1983) 750–754.
38 W.M.A. Niessen and J. van der Greef, *Liquid Chromatography–Mass Spectrometry Principles and Application*, Volume 7, Marcel Dekker, New York, 1992, pp. 157–202.
39 G.G. Jones, R.E. Pauls and R.C. Willoughby, *Anal. Chem.*, 63 (1991) 460–463.
40 T.R. Covey, R. Jong, H. Javahari, C. Liu, C. Thomson, S. Potyrala and Y. LeBlanc, in a poster presentation titled "Design Optimization of APCI Instrumentation" presented at the 19th Montreux Symposium on LC–MS, Montreux, Switzerland, November 2002.
41 J.D. Bernardin, I. Mudawar and J. Heat, *Trans.*, 121 (1999) 894–903.
42 S. Dagan and A. Amirav, *J. Am. Soc. Mass. Spectrom.*, 7 (1996) 550–558.
43 A. Amirav, U. Even and J. Jortner, *Chem. Phys.*, 51 (1980) 31–42.
44 E. Kolodney and A. Amirav, *Chem. Phys.*, 82 (1983) 269–284.
45 A.B. Fialkov and A. Amirav, *J. Chromatogr. A*, 991 (2003) 217–240.
46 T. Alon and A. Amirav, *Rapid Com. Mass. Spectrom.*, 20 (2006) 2579–2588.
47 K.F. Blom, *Org. Mass Spectrom.*, 23 (1988) 194–203.
48 V.A. Bogdanov, A.V. Vorob'ev, I. Savel'ev Yu and R.N. Shchelokov, *Bulletin Acad. Sci. USSR Division Chem. Sci.*, 40 (1991) 336–338.

49 A. Tenhosaari, *Org. Mass. Spectrom.*, 23 (1988) 236–239.
50 M.J. Karlsson, AstraZeneca, Lund, Sweden, Private communication and kind donation of several samples.
51 D. Muller, A. Levy, R. Shelef, S. Abramovich-Bar, D. Sonenfeld and T. Tamiri, *J. Forensic Sci.*, 49 (2004) 935–938.
52 X.M. Xu, A.M. van de Craats, E.M. Kok and P.C.A.M. de Bruyn, *J. Forensic Sci.*, 49 (2004) 1230–1236.
53 A. Amirav, U. Even and J. Jortner, *J. Chem. Phys.*, 75 (1981) 2489–2512.
54 S.E. Stein, *J. Am. Soc. Mass. Spectrom.*, 10 (1999) 770–781.
55 J.M. Halket, A. Przyborowska, S.E. Stein, E.G. Mallard, S. Down and R.A. Chalmers, *Rapid. Com. Mass. Spectrom.*, 13 (1999) 279–284.
56 S. Dagan, *J. Chromatogr. A*, 868 (2000) 229–247.

Achille Cappiello (Editor)
Advances in LC–MS Instrumentation
Journal of Chromatography Library, Vol. 72

Chapter 5

A case for congruent multiple ionization modes in atmospheric pressure ionization mass spectrometry

MICHAEL P. BALOGH

5.1 INTRODUCTION

This chapter examines the underlying science supporting the conditions and requirements for co-residence of discrete ionization phenomena. Electrospray ionization (ESI), atmospheric pressure chemical ionization (APCI), and atmospheric pressure photoionization (APPI) are included along with related efforts to provide indirect and non-chromatographic means of increasing analytical range for mass spectrometric analysis.

Practical and theoretical knowledge culled from the past decade in developing and applying atmospheric ionization techniques has taught a few lessons:

- Liquid, while a necessary intermediary in transporting analytes and creating ions, is also a vestigial remnant of the condensed phase liquid chromatographic process. The resulting vapor needs to be reduced but not at the expense of performance.
- The efficiency of capturing ions following creation is seen to always be in need of improvement.
- The compromise between optimal ionization producing settings for disparate modes – voltage, gas and heat – must be minimized to analyze a greater diversity of analytes from a single solution in the same instrument conformation.

Sometimes, even aided by theory, we find ourselves bound by the limits imposed by the empirical data we observe. This chapter briefly reviews the primary ionization techniques, as they relate to the three application goals above. Following that is an evaluation of combined and dual ionization strategies and the shared requirements for an atmospheric ionization source.

5.2 ATMOSPHERIC PRESSURE IONIZATION: ELECTROSPRAY

Work performed in the early 1990s has just recently yielded a more unified understanding of the mechanisms involved in ion production. The transfer mechanism of potential from the liquid to the analyte to create ions has been a topic of controversy. A balanced discussion of the two most popular theories can be found in Cole [1]. Dole [2] first proposed a charge residue mechanism in 1968 whereby as the droplet evaporates the surface tension is ultimately unable to oppose the repulsive forces from the imposed charge and a droplet explodes into many smaller droplets. These coulombic fissions occur until droplets containing a single analyte ion remain. When solvent from the last droplet evaporates a gas-phase ion is formed. Iribarne and Thomson [3] proposed an alternative model – the ion evaporation mechanism in 1976. In this theory too, small droplets are formed by coulombic fission. Here, however the electric field strength at the surface of the droplet is thought to be high enough to make it energetically favorable for solvated ions to leave the droplet surface and transfer directly into the gas-phase. It is possible, as Cole argues that the two mechanisms may actually work in concert. The charge residue mechanism dominates for masses higher than 3000 Da; the ion evaporation dominates for lower masses.

The ESI probe or device is typically a conductive capillary, usually of stainless steel, in which the condensed phase liquid-containing analyte is carried. The capillary is captured in a larger-bore cylinder, which allows a concentric flow of nitrogen to be applied to the aerosol at the outlet. The added shear forces of the gas and heat transmitted from adjacent supplemental devices and/or direct heating of the gas itself enhances the aerosol droplet formation. When the solvent leaves the ESI probe through the creation of the Taylor cone, it carries a net ionic charge. The liquid, introduced typically as the result of a liquid chromatographic process, enters the ESI probe in a state of charge balance. For ESI to be continuous, the solution must be charged by electrochemical reactions, whereby electrons are transferred between a conductive surface acting as an electrode and can lead to, among other effects, pH changes. In positive mode, it is assumed, positive droplets leave the spray and electrons are accepted by the electrode (oxidation); the reverse occurs in negative mode. The surface area of the electro-active electrode, the magnitude of the current, and the nature of the chemical species and their electrode potentials all have an effect. Overall, ESI is an efficient process. However, the activation energy varies as does the combined energy differences for individual species. In general, although the ion creation process is considered highly efficient, transferring the ions generated at atmosphere into the vacuum system, and ultimately detecting them, is not. To detect a single ion may take as many as 5000 molecules in the original condensed phase solution. The efficiency is highly dependent on the flow rate of the original solution as well as the chemical characteristics of the molecules themselves. Despite iterative and generational improvements in conserving ions once they are introduced to the instrument critical path the means of extracting ions formed at atmosphere has changed little: a larger inlet orifice complemented by a necessary increase in pumping capacity.

The solution flow rate and the current applied define limits for each droplet. Competition between molecules occurs and suppression of analytes of interest is not uncommon. When considering the prospects for combining ionization mechanisms in the same-source environment, chemical emendations that prove effective for one compound may prove inimical to others in the same mixture. The elements in common are the ability to initiate a liquid droplet aerosol that, in the presence of heat, promotes desolvation of the droplets. An excellent review of our current understanding of these mechanisms can be found in a recent article by Kebarle [4].

5.3 ATMOSPHERIC PRESSURE IONIZATION: CHEMICAL IONIZATION

APCI was not widely adopted until API technology in general was commercialized following Fenn's work in 1985 [5] (although the work was published in parallel with ESI). APCI was first introduced in 1973 by Horning [6] to analyze volatile compounds using a variety of introduction techniques, one of which was a liquid chromatography [7]. The adjunctive capability of APCI permits analytes that are not amenable to conversion to gas-phase ions by ESI, that is the less polar and more volatile ones – to be introduced to a mass spectrometer from a condensed phase or liquid stream.

Unlike ESI, APCI transfers neutral analytes into the gas-phase by vaporizing the introduced liquid in a heated gas stream. Chemical ionization relies on the transfer of charged species between a reagent ion and a target molecule to produce a target ion that can be mass analyzed. Most commonly in positive ion mode, an adduct forms between the target molecule and the small H^+ ion. The maximum number of ions capable of formation by APCI is much higher than ESI, since reagent ions are redundantly formed [8]. Yet, as with ESI, a small proportion of the ions formed are drawn through the small ion inlet orifice and into the vacuum system.

The liquid is pushed through a non-conductive (usually fused silica glass) tube around which a nebulizing gas is also directed. The resulting fine droplets subsequently collide with the inner heated wall of a tube, or probe, extending beyond the end of the non-conductive tube and are converted into the gas-phase. The desolvated analyte molecules are then ionized via chemical ionization. The reagent ions are produced by ionizing solvent molecules in a corona discharge generated by applying a high voltage ($\sim 5\,\text{kV}$) to the tip of a sharp needle.

This form of ionization is usually performed at much greater linear velocities than those provided by the high pressure liquid chromatography (HPLC) flow rates associated with electrospray (closer to 1 mL/min as opposed to decreasing the flow shown to improve electrospray performance). The ionizing potential is applied, not through the liquid as in ESI, but at the tip of a needle as a plasma or corona through which the droplets pass. In effect the mobile phase acts as an intermediary to transfer the charge to the analyte.

The aerosol exiting the non-conductive capillary tube passes through a larger-circumference, heated chamber for further desolvation before exiting the probe or device to be exposed to the APCI plasma. Unlike ESI, which produces a relatively wet aerosol, APCI requires a greater degree of desolvation before introducing the analyte-containing droplets to the plasma for charge transfer initiation. In this more traditional APCI design, initial aerosol formation and desolvation occurs via nucleate boiling or leidenfrost effect (or perhaps by a combination of both).

Nucleate boiling is more typical of contemporary designs where the heated chamber portion of the probe, extends as much as a few inches beyond the aerosol outlet. Latent heat measured by a thermocouple placed at an arbitrary point in the device can measure as much as 600°C as part of the temperature control feedback loop although the analyte encased in the aerosol droplet experiences little of this heat directly. Gas temperature measured at the outlet of such a device may be only 20% of the heat measured at input. Unlike some gases commonly used in mass spectrometry, such as helium, nitrogen has relatively low heat induction capacity and is more instrumental in providing shear forces for desolvating the droplets. The heavier nitrogen is pumped away more efficiently, allowing more latitude in vacuum design, and is generally a less expensive, easily generated commodity that is available worldwide.

The residence time on the heated surface for nucleation to occur and subsequently the time required for the droplet to travel the length of the heated chamber is relatively brief, and the "protective" liquid layer usually preserves the analyte from pyrolytic degradation. In some cases, e.g., *N*-oxide structures and some metabolites such as glucuronides and sulphates, an inherent, increased sensitivity to thermal degradation instead yields the original drug structure [9,10]. Thermal degradation can, when well characterized, be a useful analytical tool in itself [11].

Leidenfrost effects do not necessarily offer greater protection to the analyte. The Leidenfrost, or vapor film, effect achieves rapid evaporation without direct sample contact with the surface. Very high, uniform temperatures and surface finish are required to produce a vapor film, and if the proper conditions can be maintained, the vapor pressure eventually exceeds the droplet momentum toward the surface and the droplet rebounds before contact, by some estimations losing as much 50% of its volume. Few designs except those that incorporate materials capable of a high-polish, low-adsorptive finish, such as ceramics, claim this as a dominant mechanism.

5.4 ATMOSPHERIC PRESSURE IONIZATION: PHOTOIONIZATION

APPI is a relatively new development in analytical mass spectrometry. So far, it has been used to complement APCI and ESI and to extend the range of molecules that can be efficiently ionized in a liquid chromatography–mass spectrometry (LC–MS) analysis [12,13]. The basic mechanism of photoionization is $M + h\nu \rightarrow M^+ + e$. However, the ion most frequently observed by APPI is typically MH^+. This indicates involvement of a chemical reaction following the photoionization event, regardless of whether protonation occurs in the presence of reagent dopant ions or is a product of direct APPI [14,15]. In some instances, which seem to correlate with molecules of low proton affinity, the molecular ion M^+ is observed. However, it is not certain whether this process dominates at atmospheric pressure. There are two general mechanisms possible to explain the product of MH^+ from M:

$$M + R^+ \rightarrow MH^+ + R(-H)$$
(Protonation by charge carrier R^+) (5.1)

$$M^+ + S \rightarrow MH^+ + S(-H)$$
(Hydrogen atom abstraction from a protic molecule S) (5.2)

This protonation mechanism is the basis for APCI and ESI as well. Much of the work published on APPI is the product of a dual source approach, where the existing API source (usually an APCI design) generates the aerosol and a krypton, gas-filled lamp is positioned to transmit 10 eV photons at the cross-section of the aerosol in line with the ion inlet. As in APCI, the desolvation required for APPI must exceed that required for ESI, since the vapor acts as an interferent that is analogous to the high flow rate artifacts found in other forms of ionization. Photoionization has been widely used in gas chromatography applications since the 1970s. Though liquid chromatography–mass spectrometry (LC–MS) can be traced to Chen [16] in the 1980s it was not considered commercially feasible as an LC–MS application due to vapor load restrictions. Enhanced source desolvation developments and an increased need to analyze compounds not easily amenable to ESI, and those typically out of the range of APCI, like steroids and polyaromatic hydrocarbons (PAH), gave HPLC–APPI–MS a commercial impetus at about the turn of the century.

5.5 COMBINING IONIZATION MECHANISMS

5.5.1 Vapor load effects

The dominant influence on ESI behaviors is liquid physicochemical properties. Reducing liquid interference has yielded improved response linearity in mass spectrometer interfaces [17]. Through whatever mechanism ion formation proceeds, if we accept the electro-hydrodynamic theory that ions are produced when droplets are reduced to approximately a 10-micron radius, the prospect for a larger population of ions produced at any given point improves when a more homogenous population of smaller diameter droplets is initially produced. The surface properties of the liquid and the volume admitted to the mass spectrometer source enclosure determine the number of ions liberated from the liquid per unit time. Reduced surface tension and reduced flow rate are primary factors. Reducing the flow rate to the nanoliter-per-minute range has also been postulated as a means to reducing signal suppression, possibly by generating smaller, more highly charged droplets that are more tolerant of non-volatile salts [18] (Fig. 5.1).

A mass spectrometer's ion source geometry and properties related to heat propagation and gas dynamics are less easily compared between competing designs. In practice, a design is derived that suits a number of practical (often market-driven) requirements, including upper and lower liquid flow rates, tolerance to non-volatile components in the liquid, and aerosol production and maintenance. Once the API aerosol has been initiated, further desolvation must occur within the geometry of the ion source. Dispersion of droplets away from the ion inlet and incomplete droplet desolvation play a major role in ion losses. Both are controlled by fundamental gas dynamic principles, which can be studied via computational fluid dynamic models. Both are modeled analytically, which ultimately teaches us that the physical processes of gas entrainment and recirculation dominates the trajectory of the aerosols. Various equations have been derived to characterize source behaviors, for example Busman [19] for APCI space charge limitations in a spherical geometry.

A commercially viable form of ionization used in LC–MS in the early 1990s was referred to as thermospray [20]. The technique, which has since been supplanted by ESI, produced ions through desolvation of the liquid stream using the shear forces associated with largely aqueous mixtures at high flow rates (achieving high linear velocities). Heating the conducting tube or probe to just below the point where total vaporization occurred at

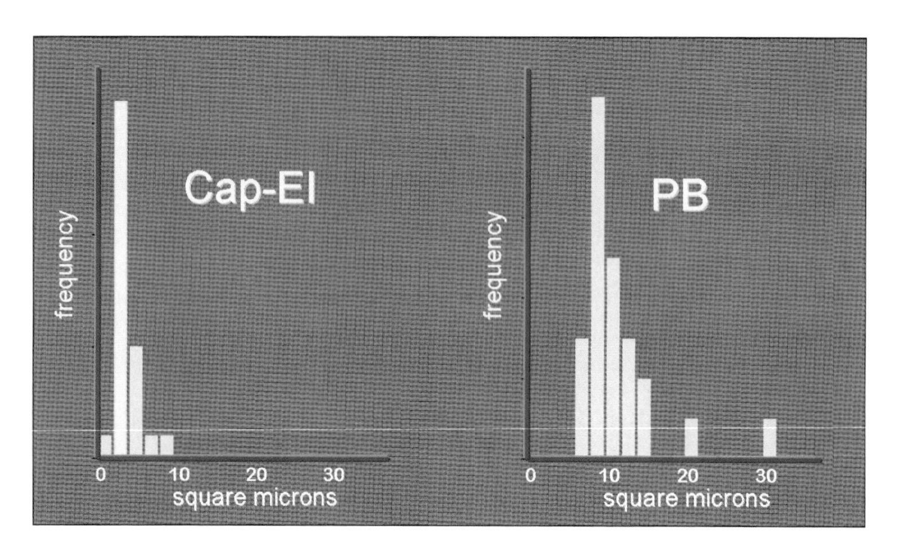

Fig. 5.1. Reduced liquid flow rates are at least one contributing factor producing improvements in response attributable to improved droplet population characteristics. (Courtesy Achille Cappiello.)

the outlet allowed final desolvation to occur in the reduced pressure region with the resulting ions produced from the droplets in position to be sampled with good efficiency. The optimal flow rate of 1 mL/min into a vacuum system required very high capacity pumping and trapping systems to remove the resulting vapor. Since this technique worked with a highly aqueous mobile phase an approximate illustration taken at atmospheric equivalent with water indicates 1 mL/min produces 1 L/min of vapor.

ESI works largely with compounds exhibiting little or no thermally achievable vapor pressure. APCI extends the utility of API instruments to enhance sensitivity for less polar and more volatile compounds where ESI is not as effective. It is not uncommon to find that some compounds produced ions in the manner of ESI without an additional form of ionization present before the development of the heated, pneumatically assisted ESI that is more common today. "Mixed mode" behaviors have been noted where, having removed any form of external ionization, such as an APCI needle when ostensibly performing APCI, resulted in thermally induced ions [21]. These ions are akin to those produced by thermospray. However they are produced at atmosphere rather than under reduced pressure (the case with thermospray) and pneumatically assisted ESI.

Publications illustrating corona- or plasma-assisted ESI employed for large molecules [22] have appeared. In recent years, however, ESI's success has generally overshadowed mixed mode's viability as a commercial entity. Instrument design, both hardware and software, has focused on the lucrative and more prevalent need.

5.6 DEVELOPING REPRODUCIBLE MECHANISM BOUNDARIES

It is not uncommon to find that adjusting solvent characteristics – that is increasing or decreasing pH to shift protonation tendencies or changing the surfactant properties of the liquid to promote the liberation of ions from the surface once formed – will promote electrospray ions from an analyte previously thought intractable in terms of the ESI mechanism of ion production.

One multimode source [23] allows separate ESI and APCI delivery to be optimized in the same source via specifically designed materials and temperature control [24]. The source is set to reestablish either full ESI or full APCI conditions after an approximately 100 ms interval following the source switchover which is defined by the time the ion signal takes to achieve 90% of the steady-state level. Theoretical measurement cycling time can only be as short as 1 s (or a switching frequency of 1 hertz) and still achieve a measuring duty cycle that approaches 90%. The design attempts to establish traditional APCI conditions in a shorter time, but the conditions remain the same: *i.e.*, high flow rate and heat for desolvation as illustrated by the presence of thermal degradation when labile compounds are used.

5.6.1 Importance of segmentation of APCI and ESI mechanisms

The importance of characterizing multiple-mode behaviors within a given mechanism is illustrated by APCI characterization work by Bajic and Bateman [25]. Analysis of mixtures containing both highly polar and moderately polar to non-polar analyte types would require different corona current settings for the optimization of analysis. Highly polar analytes can emerge from a heated nebulizer as either gas-phase ions or charged micro-droplets. In this case a higher corona current/voltage on the APCI pin while optimal for some ions in a mixture can "defocus" these ions in the vicinity of the sampling cone orifice. This leads to a decrease in ion signal and an increase in corona voltage. (Fig. 5.2 upper) shows an overlay of two separate HPLC–MS analyses of the same four analytes at two different corona current values for a standard APCI geometry. The highly polar analytes (verapamil and reserpine) are optimized at a low corona current of 0.2 μA and the moderately polar samples (corticosterone and hydroxyprogesterone) are optimized at 5 μA.

In Bajic's work, reagent ions are formed in an ancillary chamber and carried by gas flow to a reaction chamber where they mix with desolvated analyte molecules and ionize by chemical ionization. This type of API source, a dual-chamber APCI source, is similar to conventional source designs except that a collection of stable reagent ions is allowed to form at the tip. Ions are created due to the corona discharge resulting from an application of voltage at about 3 kV. Ions are then carried by the reagent gas flow through an interconnecting tunnel. There they mix with analyte molecules in the reaction chamber. The analytes then undergo gas-phase ion–molecule reactions before they exit the reaction chamber.

In the case of moderately polar analytes, the predominantly neutral analyte molecules are carried by the probe gas flow into the reaction chamber adjacent to the corona chamber. The gas flow from the corona chamber contains reagent ions, which enter the reaction chamber and ionize the analyte molecules. The analyte ions then exit the chamber with the reagent ions and nebulizer and reagent gas flows, and they enter the sampling orifice. In the case of highly polar analytes, pre-formed analyte ions that emanate from the probe reach the ion sampling orifice without influence from the corona electric field which is confined to the volume of the corona chamber. The system gas flow is such that analytes and contaminants do not directly flow past the corona tip.

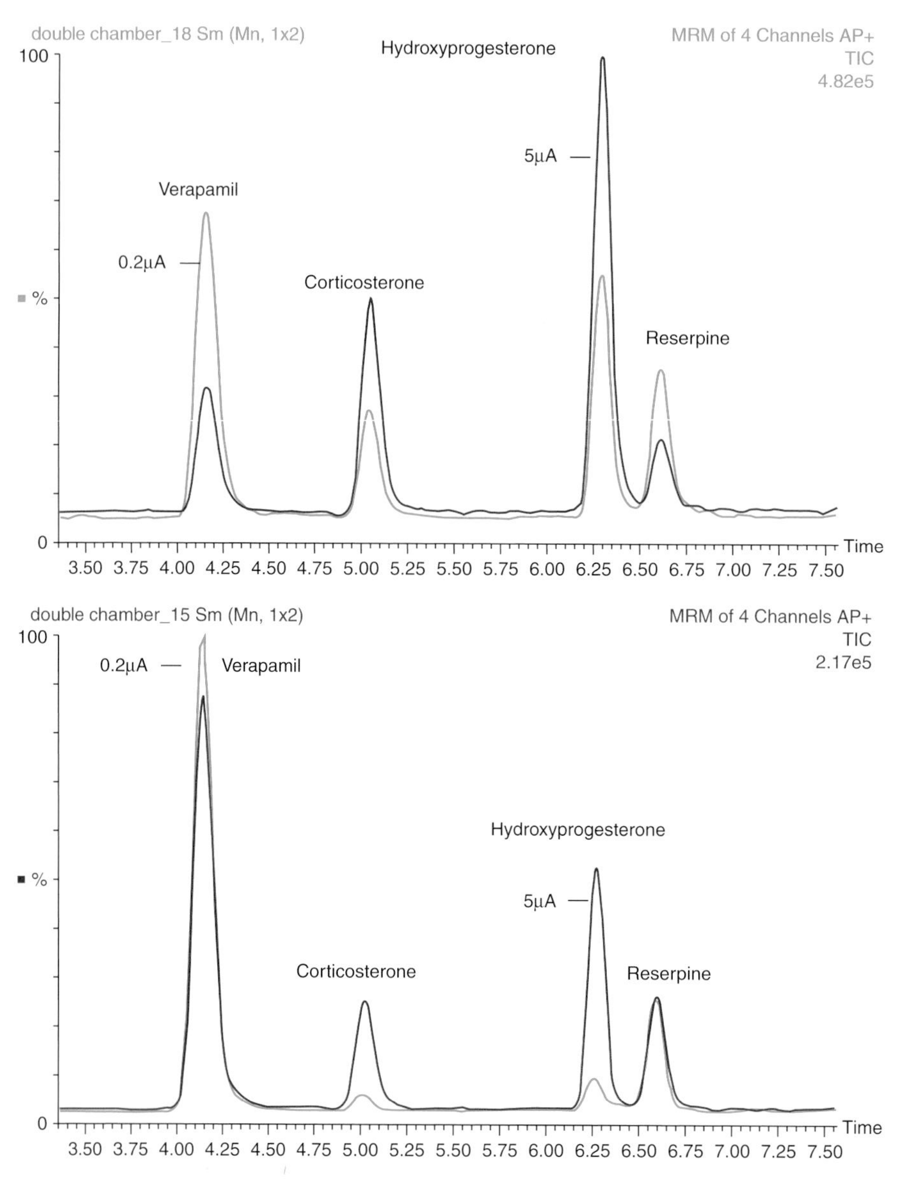

Fig. 5.2. Upper, defocusing of some compounds by a single optimal corona current setting. Lower, negligible defocusing of susceptible species while using a single optimal corona current setting in a segmented source design. (Ref. 26.)

Repeating the ion signal-versus-corona-current measurements at 0.2 and 5 μA for the four test analytes produces the responses shown in (Fig. 5.2 lower). Thus, the response for corticosterone is similar to that given by the standard geometry source. As anticipated however, the ion signals for reserpine and verapamil show no significant fall-off with increasing corona current. This would imply that the dual chamber geometry negates the effects of defocusing for highly polar compounds with a single corona current setting. In contrast, only verapamil and reserpine gave optimum sensitivity at 0.2 μA. Thus it no longer is necessary to repeat an LC–MS analysis under different experimental conditions to accommodate different analyte types [26].

Desolvation occurs in various zones within the source and its immediate environment. As described by Jarrell and Tomany [27], secondary desolvation can and does occur to differing degrees in areas in the process, including the source chamber and ion entrance, depending on the design. Coupling an appreciation of gas dynamics with desolvation requirements and being able to test the fidelity and reproducibility of ionization mechanism boundaries contributed to creating a true API multimode source.

5.7 TRUE COMBINED MECHANISM DESIGNS

The development of a high-speed ESI/APCI switching source to differentiate between ionization mechanisms was of paramount interest to answer geometry-dependent questions like how much ion residence time must be allowed when changing between ESI and APCI, and mechanistic questions like how can we be sure we are doing true APCI? Answering such questions would allow examination in real time of ion suppression. According to Kebarle and Tang [28], APCI frequently gives rise to less ion suppression than ESI. Delineating ion suppression is often done by introducing the analyte into the suspected suppressive system [29]. High-speed switching using clearly defined reproducible boundaries might allow at least partial characterization of a given chemical entity, including physicochemical properties, by characterizing its ionization mechanism. Aerosol initiation derived from the ESI process and supplemented by secondary desolvation within the ion source reduces exposure to high heat, avoiding or reducing the effects of thermal degradation.

5.8 DEVELOPMENT OF ESCi® MULTIMODE IONIZATION

The Mark II ion source design employed in the ZQ single quadrupole mass spectrometer (Waters Corporation) incorporating the existing Z-spray orthogonal design (developed in 1997 by Bajic [30]) provides two intrinsic features essential for combining and characterizing truly independent ESI and APCI mechanisms: high heat-transfer coefficients in a confined geometry that promotes efficient gas dynamics and desolvation secondary to the formation of the aerosol (Fig. 5.3).

The ZQ mass spectrometer's design provides voltage to the ESI capillary and the APCI needle simultaneously. Proof of the simultaneous utility was established by Castoro [31] in work at ASMS 2002. Subsequent work by Fischer produced a similar capability relying on IR lamps to provide the necessary additional desolvation for APCI to proceed [32]. In this similar undifferentiated process, the potential well effect continues to drive ESI ions, once formed, toward the ion inlet and mass spectrometer (as in an ESI-only design), while the remaining neutral analytes (APCI candidates) are subjected to APCI ionization further downstream. Castoro's results indicated that at the reduced flow rates typical of ESI, in this case

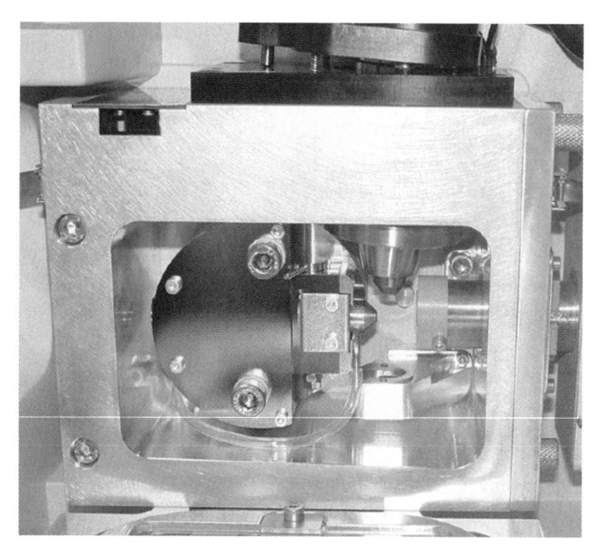

Fig. 5.3. ESCi type source (Upper, 810 cc gross calculated volume) and earlier glass housing requiring preheated nebulizing gas to perform ESI and APCI in the same source. (Lower, 1540 cc gross calculated volume.) (Courtesy Waters Corporation.)

0.1 mL/min, there is sufficient energy transfer from the heated nebulizing gas for desolvation and gas-phase ionization via an APCI mechanism (Fig. 5.4). In this study, the charged droplets do not appear to affect the APCI process. Ions formed by electro-hydrodynamic means in ESI are not seen to be affected by the APCI reagent ions.

Subsequent work in APCI development by Bajic and Bateman [25] as described earlier indicates however that more polar analytes which show both ESI and APCI response can be adversely affected (or defocused) by the increased corona current and energetic reagent ions that can be favorable to one or more other species in the mixture.

A simple set of circumstances proves the boundary and fidelity of the ionization state occurring in combined or alternating ESI and APCI [33] functions. Compounds that easily protonate in ESI mode can be added to a solution with those yielding characteristic APCI signatures. As an example, phenones, which produce radical cations only by APCI. Adjusting the duration of the internal scan delay (ISD) in alternating mode design (rather

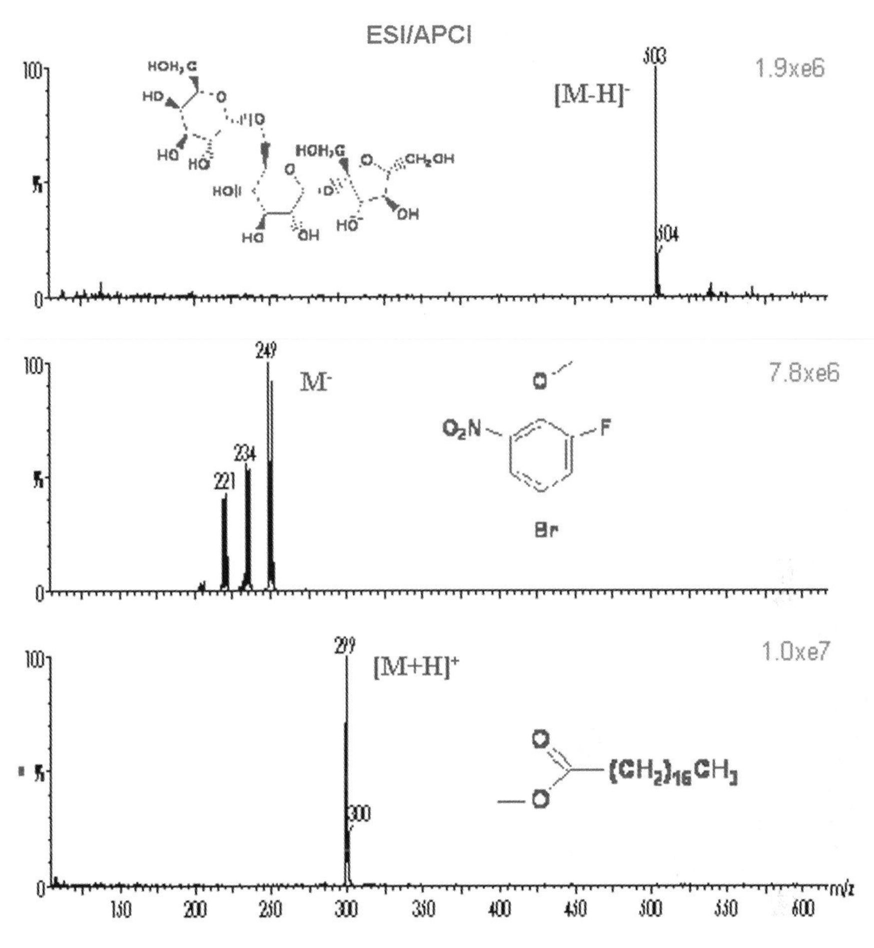

Fig. 5.4. Simultaneous ESI/APCI, using reduced flow rate ESI conditions, with alternating positive and negative detection using a conventional mass spectrometer. Upper, D-raffinose (504 Da) a polar thermally labile molecule, produced an intense deprotonated molecular ion signal, $[M-H]^-$ by ESI but not by APCI. Middle, 4-bromo-2-fluoro-6-nitroanisole and methylstearate (lower) produce an intense ion signal, M^- and a $[M+H]^+$, respectively using APCI.
The signal intensities obtained by the simultaneous ESI/APCI source are comparable to the best obtainable signals from the individual ionization technique. (Ref. 31.)

than a simultaneous mode design) as each mode is switched provides an indication of "carryover" since each channel is displayed separately.

Ultimately as many as four modes – positive and negative – in both ESI and APCI can be acquired and "mapped" against the UV signature of each using an inline UV detector. In series with MS, UV detection provides an orthogonal measurement with which to normalize response. Beers–Lambert law ensures the UV response is directly proportional to concentration and UV data points for a given wavelength are contiguous over a peak providing highly precise RSDs that are less susceptible to dispersion. MS measurement of the same eluted peak depends more on signal-to-noise response (that is, peak height versus greater noise in the signal baseline than UV) and can be subject to providing fewer data points for the same eluted band or peak. The same absolute amount can be injected and quantified under a variety of HPLC elution conditions. The same peak measured by the mass spectrometer may provide fewer data points dependent on signal-to-noise response, which is affected by dispersion of the chromatographic peak.

Various means of obtaining both ESI and APCI from the same instrument have been reported. These rely on the conditions required for desolvation using preheated nitrogen nebulizing gas flow and external switching to apply voltage to either the ESI probe and/or the APCI needle used to generate the plasma [34]. The source geometry fundamentally determines gas and heat dynamics. In this early work a Micromass Platform source – a cylindrical enclosed design that uses an inline spray, imposed a need for preheated nebulizing gas that was not needed in the later ZQ design which provided sufficient secondary desolvation as represented in the results.

5.8.1 Secondary desolvation

The cylindrical glass design gross internal volume of earlier post-Platform designs is large compared to the later ZQ source – 1540 cc in the former compared to 868 cc gross volume. In the later Premier design the internal volume is further reduced to 810 cc. The high heat-transfer efficiency of the later enclosed aluminum source housings at typical operating temperatures provided a greatly increased secondary desolvation for the aerosol before it encountered the plasma on the APCI needle (Fig. 5.5).

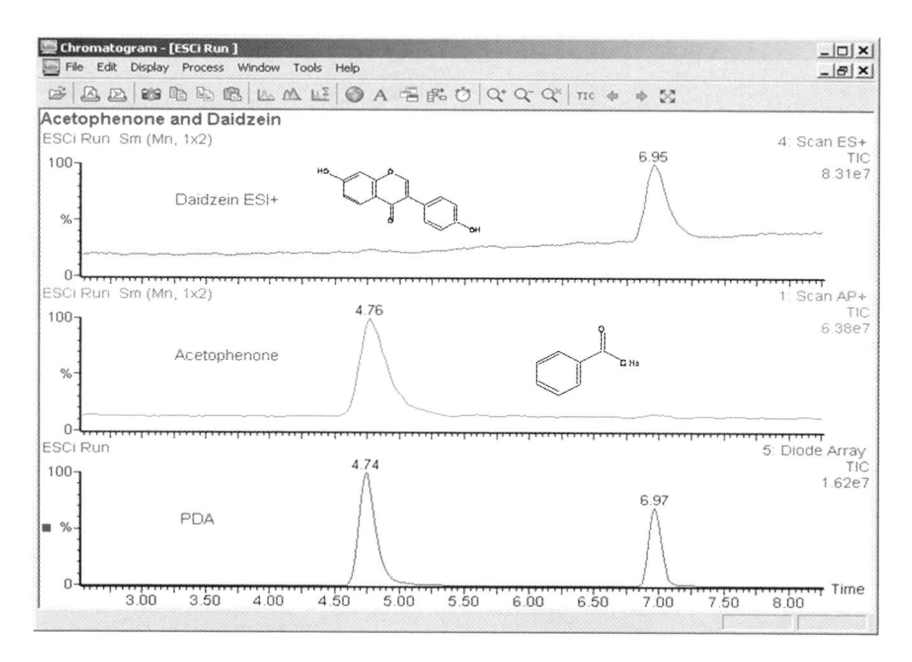

Fig. 5.5. LC–ESCi–MS characterization of an ESI/APCI candidate (daidzein) and an APCI only compound (acetophenone). Total ion chromatograms (TIC) with an accompanying UV trace. (Courtesy Mike R. Jackson, Waters Corporation)

The source of heat input to the system was unchanged from earlier designs – the source heater cartridges and the heating mantle on the ESI probe provide the same heat input to the smaller enclosure. The solvent, measured over a wide range of HPLC flow rates from 0.1 mL/min to 1.5 mL/min, was sufficiently desolvated at this point to allow true APCI mechanisms to proceed.

5.8.2 Establishing reproducible boundaries

The requirement to provide the utility for such an application required designing suitable software and a power supply as well. Designing high-speed electronics created a "same source" device capable of acquiring positive and negative ESI and APCI at millisecond intervals.

The power supply typically designed for use in mass spectrometers up to this time in, perhaps owing to the historically qualitative purpose of the instrument, required an ISD of hundreds of milliseconds to stabilize voltage before a subsequent acquisition could proceed. Thus there were no data points acquired during the ISD during which a chromatographic peak was in transition. The most arduous switching requirement placed on the design was switching polarity between positive and negative mode.

Controlling transition between modes is accomplished in MassLynx (Waters Corporation) software using bits for voltage output (V) recognition applied to the ESI capillary and current output (I) applied to the APCI needle. The same power supply could be used with the addition of solid state switching (MOSFET) to separate the two outputs and re-establish them with measurable fidelity. The power supply initially developed for what became the ESCi® multimode ionization technology project, was tested to switch positive and negative as well as voltage (for ESI) and current (for APCI) in less than 100 ms and still achieve close to 100% output (Fig. 5.6). Succeeding designs now provide power supplies that switch at $<$10 mS, reducing any non-data acquisition time [35].

Since the attempt is to switch at very high speed between inimical modes of ionization the behaviors of functional surfaces including heat, electrostatic lenses and gas flows need to be optimized in the presence of each other. The MassLynx 4.0 oscilloscope view, or tune window user interface, was redesigned to permit the operator to view ions produced by both ESI and APCI switching at high speed (so they appear simultaneous) from the same infusion (Fig. 5.7). Gas and heat can therefore be co-optimized into a single method for use with either ESI or APCI, or both in positive and negative modes as fast as every 10 ms. Also, since in a reversed phase separation weakly polar analytes requiring APCI are undoubtedly more apt to be present later in the run the method can be set up to maximize the duty cycle by invoking the mode later in the run.

5.9 APPLICATIONS OF MULTIMODE IONIZATION

5.9.1 APCI operation at ESI conditions

Some of the unique benefits of performing APCI at ESI flow rates are reducing nitrogen and solvent consumption to that of the lower flow rates common to ESI operation at low μL/min flow rates. At the same time, it maintains the column loading associated with

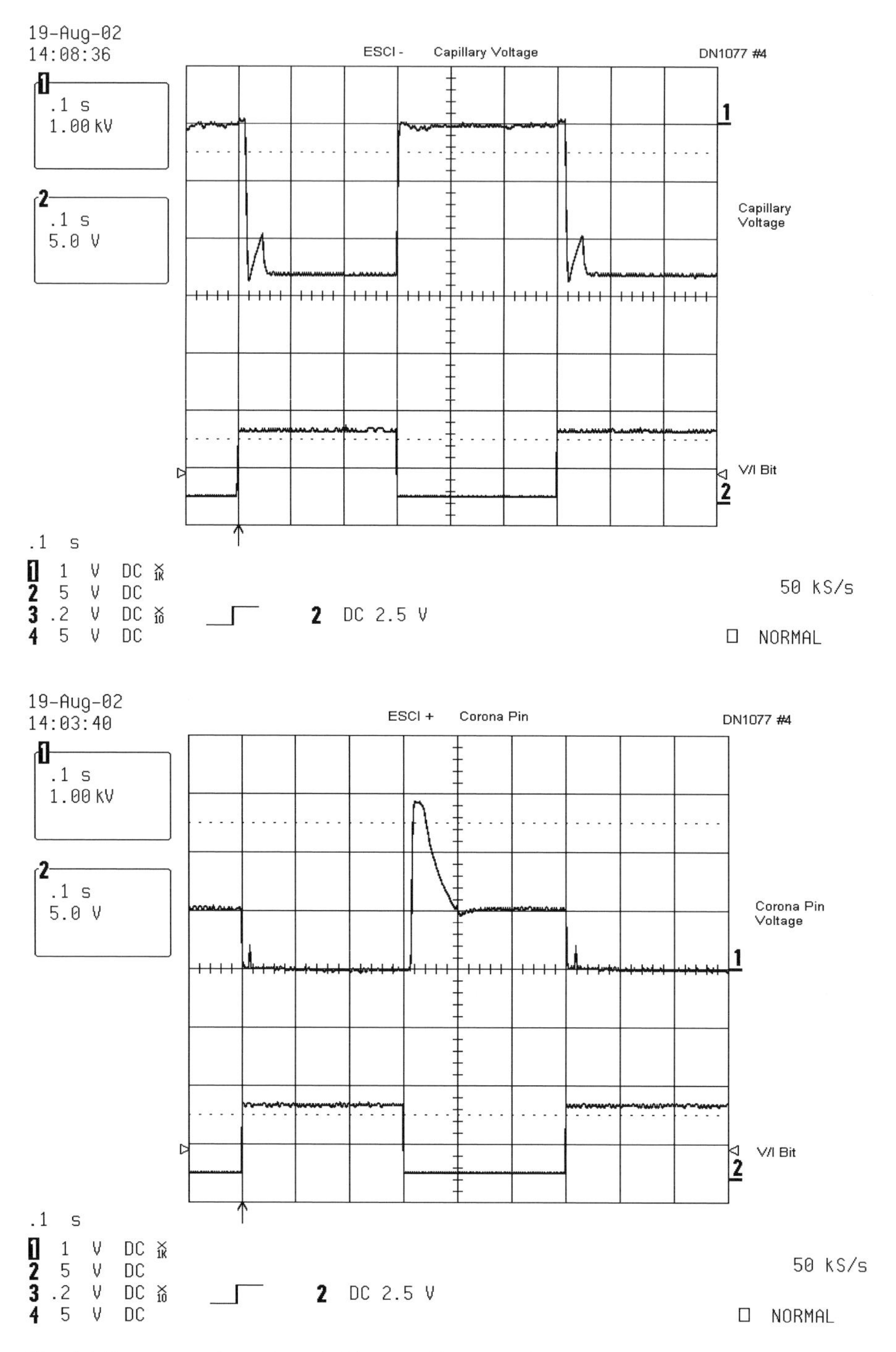

Fig. 5.6. Power supply mode switching at 100 ms: capillary voltage (*V*, left) and corona current (*I*, right). (Courtesy Gary Bertone, Waters Corporation.)

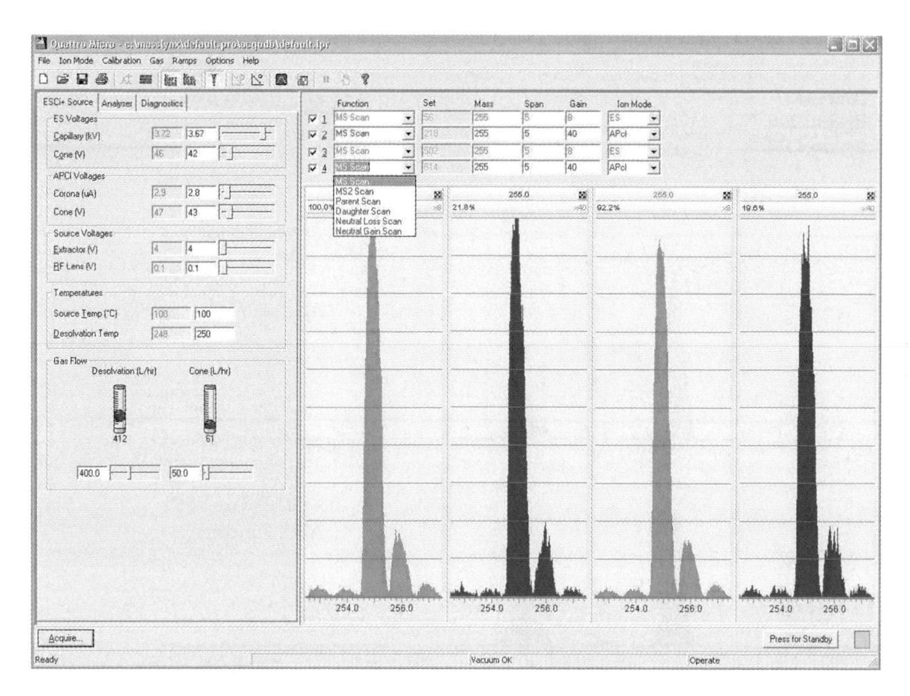

Fig. 5.7. Oscilloscope (tune window) view in MassLynx software used with ESCi incorporates the control to co-optimize gas, heat and voltage/current while infusing a test mixture. Peaks in windows 1 and 3 (left to right) have been selected as ESI while windows 2 and 4 display APCI results. (Courtesy Waters Corporation, MassLynx 4.0, Quattro Micro tandem software.)

larger-diameter formats. To preserve optimum ESI performance, it has been common practice to split the HPLC eluent as much as 10:1 before introducing it to the MS (100 μL/min or less). The mass spectrometer's relative insensitivity to the reduction in absolute mass is due to the improvements made in ion propagation at lower flow rates. The practice also preserves the optimal operating characteristics common to HPLC practice. An example of the scaling benefit is illustrated by an analysis for solution containing polymer antioxidants and additives. The traditional APCI practice using a flow rate and a column diameter appropriately chosen according to chromatographic (van Deemter) principles, typically a 4 mm (nominal) diameter coupled with a flow rate of 1 mL/min, induces more radial diffusion than the more common 2.1 mm ID column used for ESI analysis at flow rates of 0.25 mL/min (Fig. 5.8). Signal-to-noise response is significantly increased following rescaling of the separation using the smaller-diameter column since the absolute amount of material injected is conserved in narrow, less dispersed band. The UV response was used to normalize the amount injected for each trace as k' necessarily changed with scaling.

5.9.2 High throughput in a chemically diverse practice

A typical application first envisioned for this project was done in collaboration between Waters Corporation and Astra Zeneca, Macclesfield, UK. Like all pharmaceutical companies

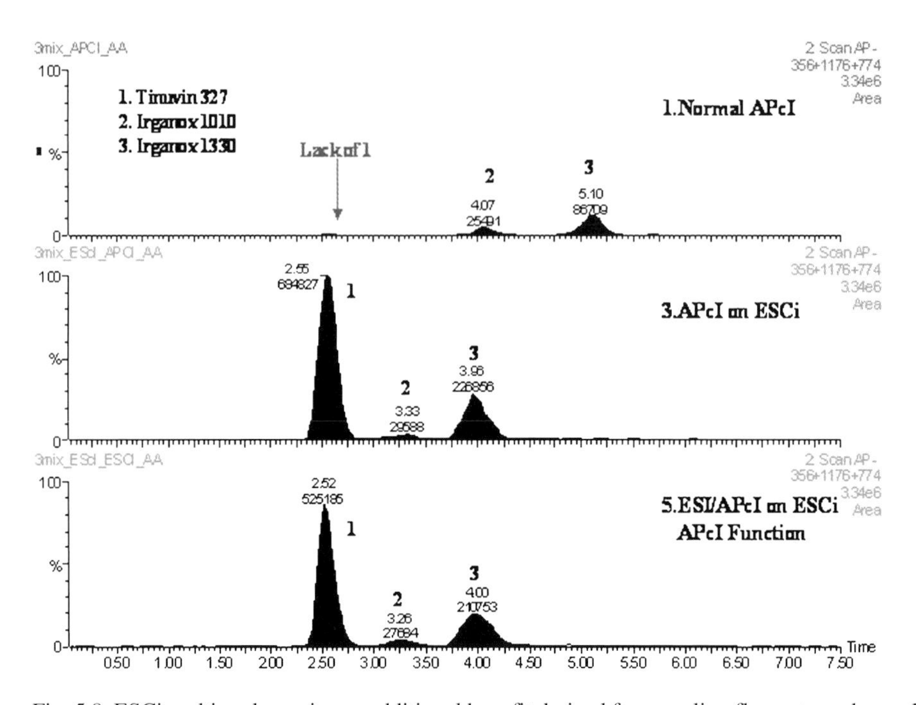

Fig. 5.8. ESCi multimode carries an additional benefit derived from scaling flow rate and sample load to improve signal-to-noise performance in LCMS when compared to dilution and dispersion effects under traditional APCI conditions. Upper trace, traditional APCI – 1 mL/min flow rate using a 4.6 mm ID column; middle trace, ESCi operating as APCI only (ESI off) – 0.25 mL/min flow rate and 2.1 mm ID column; lower trace ESCi in switching mode (ESI alternating with APCI at 100 mS ISD) same amount of material injected; separation conditions optimized for reduced flow rate. (Courtesy Jun Yonekubo, Nihon Waters).

Astra Zeneca needs to verify the contents of its libraries. A project constituting a need to analyze 500,000 compounds following the expected rule that only 80% would work by ESI left 100,000 compounds needing to be isolated and reanalyzed (Fig. 5.9). The resulting benefit for Astra Zeneca when it compared the throughput of ESI alone, APCI alone and ESCi was an increase in project efficiency of 10%. That is, the company found 50,000 more compounds using the high-speed switching method on the initial analysis. Those samples would have had otherwise required re-analyzing [36].

5.9.3 Multimode sensitivity with narrow peaks at high speed

The preservation of data acquisition with the improved duty cycle of a power supply capable of high-speed switching at 10 ms is illustrated with studies of microsomal incubates [37], which also employ an ultra high pressure chromatographic system (commercialized by Waters Corporations under the name UPLC) resulting in peak widths of 1 – 2 s, which are more commonly associated with gas chromatography (GC) separations. This added peak capacity, or more discrete peaks separated from complex mixtures in a shorter

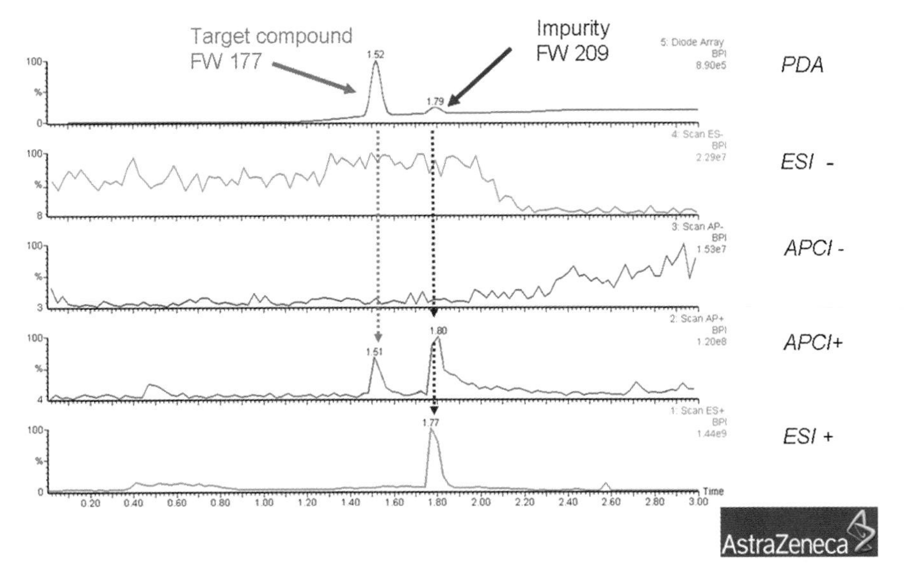

Fig. 5.9. Application of LC–ESCi–MS full switching mode: ESI+, ESI−, APCI−, APCI+ including a monitor UV trace. An additional 50,000 compounds were characterized in a 500,000 compound library using ESCi versus ESI only in a single pass. (Ref. 36.)

time span, speeds analysis times but puts an added premium on fast transition times and short ISDs. In one example, a microsomal stability assay presented large numbers of samples with a wide chemical variety.

Results for eight standards were optimized including acetophenone, which can only be analyzed by APCI in positive mode (APCI+), and Ibuprofen, which can only be analyzed by ESI negative mode (ESI−). Tolbutamide represents a common observation, as the response is more intense in ESI than APCI, and in APCI, it displays a preference for positive over negative mode. All other compounds were seen to respond preferably in positive mode with acceptable signal in both ESI and APCI. For each individual sample, the target is always a single analyte. Sensitivity at 0.01 ng/mL (injected) were routinely achieved in APCI+ exceeding expectations set by the initial sensitivity benchmark (50 ng/mL injected) established for routine confirmation work as illustrated by the work published by Astra Zeneca.

The test standards were incubated with the rat microsome and results calculated as the percent remaining for each compound, which is a typical procedure for microsome stability assays. Using first-order kinetics, the compound half-life based on a 20-min incubation assay was deduced and reported. Three additional drug standards with known microsomal activities were incubated for inclusion in the assay for QC purposes: Loperamide, Verapamil, and Zolpidem. As previously measured by the LC–MS–MS, the range of percent remaining for these three compounds were 10–15% for Loperamide, 3–9% for Verapamil, and 44–59% for Zolpidem respectively. The results of the ESCi UPLC/MS/MS microsomal stability assay are summarized in Table 5.1. According to the study, they were found in good agreement with the LC–MS–MS method. The chromatographic peak widths at the base were typically 1.8 s wide yielding good MRM statistics at 50 ms (20 data points over each peak) with an ISD of 20 ms.

TABLE 5.1

MICROSOME STABILITY ASSAY RESULTS USING ESCI IN A FAST CHROMATOGRAPHY MRM ANALYSIS. THE RESULTS COMPARED FAVORABLY WITH EXISTING TRADITIONAL LC–MS–MS DATA WHERE INDIVIDUAL IONIZATION MODES WERE RUN AS SEPARATE ANALYSES WITH TRADITIONAL LC PEAK WIDTHS. (UPPER) MODE, LINEARITY AND LOD FOR STANDARDS AND (LOWER) STABILITY ASSAY RESULTS (REF. 37)

Analyte	Ionization mode	Linear range (ng/mL)	R^2	Limit of detection (ng/mL, 5 μL injected)
Acetophenone	APCI+	5.0–1000	0.990	10.0
Corticosterone	APCI+	2.0–1000	0.991	2.0
Daspone	ESI+	0.1–1000	0.996	≪0.1
17a-Hydroxyprogesterone	APCI+	0.2–200	0.990	0.2
Ibuprofen	ESI−	0.1–500	0.990	1.0
Nortriptyline	ESI+	1–500	0.980	1.0
Sulfadimethoxine	APCI+	0.01–100	0.995	0.01
Tolbutamide	APCI−	0.1–1000	0.991	0.1

.Analyte	MW	Ionization mode	(%) Remaining	*In vitro* metabolic half-life (years)
Acetophenone	120	APCI+	55.60	18
Corticosterone	346	APCI+	56.14	18
Daspone	248	ESI+	48.19	14
17a-Hydroxyprogesterone	330	APCI+	0.54	2
Ibuprofen	206	ESI−	74.73	>30
Loperamide	476	ESI+	11.21	4
Nortriptyline	263	ESI+	0.59	2
Sulfadimethoxine	310	APCI+	87.65	>30
Tolbutamide	270	APCI−	81.88	>30
Verapamil	454	ESI+	3.79	3
Zolpadem	307	ESI+	47.12	14

5.10 DEVELOPMENTS IN PROGRESS

5.10.1 Combined ESI and APPI

APPI has been shown to expand the realm of compounds amenable to LC–MS analysis beyond that available to ESI and APCI. Limited investigations of ESI have been combined with APPI [38] and evaluated for quantitative and qualitative mass spectral analysis. A number of compound classes were examined including steroids, proteins, and polyaromatic hydrocarbons to elucidate the range of detectable compounds. The combined ESI and PI source is constructed from a standard ESI source and a Krypton (Kr) discharge lamp on a Waters/Micromass ZQ. VUV photons produced by the Kr lamp are 10.0 and 10.6 eV.

Heat of vaporization for even a small amount of water in the HPLC mobile phase at low flow rates requires a significant heat contribution. Similar to combining ESI and APCI, minimum desolvation sufficiency which allows APPI to proceed when derived from an APCI-generated aerosol must be met even though the vapor load may be significantly less at ESI flow rates than that associated with APCI. Early attempts indicate that pre-heating applied to the nitrogen gas provides insufficient desolvation capacity for APPI. However, under existing design restrictions, work at the typical ESI flow rates of 100 μL/min and 200 μL/min has displayed acceptable linearity and signal-to-noise response relative to ESI with aqueous mobile phase ($MeOH/H_20$). Attributes associated with ESCi were preserved: mode switching at 20 ms and ESI sensitivity remained intact.

A unique and possibly diagnostic feature in addition to extending the range of possible analyte compounds is found with charge-stripping APPI. A mixture of horse heart myoglobin (HHM) and testosterone provided clear examples of ESI multiple charge spectra in the same mixture as a steroid compound that is not amenable to ESI ionization. High-speed switching between the two forms of ionization yielded clear indications of both expected ESI behavior and APPI as well as simplification of the multiply charged HHM to the single MH^+ expected from APPI (Fig. 5.10).

5.10.2 Extending the existing API source utility

A further adaptation of the existing API source adds adjunctive capabilities of GCMS (CI only) plus direct solids probe utility (APCI) into a source designed for ESI. The atmospheric solids analysis probe (ASAP) [39] or in combination with the atmospheric pressure GCMS (APGCMS) [40] and ESI extends the format (solids or liquids can be taken up in a melting point capillary without sample preparation) and access to a more chemically diverse analytical capability. As McEwen observes APCI instruments are capable of ionizing samples introduced as vapors and have been used for trace analysis or organic compounds in air and less volatile compounds such as drugs of abuse and explosives aided by a hot gas directed into the APCI region.

The key to achieving acceptable results with a relatively unknown target within a mixture is the level of performance of the modern mass spectrometer. As stated in the preface to the *Methodology for Accurate Mass Measurement of Small Molecules*, "in the past the phrase "accurate mass" was interpreted very broadly and covered a wide variety of mass spectrometry measurements with varying precision. Today, most instruments used for accurate mass measurements are capable of achieving precisions of 10 ppm or better." [41]. Where candidate compositions (C_{0-100}, H_{3-74}, O_{0-4} and N_{0-4}) measure a mass-to-charge (m/z) response at 118 the error need not to exceed 34 ppm to be unambiguous [42]. A calculated mass of 118 measured by a modern mass spectrometer to within 2 millimass units (mmu) accuracy would display 17 ppm error – sufficient by today's standards for unambiguous determination of a chemical formula at that mass.

Monoisotopic calculated exact mass	= 118
Measured accurate mass	= 118.002
Difference	= 0.002 mmu
Error [difference/exact mass × 10^6]	= 17 ppm

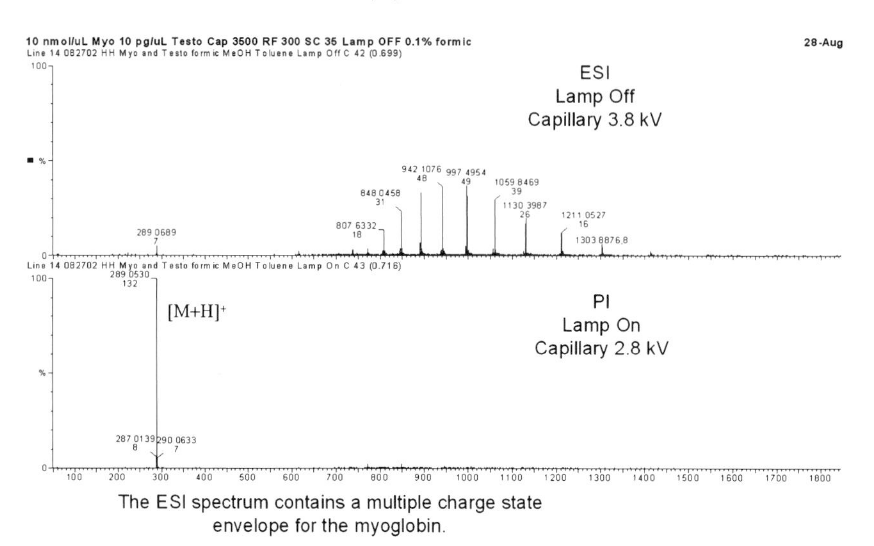

Fig. 5.10. High-speed switching between the two forms of ionization yielded clear indications of both expected ESI and APPI behavior as well as simplification of the mixture to the single MH^+ expected from APPI for testosterone. (Ref. 38.)

An instrument capable of a response at 750 *m/z* is also deficient by 2 mmu. When compared to the theoretical or calculated mass, it would be accurate to 2.7 ppm. Where the target is well characterized and the ionization event free from interference it is possible to use instruments such as quadrupoles without accurate mass capability to take advantage of these ionization techniques.

McEwen's work has shown sensitivity of the direct solids probe technique in some cases to 50 ppb. Some demonstrated applications for ASAP include: intact plant tissue for plant metabolomics, direct interrogation of cell cultures, environmental samples (especially in negative mode), and synthetic compounds and polymer additives. Food safety and surface chemistry studies (for example, drugs of abuse and explosives on currency) have been demonstrated. In addition, some well-characterized applications of GCMS are also amenable to existing API instruments (with minor modifications to the API source including interfacing to supercritical fluid chromatography (SFC)).

Automation may extend the versatile API–MS instrument beyond applications where it has been a mainstay for years, as with walk-up open-access utilities to areas of emerging interest like the process analytical technology (PAT) initiative from the FDA. Here online, or at the very least, at-line technology is required. The technology must fall easily to the operator's hands with limited or no sample preparation and be exceptionally robust. Most recent interest has been focused on developing slurry characterization tools and employs sensors and Raman and IR in addition to data reduction. An automated source for both liquids and solids based on well-characterized technologies such as ESI combined with other techniques, which expand analytical capabilities to more diverse chemicals may open possibilities for PAT interests.

There are two broad goals for the application of LC–MS – high throughput verification of suspected compounds or deconvolution of complex mixtures for identification of unknowns. Within each major sub-specialty, LC–MS technology has not changed substantially in many years. The analytical benefit of liquid chromatography coupled with MS is in its serial nature. The recent commercial development of ultra performance liquid chromatography (UPLC) based on high pressure work by Jorgenson and Lee [43,44] delivers peak capacity similar to gas chromatography with peak widths near 1s better capitalizing on the MS duty cycle. Previously one of the best means to increase throughput was employing a device that allowed multiplexing more than one HPLC stream into the same mass spectrometer source.

Acknowledging the current employment of LC–MS we could expect ESI to remain dominant with the knowledge base and conventional wisdom already in the hands of practitioners. Extending the reach of API technology means embracing a more complex understanding of the inherently multimodal processes involved. We can see the limits to conventional wisdom where multimode APCI, when the aerosol is more uniform and the vapor load is reduced, demonstrates sensitivity equal to or exceeding traditional APCI designs in some cases. Work on chip-based APPI is an example [45]. Possibly we will see an increase in ionization and introduction techniques that are not HPLC-based. Desorption ESI (DESI) [46] and direct analysis real time (DART) [47] have been highlighted recently. Without requiring sample preparation, they also promise a broader ionization range – more like ESI combined with APCI and APPI even to the compounds normally amenable to electron ionization. An energized stream is focused on the sample and either applied to an inert substrate or placed directly in the path to create secondary ions. Where evidence of known compounds can be extracted from a complex mixture or with effective support from software algorithms to provide putative structural determinations these technologies could be rapidly embraced.

ACKNOWLEDGMENTS

In addition to those cited, a few people have made significant contributions to my work and need to be acknowledged: Mike R. Jackson and Lisa Southern for their assistance leading to the development of ESCi and Marshall Siegel's early work on multimode ionization; Steve Bajic's development of Z-spray and continuing studies in atmospheric ionization; Karl Hanold's development of multimode APPI and ESPI; and David Sarro who has provided editorial and rhetorical suggestions for my work for many years.

REFERENCES

1 R. Cole, *J. Am. Soc. Mass Spectrom.*, 35 (2000) 763–772.
2 M. Dole, L.L. Mack, R.L. Hines, R.C. Mobley, L.D. Ferguson and M.B. Alice, *J. Chrom. Phys.*, 49 (1968) 2240–2249.
3 J.V. Iribarne and B.A. Thomson, *J. Chem. Phys.*, 64 (1976) 2287–2294.
4 P. Kebarle, *J. Am. Soc. Mass Spectrom.*, 35 (2000) 804–817.
5 C.M. Whitehouse, R.N. Dreyer, M. Yamashita and J.B. Fenn, *Anal. Chem.*, 57 (1985) 675–679.

6 E.C. Horning, M.G. Horning, D.I. Carroll, I. Dzidic and R.N. Stillwell, *Anal. Chem.*, 45 (1973) 936–943.
7 D.I. Carroll, I. Dzidic, R.N. Stillwell, K.D. Haegele and E.C. Horning, *Anal. Chem.*, 47 (1975) 2369–2373.
8 C.H. Bruins, C.M. Jeronimus-Stratingh, K. Ensing, W.D. van Dongen and G.J. Jong, *J. Chrom. A*, 863 (1999) 115–122.
9 W. Tong, S.K. Chowdhury, J.C. Chen, R. Zhong, K.B. Alton and J.E. Patrick, *Rapid Commun. Mass Spectrom.*, 15 (2001) 2085–2090.
10 D.Q. Liu and T. Pereira, *Rapid Commun. Mass Spectrom.*, 16 (2002) 142–146.
11 P. Siegel and T. Karancsi, In E.J. Karjalainen, A.E. Hesso, J.E. Jalonen and U.P. Karjalainen (Eds.), *Advances in Mass Spectrometry*, Elsevier Science and Technology Publications, Burlington, MA, USA 1998, p. 14, Proceedings from the 14th International Mass Spec. Conf, Tampere, Finland, August, 1997.
12 J.A. Syage, M.D. Evans and K.A. Hanold, *Amer. Lab.*, 32 (2000) 24–29.
13 D.B. Robb, T.R. Covey and A.P. Bruins, *Anal. Chem.*, 72 (2000) 3653–3659.
14 J.A. Syage, *J. Am. Soc. Mass Spectrom.,* 15 (2004) 1521–1533.
15 E. Marotta, R. Seraglia, F. Fabris and P. Traldi, *Int. J. Mass Spectrom.*, 228 (2003) 841–849.
16 H.N. Chen, W. Genuit, A.J.H. Boerboom and J. Los, *Int. J. Mass Spectrom. Ion Proc.*, 46 (1983) 277–280.
17 A. Cappiello, M.P. Balogh, G. Famiglini, F. Mangani and P. Palma, *Anal. Chem.*, 72 (2000) 3841–3846.
18 E.T. Gangl, M. Annan, N. Spooner and P. Vouros, *Anal. Chem.*, 73 (2001) 5635–5644.
19 M. Busman, J. Sunner and C.R. Vogel, *J. Am. Soc. Mass Spectrom.*, 2 (1991) 1–10.
20 C.R. Blakeley and M.L. Vestal, *Anal. Chem.,* 55 (1983) 750–754.
21 P. Brown, R. Nachi, T. Julian, A. Dzerk and P. Lin, Thermal Spray is Back: A Modified APCI Interface on the LC/MS/MS Bioanalysis of Famotidine in Human EDTA Plasma, *48th ASMS Conference on Mass Spectrometry and Allied Topics*, Long Beach, CA, June 2000.
22 K. Tang and R.D. Smith, *Intl. J. Mass Spec. Ion Proc.,* 162 (2001) 69–76.
23 S. Haider, F.-F. Alary, P. Kovarik and T. Covey, *A Dual APCI/TIS Source for LC–MS–MS Analysis*, 19th Montreux LCMS Symposium, Ithaca, NY, 2002.
24 C.S. Liu, A. Marland, F. Garofalo, H. Pang, M. Macintosh, E. Wong, M. Kennedy, T. Covey and Y. LeBlanc, Use of fast settling desolvation heaters and rapid ionization process selection for enhanced LC–MS–MS performance, *50th ASMS Conference on Mass Spectrometry and Allied Topics*, Orlando, FL, June 2002.
25 S. Bajic and R.H. Bateman, Mass Spectrometer, UK Patent GB2406703, November 30, 2005.
26 S. Bajic and R.H. Bateman, A Fast Switching APCI Source*, 52nd ASMS Conference on Mass Spectrometry and Allied Topics*, Nashville, TN, June 2004.
27 J.A. Jarrell and M.J. Tomany, *Ionization Electrospray Apparatus for Mass Spectrometry*, Patent 5,828,062, October 27, 1998.
28 P. Kebarle and L. Tang, *Anal. Chem.*, 65 (1993) 972A–986A.
29 R. Bonfiglio, R.C. King, T.V. Olah and K. Merkle, *Rapid Comm. Mass Spectrom.*, 13 (1999) 1175–1185.
30 S. Bajic, Electrospray and atmospheric pressure chemical ionization mass spectrometer and ion source, Patent 5,756,994, May 26, 1998.
31 J.A. Castoro, Investigation of a simultaneous ESI/APCI for LC/MS analysis*, 50th ASMS Conference on Mass Spectrometry and Allied Topics*, Orland, FL, June 2002.
32 S. Fischer, D.L. Gourley and R.L. Bertsch, *Multimode Ionization Source*, Patent 6,646,257, November 11, 2003.
33 M.P. Balogh, *A High Speed Combination Multi-mode Ionization Source for Mass Spectrometers*, Patent Application 10/470,648, November 11, 2004.

34 M. Siegel, K. Tabei, F. Lambert, L. Candela and B. Zoltan, *J. Am. Soc. Mass Spectrom.*, 11 (1998) 1196–1203.
35 Specification – Waters Quattro Premier^{TM} *XE Mass Spectrometer Brochure*, 720001251EN, 2005. Waters Corporation, Milford, MA.
36 R.T. Gallagher, M.P. Balogh, P. Davey, M.R. Jackson, I. Sinclair and L.J. Southern, *Anal. Chem.*, 75 (2003) 973–977.
37 K. Yu, P. Alden, S. Gaulitz, L. Di, E. Li and E.H. Kerns, High-throughput microsome stability assay for candidate profiling in drug discovery by UPLC–ESCi MS–MS, *53rd ASMS Conference on Mass Spectrometry and Allied Topics*, San Antonio, TX, June 2005.
38 K. Hanold and M.P. Balogh, Evaluation of a combined electrospray and photoionization source, 22nd Montreux LC–MS symposium, Ithaca, NY, 2005.
39 C.N. McEwen, R.G. McKay and B.S. Larsen, *Anal. Chem.*, 77 (2005) 7826–7831.
40 C.N. McEwen and R.G. McKay, *J. Am. Soc. Mass Spectrom.*, 16 (2005) 1730–1738.
41 K. Webb, A. Bristow, M. Sargent and B. Stein, Methodology for accurate mass measurement of small molecules, department of trade and industry's VIMMS program within the UK national measurement system, LGC Limited, Teddington, UK, 2004.
42 *Journal of the American Society for Mass Spectrometry*, author's guidelines, 17 (1) (January 2006) A19–A20.
43 A.D. Jerkovich, J.S. Mellors and J.W. Jorgenson, *LC–GC North America*, 21 (2003) 600–610.
44 N. Wu, J.A. Lippert and M.L. Lee, *J. Chrom.*, 911 (2001) 1–12.
45 T.J Kauppila, P. Ostman, S. Martilla, R.A. Ketola, T. Kotiaho, S. Franssila and R. Kostiainen. *Anal. Chem.*, 76 (2004) 6797–6801.
46 Z. Takats, J.M. Wiseman, B. Gologan and R.G. Cooks, *Science*, 306 (2004) 471–473.
47 R.B. Cody, J.A. Laramee and H.D. Durst, *Anal. Chem.*, 77 (2005) 2297–2302.

Achille Cappiello (Editor)
Advances in LC–MS Instrumentation
Journal of Chromatography Library, Vol. 72

Chapter 6

Atmospheric pressure laser ionization (APLI)

OLIVER J. SCHMITZ and THORSTEN BENTER

6.1 MOTIVATION

Over the past 15 years, liquid chromatography–mass spectrometry (LC–MS) has matured into a very powerful analytical tool. LC–MS is currently applied in diverse areas of chemistry including the pharmaceutical sciences, biochemistry, and environmental chemistry, to name a few. The most widely used ionization techniques in LC–MS rely on generation of charge carriers at atmospheric pressure (AP). Today, atmospheric pressure ionization (API) sources are commercially available from all major MS manufacturers in a variety of arrangements, for example, LC effluent delivery to the first MS sampling aperture in in-line, orthogonal, or z-geometry. Many more aspects are important for the performance of API sources. There are a number of excellent reviews in the literature discussing the underlying principles of API sources as well as detailing individual approaches, for example, the series of LC–MS reviews of Niessen [1–3] and also many others; the reader will also find ample general references elsewhere in this book. The most widely used AP ionization technique is electrospray ionization (ESI), followed by AP chemical ionization (APCI) [4–7] and AP photoionization (APPI) [8,9]. This ranking reflects more or less on the chemical properties of the analytes determined in API–MS: Most compounds in pharmaceutical, bioscience, and related sectors are rather polar if not ionic; thus they are efficiently ionized with ESI. Since there are vast numbers of research groups and companies active in these sectors, ESI has seen high acceptance. However, there is also considerable interest in applying API techniques for efficient ionization of less or even nonpolar compounds. Here, ESI becomes less suitable. Instead, APCI seems to be most appropriate for the intermediate analyte polarity range (Fig. 6.1); this is conceivable since proton transfer is the most frequently encountered ionization process in this case. Again, a substantial literature body exists for numerous APCI applications in LC–MS. The most recent commercially available addition to the suite of API techniques is APPI. For the purpose of the remaining discussion, we will use APPI as synonymous to direct photoionization of analytes at AP [10,11], whereas DA–APPI denotes a variation of this technique – dopant-assisted APPI [12,13], which is basically reverting to the ionization mechanisms present in APCI – insofar as proton transfer

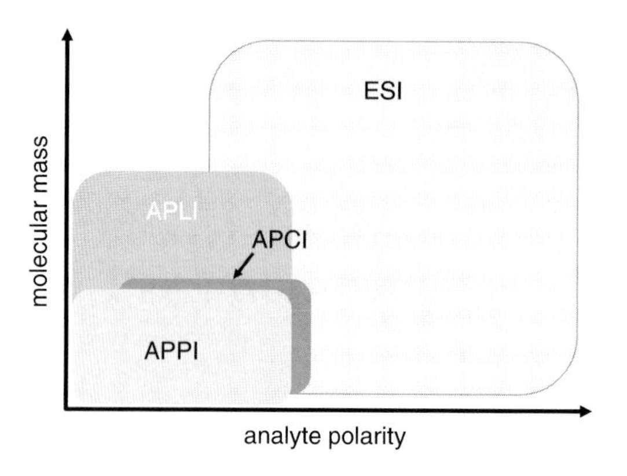

Fig. 6.1. Polarity range of analytes suitable for ionization with different API techniques. Note that the extended mass range for APLI as compared to APPI and APCI results from volatizing the analytes in an electrospray (see text for details).

through ion–molecule reactions between primary charge carriers and the analyte is the main charging process. However, the mechanism of generating primary charge carriers and their nature are of course fundamentally different in DA–APPI as against APCI. APPI and DA–APPI thus operate on fundamentally different ionization mechanisms regarding the generation of analyte ions.

APPI enables access to less polar or even nonpolar compounds (Fig. 6.1). It thus appears that a powerful suite of API methods is available, covering the entire range of analyte polarities. However, there are some aspects, in particular, with respect to APPI, which have motivated us to introduce another API technique, atmospheric pressure laser ionization (APLI). In 2005, the first paper on APLI was published [14] followed by a second, concerning LC–APLI–MS coupling stages in the same year [15]. To date, these are the two single papers representing APLI in the peer-reviewed literature, a situation quite similar to the initial release of APPI papers in the beginning of 2000 [9]. In this chapter, we will summarize the current state of the APLI development, first with respect to the fundamentals of laser ionization at AP, followed by a review of the currently operative LC–APLI–MS interface varieties established in our laboratory.

6.1.1 General principles of API source operation

Very briefly summarized, API sources generally operate as follows [1]: (i) The effluent from an LC column or from direct syringe injection is nebulized into the API source by means of either pneumatically assisted vaporization in a heated stage (APCI, APPI, and APLI) or the generation of an electrospray within a strongly nonlinear electrical field (ESI and ES–APLI). Also a combination of both, pneumatically assisted ESI with additional application if heat, sometimes called ion-spray, has been widely employed. (ii) Ions are generated either in the gas phase directly, via ion–molecule reactions (APCI, APPI, APLI), or in the liquid phase, i.e., before evaporation of the droplets formed in the nebulization process (ESI). This is the most fundamental difference in the ionization mechanisms operative in ESI vs. APCI, APPI and APLI. The ions are then sampled together with the assisting gas flows and solvent vapors into the first stage of the mass analyzer. Varying with the actual instrument design, skimmers, orifice plates, as well as transfer capillaries are frequently used as MS sampling apertures. Depending on the effective pumping speed available at the sampling aperture, a considerable fraction of the API source gas flow is directed into the first MS stage. This is particularly true for sources equipped with sampling skimmers. Any delivered gas flow exceeding the effective pumping speed at the sampling orifice is usually swept out of the API source via an exhaust line open to ambient pressure. After expansion into the first stage, usually an extraction cone further samples the gas flow. In most cases, the sampling orifice–extraction cone arrangement is in-line. For a more in-depth discussion of API interfaces, the reader is referred again to the review articles mentioned in Section 6.1.

6.1.2 ESI and APCI vs. APPI and APLI source operation

Although all mainstream API sources can be fitted with ESI, APCI, APPI, and APLI stages, there is an important aspect to be considered for source optimization using ESI/APCI vs. APPI/APLI. As already pointed out by Robb *et al.* [12], the ion formation in APPI does not require any element in the ion source at high potential other than the offset potential of the lamp, which is independently adjustable from all other source parameters; this also holds true for APLI, as explained in the following text. In contrast hereto, ESI and APCI require a highly nonlinear field gradient with absolute potentials well above 1 kV for spray formation or maintaining the corona discharge, respectively. Changing the corona needle potential in APCI affects a number of source parameters simultaneously, in particular, the total ion density in the source, regardless of whether or not analyte is present in the liquid flow delivered. Furthermore, the spatial distribution of primary and secondary ion concentrations shifts since the position and shape of the electrical equipotential lines in the source change. Similar arguments apply for ESI.

The incorporation of APLI into an existing API source requires the addition of two quartz windows of suitable size, for laser beam entry and exit. Furthermore, a repeller plate is required for ion sources in which the MS sampling region is close to ground potential. This plate has the same function as the offset potential in APPI [12] (see Section 6.4.2 for further details). It is pointed out that laser-induced ionization processes allow to completely decouple a second parameter (aside from the offset potential) from all other API source parameters: the region in which ions are generated in the source can be independently selected, since the position of the laser beam is easily manipulated using an appropriate beam delivery stage (Section 6.4.4). This is a unique feature of APLI, enabling new and exciting experiments, particularly with respect to source optimization and mechanistic studies. However, these issues are beyond the scope of this chapter and are thus not further discussed.

6.2 PRINCIPLES OF LASER IONIZATION

6.2.1 General considerations

In this section, we highlight some of the various experimental approaches in laser ionization–mass spectrometry (LI–MS), which are remotely related to the development of APLI. Since the early 1980s, numerous papers have reviewed the progress in this area [16–30]. Very diverse applications of LI–MS have evolved, but none has reached the stage of a widely applied mainstream analytical technique. At first sight, this is surprising since the promises of LI–MS, as frequently stated in the literature, are striking: extremely high selectivity upon resonant excitation of target species, outstanding sensitivity when using efficient multistep excitation schemes, and high temporal resolution. One possible reason for this situation might be that even today the reliable operation of tunable laser systems requires very skilled operators; systems capable of delivering appropriate pulse energies are still rather expensive and relatively fragile. Whereas mass spectrometers are rapidly maturing into turnkey machines, wide-range tunable laser systems are not. However, almost all the above-mentioned advantages of LI–MS rely on the availability of tunable laser radiation with sufficient pulse energy. Furthermore, resonant excitation requires detailed knowledge about the spectroscopy of the target species, but often these data are not available. There appears to be one area, however, where LI–MS has at least reached a stage of frequent application: LI–MS appears to be ideally suited for the time-resolved trace detection of aromatic hydrocarbons, i.e., compounds containing one or more individual aromatic ring systems as well as polycyclic systems. This occurrence appears to be closely related to the spectroscopic features of aromatic systems, which will be discussed in more detail in Section 6.3.1.

6.2.2 Single-photon ionization (SPI) vs. multiphoton ionization (MPI)

Single-photon ionization has a longstanding analytical use as photoionization detection (PID) [31–33] in gas chromatography. SPI at atmospheric pressure has also been widely used in ion mobility mass spectrometry (IMS) [34–37]. SPI at atmospheric pressure with mass-selective detection for LC–MS has been introduced as APPI [10,11] and DA–APPI [12,13]. Fig. 6.2A shows schematically a one-step photoionization process. The electronic transition time between the ground and ionic state of the analyte is on the order of femtoseconds. Generally, the absorption (i.e., ionization) cross sections for this process are rather high for virtually all organic compounds, provided that the photon energy is not significantly larger than the vertical ionization potential (IP). This is the case for APPI (and other SPI applications) when krypton discharge lamps are employed. Commercially available lamps emit photons with energies of 10.0 eV and to a lesser extent 10.6 eV, respectively; shorter wavelengths are cut off by the window material. The ionization potentials of the majority of organic compounds are in the range 7–10 eV [38]. (It is noted, though, that other parameters play important roles in this regard, particularly the equilibrium geometries of the molecular and ionic ground state.)

The picture changes fundamentally when multiphoton ionization (MPI) is employed. Before we discuss these aspects, we would like to define some terms that are frequently

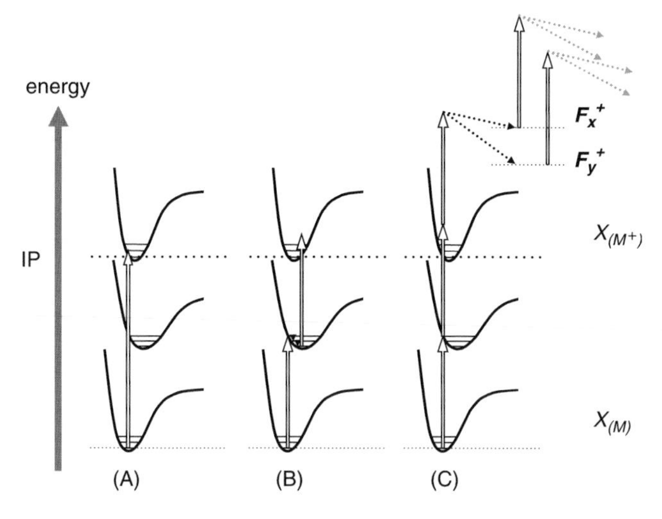

Fig. 6.2. Sketch of electronic transitions leading to ionization. (A) Single-photon ionization process. (B) $1 + 1'$ REMPI process with low laser-power density; note the intramolecular relaxation process depicted by solid arrows. (C) Same as in (B) but with substantially increased power density ($>10^8$ W/cm^2) leading to fragmentation via ladder step/switch mechanism. IP: ionization potential, X: electronic ground state, F: fragments.

used throughout this chapter. Here, MPI stands for a process in which more than one photon is absorbed by the molecule leading to the ejection of an electron. In case that the energy of one of the absorbed photons is resonant with a particular state within the ro-vibrational manifold of a molecular electronic system below the IP, the overall process is termed a resonant enhanced multiphoton ionization (REMPI) process. The number of photons absorbed before reaching this state and the number of photons required to induce ionization from this state is often indicated in brackets. An example is shown in Fig. 6.2B, which is the schematic representation of a $(1 + 1)$ REMPI process. If two or more different wavelengths are used for excitation, then a prime is added to one of the numbers, e.g., $(1 + 1')$ REMPI, which stands for a one-photon resonant, two-photon ionization process using two wavelengths. Finally, APLI stands for REMPI at AP; in virtually all applications of APLI the processes involved are either $(1 + 1)$ or $(1 + 1')$ REMPI (as shown below). In the present chapter, we focus our attention on these two types of REMPI processes.

It is worth mentioning that nonresonant excitation is also easily achievable with pulsed laser radiation. This is a common approach for resonant excitation of electronic states in the range 6–10 eV and higher, such as Rydberg states. In this case, a $(2 + 1)$ REMPI process might result in very efficient ion production [39]. However, the laser power density has to be on the order of 10^9–10^{10} W/cm^2 since the nonresonant step basically requires two photons to be present in the molecular system within a time frame of a few femtoseconds. It is straightforward to generate such power densities by tightly focusing a pulsed laser beam with pulse energies in the millijoule range, but two constraints have to be considered. First, the ionization volume becomes extremely small, on the order of 0.1 mm^3 and less; thus the amount of accessible analyte molecules is strongly reduced, leading to poor detection limits.

Second and equally important, the focused laser light is also absorbed via nonresonant multiphoton excitation by matrix molecules, which are always present in large excess. This issue is further discussed in section 6.3.3.

With (1 + 1) and (1 + 1′) REMPI, the time required for reaching the ionization region from the ground state of the molecule can be orders of magnitude longer than in SPI. The power density in APLI is generally about 10^6–10^7 W/cm^2. At this level, the lifetime of the electronically excited analyte needs to be on the order of 10^{-9} s for efficient two-photon absorption. Much shorter lifetimes lead to extensive loss of electronically excited analytes at this power level. Longer lifetimes, although strongly favoring the ionization step in collision-free environments, are not of benefit in APLI due to high collision frequency of the gas molecules. At atmospheric pressure, the mean free path of a molecule is calculated from simple kinetic gas theory to be about 0.1 μm. At average molecular velocities of 300 ms^{-1}, the time between subsequent collisions of individual molecules is then $<$1 ns [40]. Assuming that electronic energy is efficiently removed in collisions with the bath gas, the time window for absorption of a second photon of an electronically excited molecule is thus on the order of 0.1–1 ns.

Both parameters, short lifetimes in the excited state of the analyte as well as collisionally induced deactivation processes, become less important for efficient ionization when high laser power densities are employed. However, as stated above, nonresonant photon absorption, particularly, in the matrix, may occur. The situation then becomes distantly comparable to direct APPI, and may lead to unfavorable conditions (Section 6.33).

The energy range for (1 + 1) or (1 + 1′) REMPI is about 3.5–6 eV/photon, corresponding roughly to a wavelength range of 350–200 nm. An extension of this wavelength range further to the red is not useful since only a limited number of compounds have ionization potentials below 7 eV [38]. At the other extreme, generation of tunable laser radiation far below 200 nm is rather difficult and requires sophisticated instrumentation, and is thus not a useful approach for analytical applications.

6.3 SPECTROSCOPIC CONSIDERATIONS

6.3.1 Spectroscopic features of aromatic compounds

Numerous aromatic hydrocarbons are very efficiently ionized via (1 + 1) REMPI (see, e.g., [19,29] and references therein). The main reasons are high molecular absorption coefficients in the near UV, relatively large lifetimes of the intermediate S_1 or S_2 states, and favorable location of these states: Since the IP is often lower than the energy of two photons, ionization occurs readily from the resonantly pumped intermediate states. It follows that even unfocused laser beams can be used, making this compound class ideally suited for REMPI analysis. A substantial body of reports on the application of REMPI for measurements of aromatic hydrocarbons exists in the literature (see [29] for a recent review). The majority of these papers is concerned with the detection of toxic aromatic substances in exhaust gases, such as from municipal waste incinerators [41–43], but also analyses of process gases [44–46] as well as vehicle exhaust measurements [47–49], besides many others, have been reported.

With a few exceptions, other compound classes are rather difficult to analyze with (1 + 1) REMPI mainly due to unfavorable location or unfavorable properties of the involved electronic states; e.g., the energy level of the first electronic absorption bands is beyond the blue wavelength limit (e.g., aliphatic hydrocarbons), or predissociative or repulsive molecular states lead to efficient loss of excited molecules at intermediate energy levels via ultrafast dissociation processes (e.g., nitro-compounds).

In summary, (1 + 1) and (1 + 1′) REMPI processes appear to be ideally suited for analytical applications. Less sophisticated tunable laser sources are rather easy to operate, relative low power density is required for efficient ionization, and the spectroscopy of matrix molecules does not interfere with primary analyte ionization.

6.3.2 Impact of elevated temperatures on spectroscopic features

One of the main advantages of REMPI is the very high spectroscopic resolution attainable under certain experimental conditions. Many aromatic compounds exhibit dense and congested spectral absorption features in the UV/VIS range; this is particularly true for polycyclic systems. Upon adiabatic cooling in the gas phase, these dense systems become far less congested since the molecular population density is quantitatively "condensed" into the vibrational ground state and further into very low rotational states. Adiabatic cooling in supersonic gas expansions or "jets" is thus a common approach in REMPI spectroscopy. However, this approach is hardly applicable in mainstream LC–MS since a jet-expansion set-up is basically incompatible for direct coupling with an LC-stage; a detailed discussion is beyond the scope of this chapter. One principal aspect, though, is worth mentioning: if the LC effluent expands under supersonic conditions along with the analyte, extensive clustering of the solvent compounds occurs, at least in case of the rather polar reverse-phase solvents.

In contrast, in all major LC–API–MS coupling stages, heat is applied to the bulk gases or to the liquid matrix in the vaporization stage. Thus, the spectral features of the analytes develop even further into broad and basically unresolved absorption bands. At first sight, this seems to be highly counterproductive in light of the previous discussion of the high selectivity of REMPI and thus APLI. Under these conditions, however, APLI becomes selective to the *compound class* of aromatic molecules. All other constraints as laid out above still exist, i.e., unfavorable spectroscopic features of the majority of nonaromatic compounds. Furthermore, fixed-frequency lasers can be employed for ionization since there is basically no need for any narrow-bandwidth single-line excitation. By fortunate coincidence, the emission lines of three common excimer laser systems cover the entire wavelength range of interest: 308 nm (XeCl*), 248 nm (KrF*), and 193 nm (ArF*) (Section 6.4.3). As an example, the relative sensitivities for REMPI detection of a number of PAH at elevated temperatures is given in [50].

6.3.3 Spectroscopic features of typical solvents used in LC

Among the typical solvents used in normal and reversed-phase LC are H_2O, CH_3OH, CH_3CN, acetone, *n*-hexane, CCl_4, and many more. Upon coupling the LC stage with MS via an API source, N_2 is most frequently used as nebulization, desolvation, and buffer gas.

With pneumatically assisted vaporization, the N_2 gas flow is rather high, frequently upto several 100 L/h^{-1}. Usually, high-flow nitrogen generators are employed for continuous MS operation; in this case oxygen is also introduced into the system at appreciable concentrations. According to various data sheets of manufacturers, a typical value for the O_2 content appears to be 0.05–0.1% in the N_2 flow delivered. Thus, a typical LC matrix in the API source consists mainly of N_2, O_2, and the above-mentioned solvents. Most of these matrix compounds exhibit ionization potentials well above 10 eV. From this perspective, this gas-phase matrix appears to be ideally suited for APPI and APLI analysis. In APPI one-step SPI of ground-state matrix compounds is energetically not possible; in APLI two-step multiphoton ionization is highly unlikely because all matrix compounds are virtually transparent in the one-photon wavelength range 350–200 nm. Fig. 6.3 shows UV/visible absorption spectra of three widely used LC solvents, in addition to the spectrum of N_2 and O_2 [51–55].

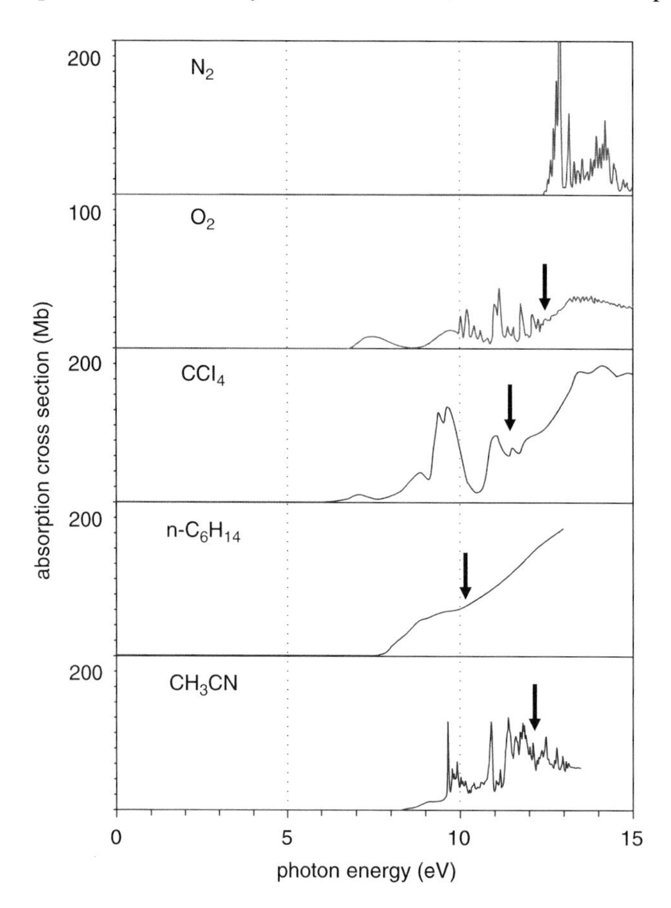

Fig. 6.3. Gas phase VUV absorption spectra of common matrix compounds present in LC–MS. (N_2: adapted from [51]; O_2: adapted from [52]; CCl_4: adapted from [53]; *n*-C_6H_{12} adapted from [54]; CH_3CN adapted from [55]). 1 Mb = 1×10^{-18} molecule/cm^2. Vertical lines at 5 and 10 eV indicate the wavelengths used in APLI and APPI, respectively. Solid arrows indicate the respective ionization potentials [38]; the value for N_2 (15.6 eV) is off scale.

The wavelength range extends into the vacuum ultra violet (VUV). The vertical line at 10.0 eV (~124 nm) indicates the photon energy of typical discharge lamps used in APPI. As can be seen, the absorption cross sections of all three solvents are extremely high and within the same range as the respective photoionization cross sections. The ionization potentials [38] for the individual solvents are indicated by arrows. Some conclusions, which have been already briefly addressed above, can be drawn from this picture:

1. The penetration depth of VUV photons in APPI is severely reduced to a couple of millimeters when solvent flow rates in the range 100 μL/min are introduced into the API source.
2. A large extent of the VUV photon energy is coupled into the matrix. It appears to be unclear in which way this energy is redistributed within the matrix. One possible fate of the highly excited matrix molecules may be ultrafast isomerization followed by subsequent ionization, as was discussed in the literature [56].
3. In APLI, virtually no electronic energy is coupled into the matrix, since the absorption cross sections in the corresponding one-photon energy range (308, 248, and 193 nm) are negligible. Thus, the photon flux fully penetrates the API source.
4. Any focusing of the laser radiation leading to power densities largely exceeding 10^8 W/cm^2 in APLI might induce resonant two-photon absorption of matrix molecules; this would potentially create the same unfavorable situation as described in point (2).

6.4 FROM JET-REMPI TO APLI

6.4.1 Typical inlet stages used in REMPI mass spectrometry

The commonly used inlet systems for LI–MS are shown in Fig. 6.4. Panel (A) shows the traditional supersonic jet expansion method using a skimmed and further collimated molecular beam. Ionization occurs far downstream of the high-pressure nozzle, usually at a distance of 5–20 cm, well behind the sudden-freeze surface [57] of the gas jet. The ro-vibrational temperatures of the analytes present in the ionization volume are usually well below 10 K, allowing high-resolution spectroscopic investigations. However, due to the issues already discussed, the method is not well suited for highly sensitive analytical applications. A variation of this technique is jet-REMPI (Fig. 6.4B) as introduced by Oser *et al.* [58] employing orthogonal beam geometry but without using skimmers. The ionization region is located just downstream of the sudden-freeze surface of the gas jet. Ro-vibrational cooling is thus still efficient, but the ionization volume is greatly increased leading to more favorable detection limits. However, the same restrictions as already discussed remain for LC–MS applications. Using medium pressure laser ionization (MPLI) (Fig. 6.4C), the ionization region is located as close as 0.5 mm from the nozzle orifice and thus within the continuous region of the jet, approaching local pressures up to 10 mbar. Ro-vibrational cooling is limited, which favors ionization with tunable broad bandwidth laser sources. The high analyte density allows for selective ultrasensitive measurements with high temporal resolution (see [59–62] for details). Nevertheless, substantial clustering is to be expected further downstream of the orifice in case that polar matrix compounds are present. It follows that

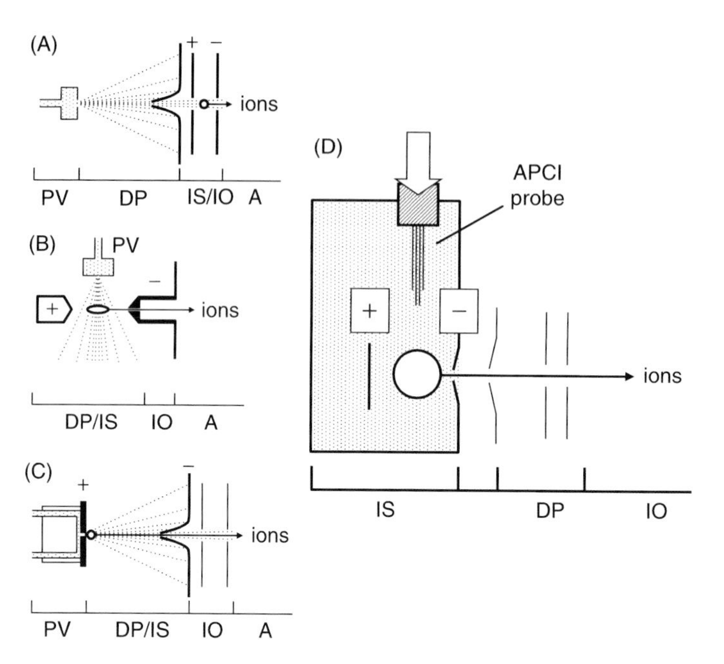

Fig. 6.4. Schematic diagrams of inlet stages frequently used in laser ionization mass spectrometry. (A) Conventional skimmed molecular beam sampling of a supersonic jet expansion. (B) jet-REMPI approach with unskimmed beam sampling. Ionization occurs shortly behind the sudden-freeze surface of the jet. (C) MPLI. Ionization occurs very close to the nozzle orifice within the continuous region of the jet. (D) APLI. Ionization occurs upstream of the MS sampling orifice; ions are extracted by electrical fields and gas flow. The dotted area indicates ambient or higher pressure. PV: pulsed valve; DP: differential pumping stage; IS: ion source; IO: ion optics; A: analyzer. ± symbolize an electrical field gradient.

none of the above stages is readily applicable to LC–MS analysis. Upon moving the ionization region upstream of the sampling orifice, however, the situation changes entirely. The APLI stage (Fig. 6.4D; [14,15]) does not feature any ro-vibrational cooling and operates directly at AP. The generated photo-ions are directed toward the sampling orifice of the MS by an external electrical field and drawn into the differential pumping stage mainly through collisions with the bath gas. This setup is discussed in more detail in the following section.

6.4.2 Design and construction of APLI stages

In principle, every commercial API source can be retrofitted for APLI operation. There are only a couple of technical prerequisites. As an illustration for the following discussion, Fig. 6.5 shows photographs of three APLI sources that were retrofitted or newly constructed in-house at the University of Wuppertal (see caption for details). First, the source enclosure is fitted with high-quality quartz windows, preferentially with an entrance and exit window. It is much easier to align the laser beam by monitoring the beam position at the entrance and exit window simultaneously. Furthermore, for ultimate performance of APLI, future applications will include two-color $(1 + 1')$ REMPI excitation schemes. An angled

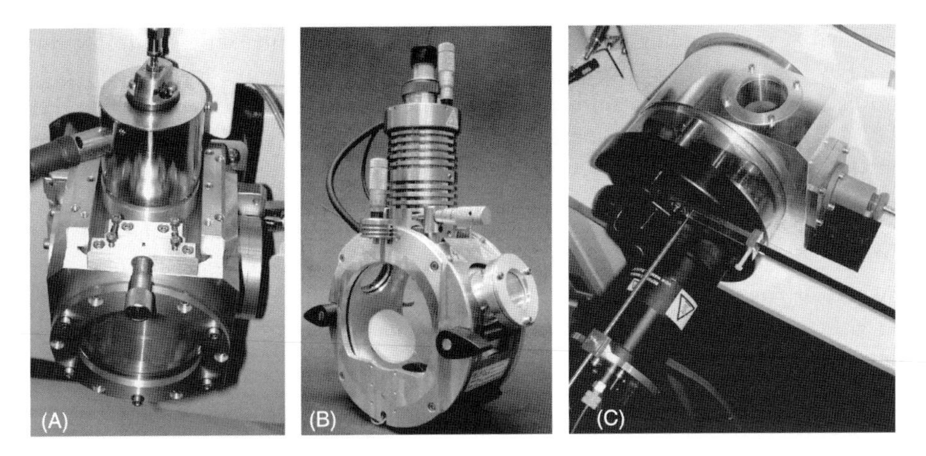

Fig. 6.5. Photographs of three APLI sources. (A) Newly designed APLI source replacing the original Bruker Apollo source. The exit window is visible on the right side; note the large front window for spatially resolved measurements. (B) Modified SCIEX API source for combined APPI/APLI operation on the SCIEX API 5000 MS. The laser beam enters diagonally from the right side, mounted inside the source enclosure is a beam dump instead of an exit window. The APPI lamp unit and the front window are removed for clarity. (C) Newly constructed APLI source for a Waters QTOF Ultima. The laser beam enters through the top window, also seen is the repeller on the right, which can be optionally replaced with a Syagen APPI lamp unit for combined APPI/APLI operation.

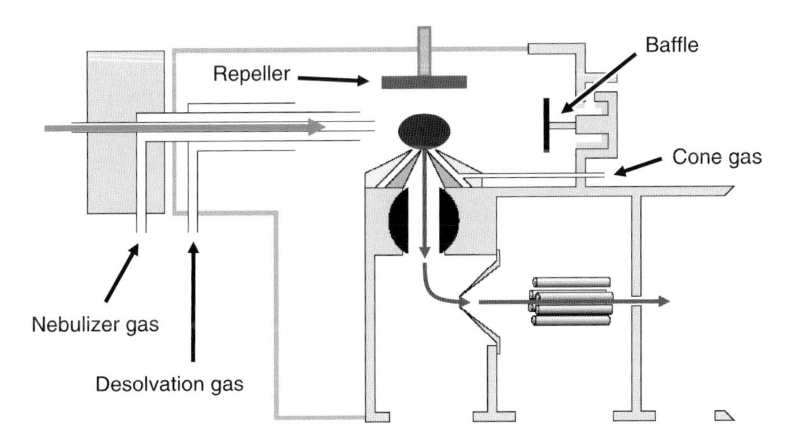

Fig. 6.6. Schematic set-up of an APLI source based on the MicroMass Z-spray interface. The dark grey area indicates the laser beam position (from top). Light grey bold arrow: liquid flow into vaporization stage, dark grey light arrows: ion travel path.

window configuration, preferably in orthogonal geometry, is of benefit in this case (Fig. 6.5B). However, this setup may need a complete reconstruction of the source enclosure due to mechanical constraints of the original design. For selected applications, focused laser radiation might be required. In this case, the exit window should be covered with a beam dump plate, fabricated, for example, from ceramic material. If the MS operates with potentials close to ground at the sampling orifice (e.g., the MicroMass Ultima series with Z-spray source, see Fig. 6.6 for a schematic diagram), a repeller plate is required for optimum performance. This plate should be positioned opposite the sampling orifice. In contrast to ESI and APCI, APLI sources are not space-charge dominated. Thus, the application of an external field has a pronounced effect on the overall ion transmission factor, as has also been shown for APPI [12]. The potentials applied to the repeller plate vary with source geometry and nozzle–repeller distance. Generally, a gentle field gradient on the order of 50–200 V/cm appears to be sufficient. MS equipped with grounded probe tips (e.g., the Bruker micrOTOF) may not require any additional means of forcing the ions toward the sampling orifice. At this stage, all conclusions are rather speculative, since these issues are subject to ongoing research in our laboratory.

6.4.3 Laser sources for APLI

As already discussed, APLI does not require tunable laser radiation to perform well. The use of pulsed fixed-frequency lasers appears to be sufficient for highly sensitive detection of numerous aromatic compounds, as will be shown further below. For the majority of APLI measurements we have carried out so far (>95%), 248-nm radiation from a KrF laser was employed. In some experiments, 193-nm radiation from an ArF laser leads to superior performance, in particular, when the electron density in the aromatic system is considerably lowered due to the presence of functional groups. The same arguments apply for electron donating groups, leading to a red shift of the absorption spectrum. In this case, 308-nm radiation from an XeCl laser may lead to optimum APLI performance.

Frequency quadrupled Nd:Yag lasers emitting 266-nm may also work with APLI but have not been employed yet. There are two main reasons. First, the laser repetition rate in APLI is directly proportional to the signal strength, as shown in Fig. 6.7. Second, the output

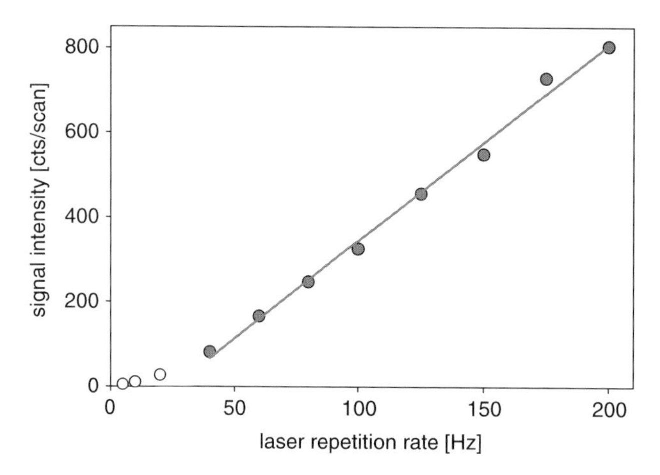

Fig. 6.7. Signal intensity recorded at $m/z = 252$ upon vaporizing a 1-μm/L benzo[*a*]pyrene solution with CH_3CN as solvent in the APLI source as function of laser repetition rate. A linear relationship is observed for rates >50 Hz. At lower rates, the effect of the currently missing synchronization of the laser and orthogonal TOF push/pull pulse becomes noticeable. This is changed in the near future.

pulse energy of the laser should be on the order of 10–50 mJ, leading to power densities around 10^6 W/cm^2 or higher. Small footprint table-top excimer lasers readily provide these pulse energies at upto 200 Hz repetition rates. Flash lamp pumped Nd:Yag lasers with comparable specifications are available but outrun by far the purchase and operational costs of the excimer system. As an example, typical flash lamp lifetimes are on the order of 10^6–10^7 pulses. Modern excimer lasers just require a simple laser gas refill after the same amount of laser pulses have been fired. Recently, bench-top excimer systems with repetition rates of upto 1 kHz became commercially available. The market need for compact high-repetition rate UV lasers is driven by fast growing industry sectors (e.g., micro-machining, lithography) as well as medical applications (e.g., eye surgery, skin disorders). Future applications of APLI will certainly benefit from the rapid further development of such laser systems.

6.4.4 Laser safety issues

A very important subject when working with class 4 laser systems is laser safety [63]. Although even the 193-nm ArF beam travels several meters through ambient air without severe attenuation, it is highly dangerous and thus working without a proper UV laser-beam enclosure is strongly discouraged. One possible way of minimizing laser radiation hazards is to curtain off the entire laser area, i.e., the laser, beam travel path, and part of the MS including the API source. However, this renders the routine work around the MS copious. As a more versatile alternative, a flexible laser-beam delivery stage has been employed, consisting of a laser mount, as many optical cubes as required – which turn the beam by 90° using precision mounted dielectric mirrors with a reflectivity of >99.8%, a beam

Fig. 6.8. Photograph of the movable parts of a beam delivery stage as used in APLI. The cubes holding the mirrors can be attached either directly to each other or via tubes. Full 360° rotation is possible. At the top left, an optional PVC tube connection is shown; at the bottom right the precision screws for a tilting mirror holder are visible. The entire stage can be purged with inert gas for deep UV operation.

manipulation stage for beam positioning and shaping, and the ion-source mount. The cubes can be attached directly to each other (resulting in a knuckle), allowing full 360° rotation of the beam. Alternatively, any type of tubing with appropriate diameter can be attached. If stiff connectors with precision-machined end pieces, for example, aluminum tubing with at least 3-mm wall thickness, are used to connect the cubes or knuckles, only minimal off axis beam deflection is observed. Should off-axis beam deflection become an issue upon use of more flexible connectors (e.g., PVC tubing), optical cubes equipped with a tilting steering plate can be employed (Fig. 6.8).

6.4.5 Currently installed APLI MS instruments

Fig. 6.9 presents photographs of two APLI MS systems currently installed in our laboratory. The Waters QTOF Ultima features a newly designed combined APCI/APPI/APLI stage, which is based on the original MicroMass APPI Z-spray source. This version utilizes the APPI lamp offset voltage, as supplied to the face plate of the lamp mount, also for APLI operation. Thus, an additional repeller plate becomes obsolete. Upon turning off the lamp driving voltage and adjusting the offset potential appropriately, rapid switching between APPI and APLI operation is possible. Also, both radiation sources can be turned on for simultaneous operation. It is noted, though, that for both techniques rather different operational parameters exist; in particular, the offset potential and even more pronounced the inlet probe position.

The Bruker micrOTOF is currently equipped with a preliminary version of a newly designed APLI source replacing the original Bruker Apollo source. Since the Apollo source is mounted on two hinges on the MS body and locked into position with a simple spring loaded latch, the two source types can be rapidly changed. Again, simultaneous or switched operation of APLI and APPI is possible.

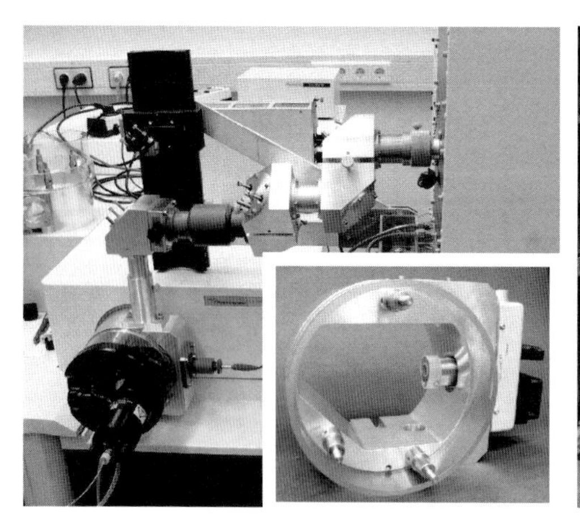

Fig. 6.9. Photographs of two APLI MS systems currently installed in our laboratory. Left: waters QTOF Ultima equipped with a combined APPI/APLI source as shown in the insert. The laser is partly visible on the top right; the laser beam delivery stage is just shown for demonstration of its flexible use. Right: Bruker micrOTOF equipped with an APLI stage as shown in Fig. 6.5A.

6.5 COUPLING STAGES FOR APLI MS AND EXPERIMENTAL RESULTS

This section describes different coupling stages for the APLI interface. In addition, for each setup selected examples for APLI mass spectra of different compound classes are given.

6.5.1 Direct syringe injection

In this case, the samples were injected directly into the pneumatically assisted heated probe with a syringe pump at flow rates around 300 μL/min. The ion-block temperature was generally maintained at 200°C. High-purity nitrogen was used as nebulizing, cone, and desolvation gas with flow rates of approximately 0.3, 0.16, and 2.5-L/min, respectively. The REMPI light source employed was a KrF* excimer laser (λ = 248 nm). Typical operating conditions were 100 Hz repetition rate, 15-mJ pulse energy, and 8-ns pulse duration. The laser beam was delivered unfocused to the APLI interface as discussed above. In all cases, the MS was operated in full scan mode with a scan time of 1 s.

Many heterocyclic aromatic compounds are readily ionized by APLI to predominantly yield the radical cation $M^{\bullet+}$. Example spectra are shown in Fig. 6.10A,B. Generally, the soft ionization process strongly suppresses fragmentation, simplifying the interpretation of the spectra. If structure determination is desired, however, in-source fragmentation can be achieved by increasing the photon density of the laser beam.

A comparison of the performance of APLI vs. APCI is presented in Fig. 6.11. A solution containing benzo[*a*]pyrene in CH_3CN was directly injected through the heated probe, as described above. For APCI operation, the corona needle was installed, and the source parameters were then optimized. For APLI, the source geometry was identical, but the needle was removed and the repeller electrode (see above) was installed. The APLI method generates only parent radical cations $M^{\bullet+}$ (along with background noise), whereas APCI generates exclusively quasimolecular ions, as expected. However, the relative signal intensity $M^{\bullet+}$ vs. $[M + H]^+$ is more than 500:1 with simultaneously greatly reduced noise.

Fig. 6.12 shows a sensitivity comparison for APPI vs. APLI. To exclude the influence of source parameters, for example, flow conditions, this experiment was carried out with the combined APPI/APLI source shown in Fig. 6.5A. The major reason for the significantly decreased sensitivity in APPI as compared to APLI may be severe absorption of the VUV photons by the solvent (CH_3CN). The more numerous background signals in the APPI spectrum indicate a more complex ionization process than with APLI. Nevertheless, the APLI spectrum clearly shows that not all PAHs are ionized with comparable efficiency. Besides MS settings that lead to a discrimination of ions toward the low mass region, the spectroscopic features of the smaller PAH are mainly responsible for the lower sensitivity. A discussion of these issues is beyond the scope of this chapter; the reader is referred to [50] for an indepth spectroscopic analysis.

The compound shown on the left in Fig. 6.13A was the target of a recent synthesis effort at the University of Wuppertal. The synthesis product was investigated by APLI. The right panel in Fig. 6.13A shows an APLI spectrum obtained at an intermediate stage of the work. The signals in the two individual regions of the survey mass spectrum, as shown in Fig. 6.13B, are fully compatible with the proposed structures, as judged by the isotopic pattern of the

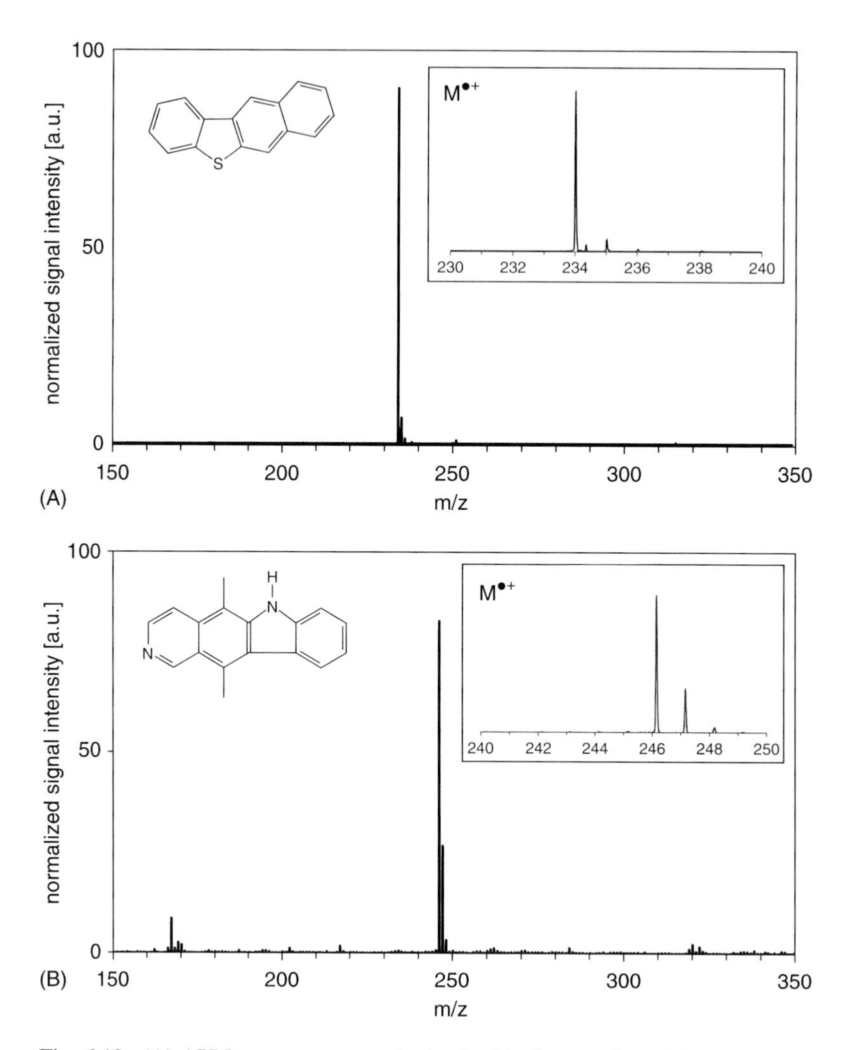

Fig. 6.10. (A) APLI mass spectrum obtained with direct syringe injection of a 1-μm/L solution of benzo[*b*]naphtho[2,3-*d*]thiophene dissolved in CH_3CN. MS: Full scan mode (1-s scan time). (B) APLI mass spectrum obtained with direct syringe injection of a 1-μm/L solution of the pyridine derivate shown in the upper left corner of the graph. Same conditions as in A.

signals recorded. As can be seen, the synthesis procedure was not yet fully worked out, but the experimental approach appeared to be promising. State-of-the-art analysis of such building blocks for novel materials synthesis and their reaction products includes field desorption (FD) MS. However, the sample preparation time for FD–MS is significantly longer than for APLI–MS. The latter method requires only a dilution step and loading of the syringe pump.

APLI is particularly suited to the ionization of nonpolar aromatic analytes. For this compound class, APLI is thus regarded as an optimum complement to existing API techniques, as long as the analytes are soluble in a polar (e.g., methanol, acetonitrile), aliphatic (e.g., hexane,

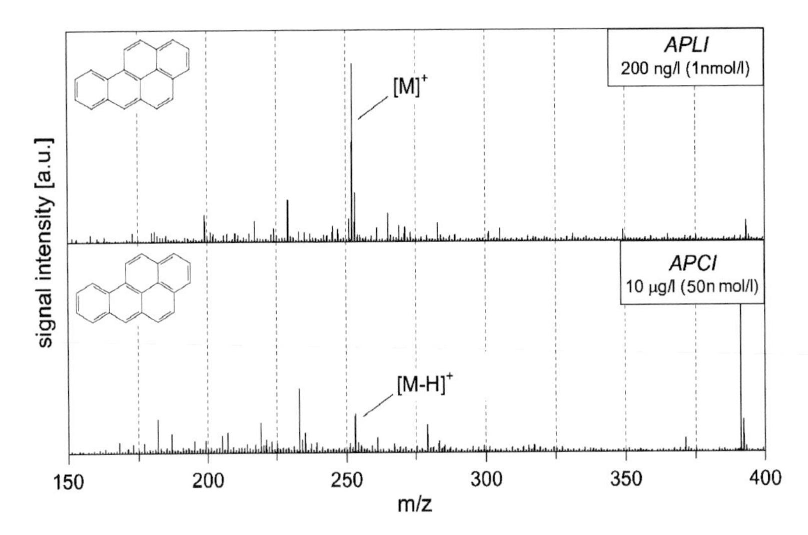

Fig. 6.11. Comparison of two raw mass spectra of benzo[*a*]pyrene dissolved in CH_3CN obtained with APLI (top trace) and APCI (bottom trace). For APLI the concentration was a factor of 50 lower. Direct syringe injection was used. The background signals originate from contamination from previous experiments.

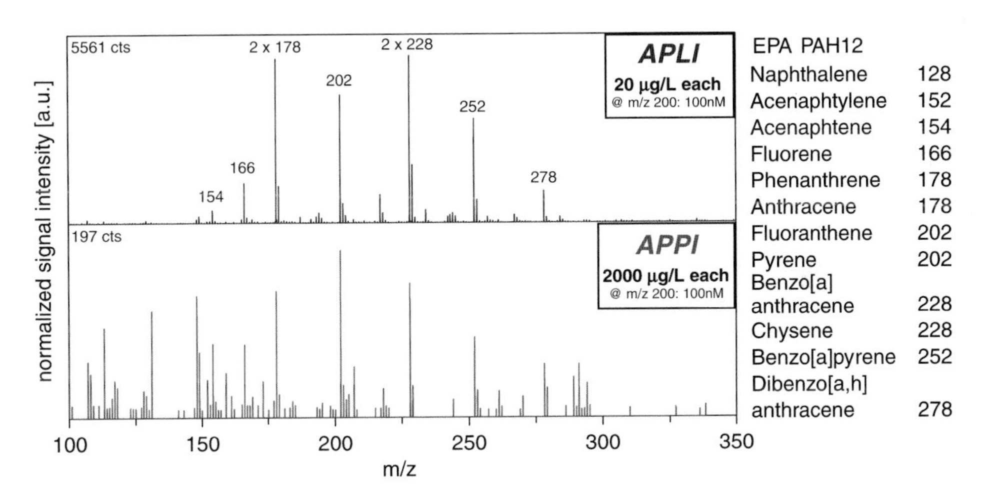

Fig. 6.12. Comparison of two raw mass spectra of a standard PAH mixture dissolved in CH_3CN obtained with APLI (top panel) and APPI (bottom panel). For APLI, the concentration was a factor of 100 lower and the maximum signal intensity in APLI was more than 25 times higher than in APPI (cf. max. count rate shown on top left corner of each panel).

heptane), or chlorinated (e.g., dichloromethane) organic solvent. If toluene is used as solvent, however, extensive laser ionization of this compound leads to a situation comparable to dopant-assisted APPI. In other words, if desired, dopant-assisted APLI is readily invoked by the addition of toluene to the liquid flow containing the analytes. With regard to both

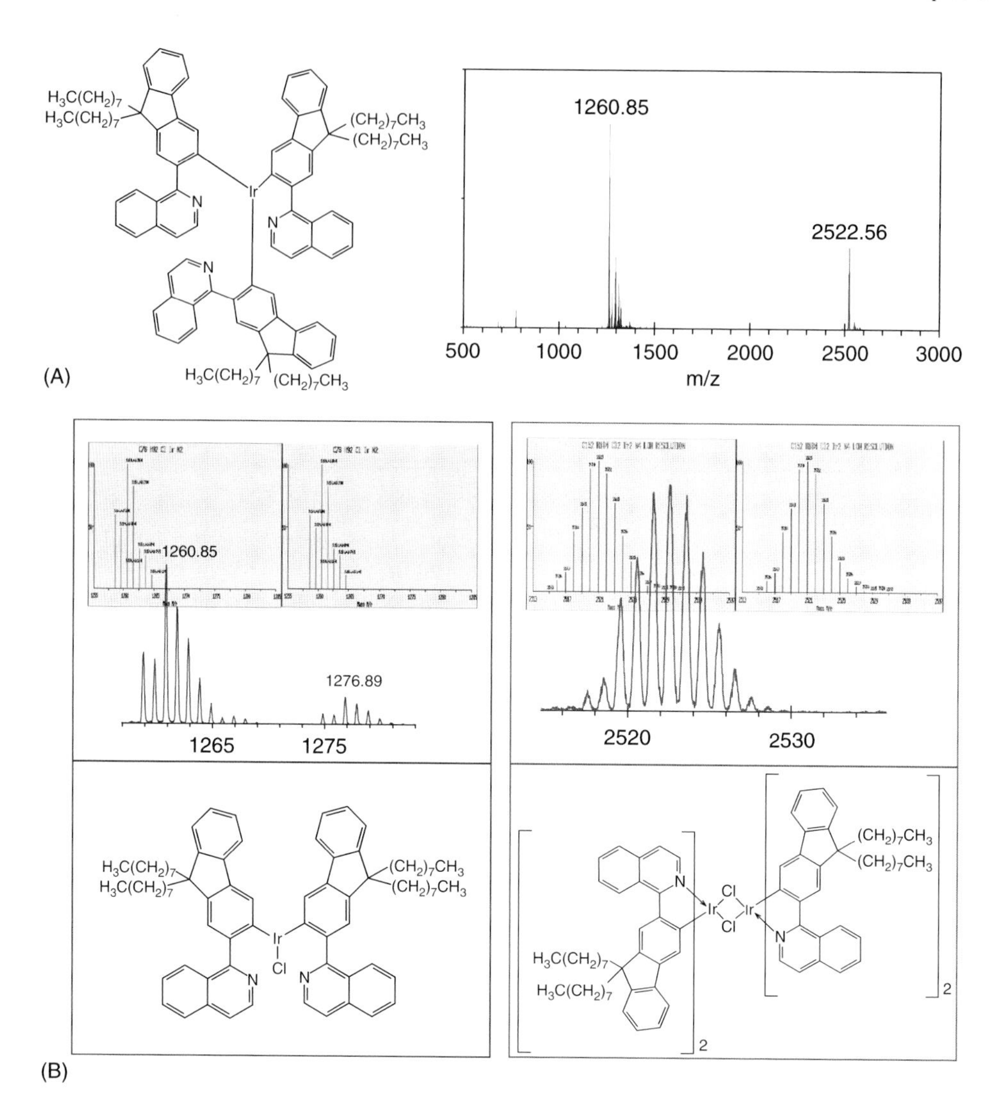

Fig. 6.13. (A) Target compound of a synthesis procedure (left) and the resulting APLI mass spectrum of the raw product (right) upon pneumatically assisted vaporization of a CH_2Cl_2 solution with a heated probe. (B) Close-up of the corresponding mass regions in the survey spectrum shown in a along with calculated stick spectra for the molecular ions of the proposed structures shown at the bottom left and right.

sensitivity and applicability, no difference has been observed between the two dopant-assisted ionization techniques. Future research will address the further improvement of DA–APLI, since we have preliminary evidence of superior performance of DA–APLI as compared to conventional DA–APPI. The most striking difference is the capability of spatially resolved generation of dopant ions within the API interface. However, this technique is regarded as a variation of AP chemical ionization and thus not the focus of our current work.

6.5.2 HPLC–APLI–MS

The HPLC stage was coupled via the pneumatically assisted heated probe to the APLI source. Four PAHs (benzo[*a*]pyrene, fluoranthene, anthracene, and fluorene) were dissolved in CH_3CN and then separated chromatographically for mass-selective detection. For comparison, the same mixture was analyzed using APLI and APCI under otherwise identical chromatographic conditions. With APLI, the parent ion $M^{\bullet+}$ was monitored, with APCI, the $[M + H]^+$ ion. The signals from the UV-diode-array detector of the HPLC were monitored simultaneously. The results are summarized in Fig. 6.14 and clearly show the advantages of APLI for PAH analysis. First, the spectra obtained with APLI are distinguished by strongly reduced baseline noise. Second, the ionization efficiency for the various PAHs is generally orders of magnitude higher with APLI. In APCI, only in the case of benzo[*a*]pyrene comparable signal intensities were obtained, while for anthracene significantly lower signal intensities were recorded and fluoranthene as well as fluorene were even below the APCI detection limit. Both effects, baseline-noise reduction and much greater ionization efficiency, lead to detection limits that are at least two orders of magnitude lower with APLI than with APCI.

6.5.3 Capillary electrochromatography

The interface described in the preceding chapters was based on pneumatically assisted heated AP probe vaporization of the liquid effluent. This setup, however, does not permit

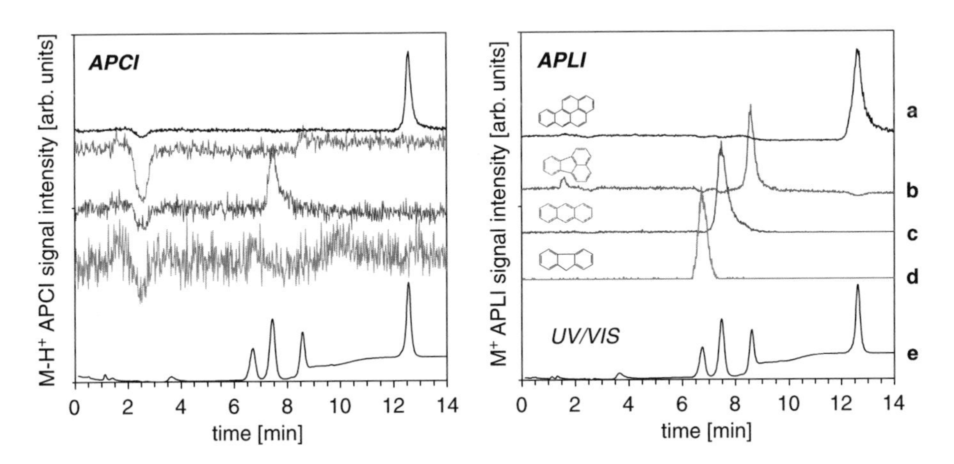

Fig. 6.14. Comparison of mass selectively recorded HPLC chromatograms of a PAH mixture containing benzo[*a*]pyrene (trace a), fluoranthene (trace b), anthracene (trace c), and fluorene (trace d), with APCI (left panel) and APLI (right panel). Additionally, the signal from the UV diode array detector mounted at the end of the HPLC column is shown (trace e in both panels). The samples were injected into the heated probe with a 30-μL loop at a flow rate of 800 μL/min. RP18e (LiChrosphere 100; 250 × 4 mm; 5 μm) was used as a separation column and the concentration of each PAH in the mixture was 20 mg/L. The gradient program for the HPLC analysis was started with MeOH/Water (88/12) and finished at MeOH/Water (100/0). (APLI and MS parameters as described in Section 6.5.1).

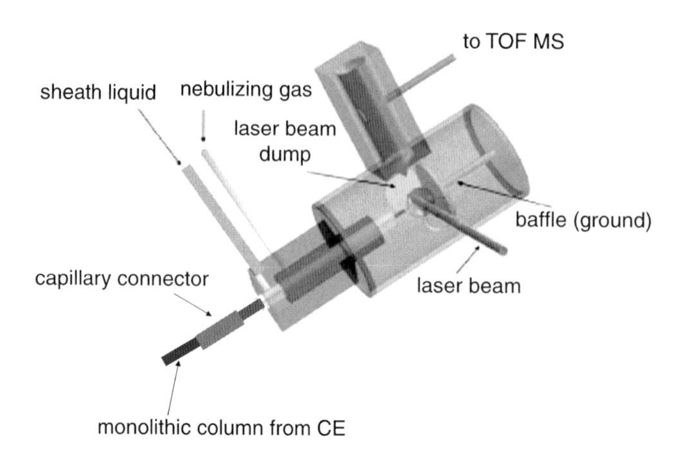

Fig. 6.15. Schematic drawing of an ES APLI-source.

the coupling with CEC. The high temperature in the probe leads to evaporation of the electrolyte in the capillary and thus to the formation of air bubbles and disruption of the separation current. Instead, the laser beam was directed onto the sprayed effluent from a micro ESI source coupled to the CEC stage (Fig. 6.15). A silica monolith CapROD–RP–18 (Merck, Germany) was used as capillary. However, the construction of the interface and the outer diameter of this monolith (360 μm) did not allow pushing of the capillary all the way up to the tip of the ESI interface. For the present feasibility study, the separation capillary was thus coupled with a transfer capillary (10 cm long, 192 μm external diameter, 75 μm internal diameter), which resulted in a reduced field strength for the CEC separation and, consequently, the analysis time was increased.

The analyte solution was vaporized by electrospray. While most polar compounds are already ionized in the liquid phase, the nonpolar aromatic analytes are ionized by the laser beam after spray evaporation (electrospray-atmospheric-pressure laser ionization, ES–APLI). In contrast to the former set-up, the laser beam was gently focused onto the spray, leading to a total ion yield increase of a factor of ten as compared to measurements with an unfocused laser beam delivery. This is readily rationalized by assessing the spatial overlap between the laser beam and microspray, which is optimized upon focussing the laser beam.

Fig. 6.16 shows the simultaneous application of the two ionization methods, ESI and APLI, to a mixture of fluorene and catechine in acetonitrile, which was forced through a fused-silica capillary by pressure. Fluorene is efficiently ionized by APLI, but not by ESI. For catechine, the exact opposite is true: with the laser turned off, only the quasimolecular ion of catechine ($m/z = 291$) was detected; when the laser is switched on, fluorene also appears in the spectra, detected as a radical cation ($m/z = 166$). This experiment shows clearly that the two ionization techniques operate completely independent of each other.

In another set of experiments, a mixture of PAHs consisting of fluorene, anthracene, fluoranthene, pyrene, and benzo[*a*]pyrene was separated by CEC on the monolithic column, and the PAHs were detected by ES–APLI–MS (Fig. 6.17). Since fluoranthene and pyrene have identical molecular weights, the parent-ion signals obtained with APLI did not allow an unambiguous assignment; this was achieved by the addition of fluoranthene and reanalysis.

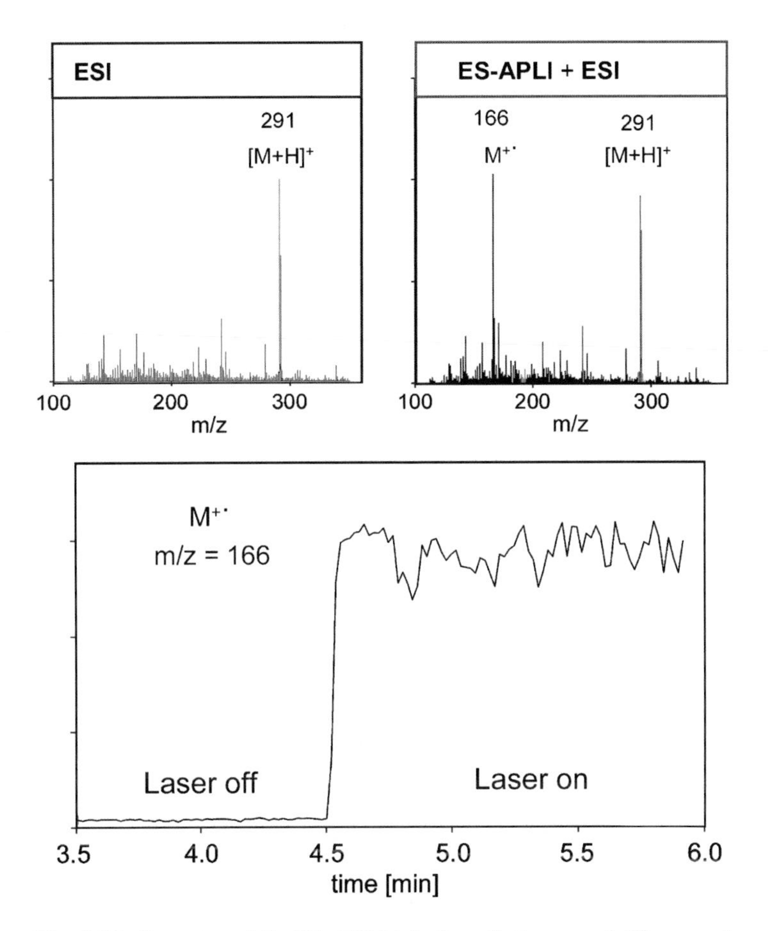

Fig. 6.16. Operation of the ES–APLI interface. Bottom panel: The mass trace at $m/z = 166$ is shown with the laser turned off and on. Without the laser beam, only the polar catechine is ionized by electrospray (mass spectrum top left panel), whereas the ionization of the nonpolar fluorene is only observed in combination with APLI (mass spectrum top right panel).

The setup of the present microspray source is far from an optimum design for separation by capillary electrophoresis and is presented as merely a study of the feasibility of the approach. As an example, the analysis time was considerable reduced on a shortened column using UV analysis. The mixture was analyzed on a 25-cm capillary within 15 min.

We are aware of the fact that only a few real-world analytical problems may exist in which polar and nonpolar substances need to be analyzed by AP–MS in a single run. This is not the goal of our efforts of coupling ES with APLI. Rather, the generation of a "cold" vaporization stage is of interest to us in many aspects: ES–APLI allows the detection of low-polar thermally labile compounds or compounds with low vapor pressure. Another advantage is that the laser energy can be efficiently coupled into micro- and nano-flow sprays by optically reshaping the laser-beam geometry.

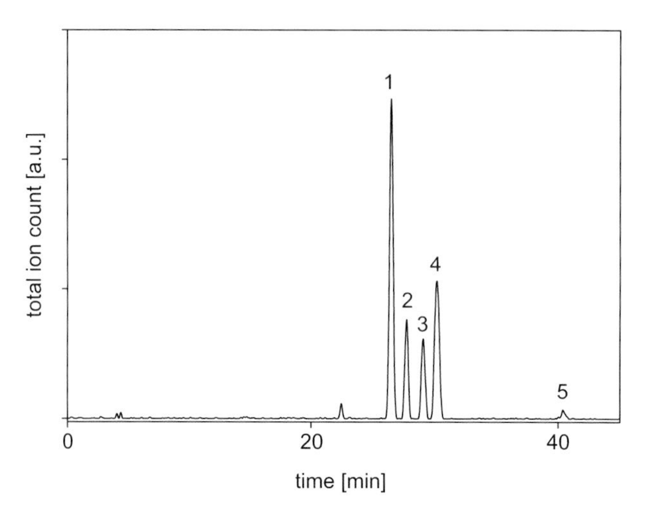

Fig. 6.17. Analysis of fluorene (1), anthracene (2), fluoranthene (3), pyrene (4), and benzo[*a*]pyrene (5) with CEC–ES–APLI–MS. Experimental conditions:
CE: Samples were injected electrokinetically at 10 kV for 5 s; separation voltage set to 30 kV; buffer mixture consisting of 20% 5 mM ammonium acetate at pH 8.0 and 80% acetonitrile; monolithic capillary with 100 μm inner diameter and 450 mm length.
MS: capillary voltage 3 kV; cone voltage 40 V; source temperature 80°C; nebulizing gas N_2 at 22 psi; sheath flow 1 μL/min (99% MeOH:H_2O = 1:1 + 1% acetic acid).

6.6 CONCLUSIONS AND OUTLOOK

In this chapter, we have tried to summarize the current state of the development of APLI–MS as presented in the literature and at international conferences. This research is rapidly advancing and the content of this chapter may have become outdated by the time this chapter is read. APLI is a complementary technique to the existing AP ionization methods. The benefits of APLI are as follows:

1. The location of the ionization region, electrical field gradient, and gas flow can be independently manipulated. This is a unique feature of APLI and extremely useful for efficient and reproducible source optimization.
2. The ionization method is soft. Molecular ions are generated in high yield. In-source fragmentation can be achieved upon increasing the laser power density.
3. Aromatic hydrocarbons are selectively ionized, including nonpolar species, such as PAH. For these compounds, APLI yielded the lowest detection limits by orders of magnitude compared to all other AP ionization methods.
4. Using APLI, virtually no electronic energy is coupled into the matrix, leading to much better-defined boundary conditions concerning investigations of secondary ion–molecule chemistry.
5. Upon addition of a dopant (e.g., toluene or anisole), DA–APLI is established. DA–APLI currently shows the same performance as DA–APPI. Thus, nonaromatic compounds become accessible using the same API source. It is envisioned that further

enhancement, for example, the spatially resolved preparation of dopant ions, will result in a much better performance as currently reported.

To further highlight the potential of APLI, two approaches, which have not been published yet, are very briefly touched on in the concluding two paragraphs.

6.6.1 GC APLI MS: A one-step approach to turning an LC–MS into a GC–MS setup

This book is about advances in LC–MS. Nevertheless, we would like to note that an APLI source is easily adapted for ultrasensitive GC–MS measurements. The pneumatically assisted heated probe used for LC coupling is simply replaced with a heated GC probe of similar design but with optimized gas flow and temperature-control characteristics. The transfer capillary from the GC attaches directly to the GC probe. First results obtained for PAH analysis show that the GC–APLI–MS method outperforms conventional GC–EI–MS by an order of magnitude with respect to sensitivity. The temporal resolution is at least as good.

6.6.2 Leaving the aromatic compound class: derivatization with REMPI tags

In routine LC analysis, the sensitivity and selectivity of the detection stage is often drastically enhanced by "tagging" or derivatization of the analyte(s) to generate compounds with specifically enhanced properties, for example, high fluorescence intensity. This approach allows the sensitive detection of compounds that do not exhibit any native fluorescence with a fluorescence detector. A number of books with excellent collections of lists of specifically designed tags for various analytical detection methods are published (see, e.g., [64,65]). We have just begun to investigate a similar approach for APLI; here, the tag or label exhibits strong overall two-photon ionization cross sections at, for example, 248 nm. One such label was recently synthesized in our laboratory. The label itself ionizes very efficiently, as do analytes that have been tagged. This approach allows access to a vast number of nonaromatic compounds for selective, ultrasensitive APLI analysis and might stimulate very exciting new research.

REFERENCES

1 W.M.A. Niessen, *J. Chromatogr. A,* 1000 (2003) 413–436.
2 W.M.A. Niessen, *J. Chromatogr. A,* 856 (1999) 179–197.
3 W.M.A. Niessen, *J. Chromatogr. A,* 794 (1998) 407–435.
4 R.B. Cole (Ed.), *Electrospray Ionization Mass Spectrometry: Fundamentals, Instrumentation and Applications*, John Wiley & Sons, New York, 1997.
5 N.B. Cech and C.G. Enke, *Mass Spectrom. Rev.,* 20 (2001) 362–387.
6 P. Kebarle, *J. Mass. Spectrom.*, 35 (2000) 804–817.
7 W.M.A. Niessen (Ed.), *Liquid Chromatography–Mass Spectrometry*, Marcel Dekker Inc., New York, 1999.
8 A. Raffaelli and A. Saba, *Mass Spectrom. Rev.,* 22 (2003) 318–331.

9 S.J. Bos, S.M. van Leeuwen and U. Karst, *Anal. Bioanal. Chem.,* 384 (2006) 85–99.
10 J.A. Syage and M.D. Evans, *Spectroscopy,* 16 (2001) 15–21.
11 J.A. Syage, K.A. Hanold, M.D. Evans and Y. Liu, Patent no. WO0197252, 2001.
12 D.B. Robb, T.R. Covey and A.P. Bruins, *Anal. Chem.,* 72 (2000) 3653–3659.
13 D.B. Robb and A.P. Bruins, Patent no. WO0133605, 2001.
14 M. Constapel, M. Schellenträger, O.J. Schmitz, S. Gäb, K.J. Brockmann, R. Giese and T. Benter, *Rapid Commun. Mass Spectrom.,* 19 (2005) 326–336.
15 S. Droste, M. Schellenträger, M. Constapel, S. Gäb, M. Lorenz, K.J. Brockmann, T. Benter, D. Lubda and O.J. Schmitz, *Electrophoresis,* 26 (2006) 4098–4103.
16 S.D. Kramer, G.S. Hurst, J.P. Young, M.G. Payne, M.K. Kopp, T.A. Callcott, E.T. Arakawa and D.W. Beekman, *Radiocarbon,* 22 (1980) 428–434.
17 J.D. Fassett, L.J. Moore, J.C. Travis and J.R. Devoe, *Science,* 230 (1985) 262–267.
18 D.H. Smith, J.P. Young and R.W. Shaw, *Mass Spectrom. Rev.,* 8 (1989) 345–378.
19 D.M. Lubman (Ed.), *Lasers and Mass Spectrometry*, Oxford University Press, New York, 1990.
20 U. Boesl, *J. Phys. Chem.,* 95 (1991) 2949–2962.
21 L. Goodman and J. Philis, Multiphoton absorption spectroscopy, In D.L. Andrews (Ed.), *Applied Laser Spectroscopy: Techniques, Instrumentation and Applications*, VCH Publishers, Inc., New York, 1992.
22 U. Boesl, R. Zimmermann, C. Weickhardt, D. Lenoir, K.W. Schramm, A. Kettrup and E.W. Schlag, *Chemosphere,* 29 (1994) 1429–1440.
23 M.G. Payne, L. Deng and N. Thonnard, *Rev. Sci. Instrum.,* 65 (1994) 2433–2459.
24 M.N.R. Ashfold and J.D. Howe *Ann. Rev. Phys. Chem.*, 45 (1994) 57–82.
25 J. Pfab, Laser-induced fluorescence and ionization spectroscopy of gas phase species. In R.J.H. Clark and R.E. Hester (Eds.), *Spectroscopy in Environmental Science,* Wiley, New York, 1995.
26 K.W.D. Ledingham and R.P. Singhal, *Int. J. Mass Spectrom. Ion Proc.,* 163 (1997) 149–168.
27 U. Boesl, Multiphoton excitation in mass spectrometry. In C. McNeil (Ed.), *Encyclopedia of Spectroscopy and Spectrometry*, Academic Press, New York, 1999.
28 C. Montero, B. Bescos, J.M. Orea and A.G. Urena *Rev. Anal. Chem.*, 19 (2000) 1–29.
29 U. Boesl, *J. Mass. Spectrom.,* 35 (2000) 289–304.
30 K. Wendt and N. Trautmann, *Int. J. Mass Spectrom.,* 242 (2005) 161–168.
31 J.N. Discroll, *Am. Lab.,* 8 (1976) 71–75.
32 J.N. Discroll, *J. Chromatogr.,* 134 (1977) 49–55.
33 D.C. Locke, B.S. Dhingra and A.D. Baker, *Anal. Chem.,* 54 (1982) 447–450.
34 M.A. Baim, R.I. Eartherton and H.H. Hill Jr., *Anal. Chem.*, 55 (1983) 1761–1766.
35 C.S. Leasure, M.E. Fleischer, G.K. Anderson and G.A. Eiceman, *Anal. Chem.,* 58 (1986) 2142–2147.
36 G.E. Spangler, J.E. Roehl, G.B. Patel and A. Dorman, US. Patent no. 5338931, 1994.
37 T.J. Kauppila, T. Kuuranne, E.C. Meurer, M.N. Eberlin, T. Kotiaho and R. Kostiainen, *Anal. Chem.,* 74 (2002) 5470–5479.
38 S.G. Lias, Ionization energy evaluation, In P.J. Linstrom and W.G. Mallard (Eds.), *NIST Chemistry WebBook, NIST Standard Reference Database Number 69*, National Institute of Standards and Technology, Gaithersburg MD, 20899, March 2003 (http://webbook.nist.gov).
39 T.P. Softley, *Int. Rev. Phys. Chem.,* 23 (2004) 1–78.
40 P.W. Atkins, *Physical Chemistry* (6th edn.), W.H. Freeman and Co, New York, 1999.
41 M. Nomayo, R. Thanner, H. Pokorny, H.H. Grotheer and R. Stutzle, *Chemosphere,* 43 (2001) 461–467.
42 H. Oser, K. Copic, M.J. Coggiola, G.W. Faris and D.R. Crosley, *Chemosphere,* 43 (2001) 469–477.
43 R. Zimmermann, H.J. Heger, A. Kettrup and U. Nikolai, *Fresenius J. Anal. Chem.,* 366 (2000) 368–374.

44 L. Cao, F. Mühlberger, T. Adam, T. Streibel, H.Z. Wang, A. Kettrup and R. Zimmermann, *Anal. Chem.,* 75 (2003) 5639–5645.
45 R. Dorfner, T. Ferge, A. Kettrup, R. Zimmermann and C. Yeretzian, *J. Agric. Food Chem.,* 51 (2003) 5768–5773.
46 R. Zimmermann, H.J. Heger, C. Yeretzian, H. Nagel and U. Boesl, *Rapid. Commun. Mass Spectrom.,* 10 (1996) 1975–1979.
47 L. Oudejans, A. Touati and B.K. Gullett, *Anal. Chem.,* 76 (2004) 2517–2524.
48 J. Franzen, R. Frey and H. Nagel, *J. Mol. Struct.,* 347 (1995) 143–152.
49 U. Boesl, H. Nagel, E.W. Schlag, C. Weickhardt and R. Frey, Vehicle exhaust emission time-resolved multicomponent analysis by laser mass spectrometry. In R.A. Meyers (Ed.), *The Encyclopedia of Environmental Analysis and Remediation,* Vol. 8, Wiley, New York, 1998.
50 P.O. Haeflinger and R. Zenobi, *Anal. Chem.,* 70 (1998) 2660–2665.
51 W.F. Chan, G. Cooper, R.N.S. Sodhi and C.E. Brion, *Chem. Phys.,* 170 (1993) 81–97.
52 W.F. Chan, G. Cooper and C.E. Brion, *Chem. Phys.*, 170 (1993) 99–109.
53 G.R. Burton, W.F. Chan, G. Cooper and C.E. Brion, *Chem. Phys.,* 181 (1994) 147–172.
54 J.W. Au, G. Cooper, G.R. Burton, T.N. Olney and C.E. Brion, *Chem. Phys*., 173 (1993) 209–239.
55 S. Eden, P. Limao-Vieira, P. Kendall, N.J. Mason, S.V. Hoffmann and S.M. Spyrou, *Eur. Phys. J. D,* 26 (2003) 201–210.
56 E. Marotta, R. Seraglia, F. Fabris and P. Traldi, *Int. J. Mass Spectrom.,* 228 (2003) 841–849.
57 G. Scoles, (Ed.), *Atomic and Molecular Beam Methods,* Oxford University Press, New York, 1988.
58 H. Oser, R. Thanner and H.H. Grotheer, *Combust. Sci. Technol.,* 116 (1996) 567–582.
59 S. Schmidt, M.F. Appel, R.M. Garnica, R.N. Schindler and T. Benter, *Anal. Chem.,* 71 (1999) 3721–3729.
60 R.M. Garnica, M.F. Appel, L. Eagan, J.R. McKeachie and T. Benter, *Anal. Chem.,* 72 (2000) 5639–5646.
61 J.R. McKeachie, W.E. Van der Veer, L.C. Short, R.M. Garnica, M.F. Appel and T. Benter, *Analyst,* 126 (2001) 1221–1228.
62 M.F. Appel, L.C. Short, T. Benter, 15 (2004) 1885–1896.
63 R. Handerson and K. Schulmeister, *Laser Safety*, Taylor & Francis, New York, 2004.
64 G. Lunn and L.C. Hellwig, *Handbook of Derivatization Reactions for HPLC*, Wiley Interscience, New York, 1998.
65 T. Toyo'oka, *Modern Derivatization Methods for Separation Sciences*. Wiley Interscience, New York, 1999.

Achille Cappiello (Editor)
Advances in LC–MS Instrumentation
Journal of Chromatography Library, Vol. 72

Chapter 7

LC–ICP-MS – a primary tool for elemental speciation studies?

KATARZYNA WROBEL, KAZIMIERZ WROBEL and
JOSEPH A. CARUSO

7.1 INTRODUCTION: TARGET ELEMENTS AND SAMPLES IN SPECIATION ANALYSIS

It is widely accepted that the mobility, bioavailability, retention and specific biological functions of an individual element depend on its physicochemical form. Furthermore, the variety of elemental forms occurring in the environment and in living organisms depend on the natural composition of earth crust, certain anthropogenic activities, as well as different degradation and/or biotransformation processes. Thus, the information of element speciation is of great importance in all research areas related to human health and environmental pollution. According to the IUPAC recommendations, the term "chemical species" refers to a specific and unique molecular, electronic or nuclear structure of an element and "speciation analysis" is the measurement of the quantities of one or more individual chemical species in a sample [1]. Analytical results may help in better understanding of the mechanisms responsible for element transport and/or degradation in the environment and in elucidation of the element pathways in living organisms [2–4]. Such results are also useful in studies on possible impact of elements in certain pathological processes (protective role versus toxicological aspect) [3,5]. Within this context, the current trend is often termed metallomics, referring to studies on the entirety of metal and metalloid species in a defined biological system (cell, tissue, organ or even entire living organism) [2,6–10]. Other relevant topics include more efficient elimination and/or recovery of elements from industrial and mining effluents (Hg, As, Cd, etc.), refining certain industrial processes (tanning with chromium salts, metal-based catalytic mixtures, etc.), development of new metal drugs (Pt, Ru, Ti) as well as promoting necessary regulations for the commercial use of metal/metalloid based biocides (As, Sn) [11–16].

Several elements are of key importance in speciation analysis, mainly because their species present drastically different biological behavior. Thus, lower oxidation states of arsenic and of inorganic selenium are more toxic than their higher oxidation states [3]. In contrast, hexavalent chromium (Cr(VI)) presents higher toxicity with respect to trivalent form (Cr(III)), which at lower concentrations exhibits beneficial effect in glucose metabolism [17]. Furthermore, depending on the element, the organic compounds can be more or less harmful than inorganic forms (mercury and arsenic respectively) [18]. Selenium is one of the most studied elements because of its species-dependent essentiality observed in a very narrow concentration band [19]. The target elements together with their primary species and samples typically analyzed are given in Table 7.1. As can be observed, element species of interest include: different oxidation states, alkylated metal and/or metalloid compounds, selenoaminoacids and selenopeptides [20]. In addition, applications in studies on the pharmacokinetics of metal-based drugs (Pt, V, Au), metalloporphyrins (Ni, V, Fe, etc.), heavy metals in phytochelatins (Cd, Cu, Zn, Hg, etc.) and in humic substances should be mentioned [3,21]. Recently, the research interest has been expanded toward molecular forms of few non-metals (P, I, Br, S) that are important in biological systems and also in the environmental studies. In the living organisms, phosphorus is associated with different biomolecules, affecting their biological activity. Thus, protein phosphorylation, as the primary post-translational modification, is essential for signal transduction; phosphorus provides the backbone in the RNA or DNA chain and phospholipids are present in all cell membranes being involved in the transport processes [22–25]. In relationship to the environmental studies, organic phosphorus compounds are widely used as pesticides, surfactants and flame retardants [26,29]. Iodine is another essential non-metal, which is required for the synthesis of thyroid hormones (T4 – tetraiodothyronine, T3 – triiodothyronine) [30–32]. Since the information on iodine species (iodide, monoiodothyrosine (MIT), diiodothyrosine (DIT), T3, T4) in clinical samples may explain metabolic abnormalities related to thyroid metabolism, speciation analysis is needed [20,33]. Pharmaceutical formulations, food products and sea water have also been analyzed [31,34–36]. Sulfur atom is present in two proteinogenic aminoacids (methionine and cysteine), so in this case speciation studies have been focused principally on metallothioneins [2]. The relation between metals (Mn, Fe, Cd, Zn) or metalloids (As, Se) and sulfur has provided useful information on element binding to proteins [9,10,37–39]. The interest in bromine-containing compounds in environmental samples has arisen because their use as preservatives and flame retardants [40–42].

The challenge of element speciation analysis is to identify and/or quantify relatively low concentrations of few to several target species (concentration values far below the total element content) in a complex chemical matrix of the real-world samples. Very often not all element species are known, they may exhibit similar physicochemical properties and usually present certain chemical lability [1]. Thus, the important features of an analytical tool suitable for speciation analysis are excellent selectivity and high sensitivity. A common speciation scheme after sample preparation involves a fractionation step followed by the quantification of element in the obtained fractions [1,4,18,43–48] Using hyphenated techniques that combine a separation and detection steps into one operating on-line system, the selectivity is achieved by application of powerful chromatographic or electrophoretic techniques and high sensitivity is assured by atomic spectrometric detection. The important advantages of hyphenated techniques are minimized contamination risk, enhanced precision, short time of analysis and relatively easy data handling [1,17,44,46]. It should be stressed, however, that hyphenated techniques with element-specific detection do not provide structural information for the species. If the appropriate standards are available, the species identity can be

TABLE 7.1

TARGET ELEMENTS, THEIR SPECIES AND TYPICAL SAMPLES IN THE SPECIATION ANALYSIS

Element	Species of interest	Chemical matrices	Relevance
As	Stable oxidation states: As(III) and As(V) Monomethylarsonic acid (MMA) Monomethylarsonous acid (MMA(III)) Dimethylarsinic acid (DMA) Dimethylarsinous acid (DMA(III)) Trimethilarsineoxide (TMAO) Tetramethylarsonium ion (TETRA) Arsenocholine (AC) Arsenobetaine (AB) Arsenosugars (AS)	Aquatic samples, soils, sediments, marine organisms, plants and clinical samples	Evaluation of contamination and/or toxicological risk, studies on detoxification routes and on biomethylation pathways
Cr	Stable oxidation states: Cr(III) and Cr(VI)	Natural and waste waters, soils, sediments, microbiological samples, tanning liquors	Environment pollution, bioremediation, efficiency of tanning process
Hg	Inorganic (Hg(II)) Mono- and dimethylmercury (MMM, DMM) Mono- and diethylmercury (MEM, DEM) Monophenylmercury (MPM)	Sea water, natural and waste waters, sediments, food products, fish tissues, biological fluids	Evaluation of toxic organomercuric compounds – products of biomethylation processes
Pb	Inorganic Pb(II) and alkyllead species: $R'_nR''_mPb^{(4-m-n)+}$ (R′-methyl, R″-ethyl)	Air, sea water, natural waters, plants, food products and clinical samples	Determination of toxic and persistent lead species in biological matrices
Pt	Anticancer drugs and their metabolites: Cisplatin, Carboplatin, Oxaplatin, Lobaplatin, Nediplatin, JM216	Drug formulations, clinical materials	Pharmacokinetics and toxicological studies

(Continued on next page)

TABLE 7.1 (*contd.*)

Element	Species of interest	Chemical matrices	Relevance
Se	Stable oxidation states: Se(IV) and Se(VI) Organic selenium species: Mono- and dimethylselenide (MMSe, DMSe) Mono- and diethylselenide (MESe, DESe) Trimethylselonium ion (TMSe) Selenomethionine (SeMet) Methyl-selenomethionine (MeSeMet) Selenocysteine (SeCys or oxidized SeCis) Methyl-selenocysteine (MeSeCys) Selenohomocystine (SeHcy) Selenoethionine (SeEt) Adenosylselenomethionine (AdoSeMet) Adenosylselenohomocysteine (AdoSeHcy) γ-Glutamyl-methylselenocysteine *N*-acetylselenohexosamines	Environmental samples, yeast, plants from allium family and Se-accumulating plants, clinical samples	Elucidation of biological pathways and specific role of Se species, phytoremediation
Sn	Inorganic forms: Sn(IV) and Sn(II) Organotin compounds: $R_nSn^{(4-n)+}$ ($1 < n < 4$; R-methyl, ethyl, butyl or phenyl)	Sea water, natural waters, sediments, soils, marine organisms	Harmful effects of organotin pesticides, anti-fouling paints and their degradation products

established based on its retention time, the shape of chromatographic peak and by using the method of standard addition. On the other hand, the identification of unknown forms and/or ultimate confirmation of unexpected compounds observed in the sample require the use of complementary techniques (molecular mass spectrometry or NMR, if the concentration is high enough) [6,49].

The primary separation techniques in speciation analysis have been liquid chromatography (LC) [2,17,20,44,50–52], gas chromatography (GC) [4,39,48,53] and to less extent capillary electrophoresis (CE) [46,54,55]. Since many elemental species of biological and clinical relevance are not volatile, LC has been widely used. This chromatographic mode offers exceptional versatility based on the convenient selection of column type and mobile phase composition [20,50]. On the other hand, among different element-specific detectors available, inductively coupled plasma mass spectrometry (ICP-MS) is today a method of choice. In addition to the unique features of this technique observed in the analysis of total element content, it offers the important advantage of relatively easy coupling with fractionation/separation step [1,6,21,43,44,56–58]. In this chapter, recent progress in LC–ICP-MS instrumentation is reviewed and a survey over key applications in the real-world samples is presented.

7.2 INDUCTIVELY COUPLED PLASMA MASS SPECTROMETRY (ICP-MS) AS ELEMENT-SPECIFIC DETECTOR

Inductively coupled plasma (ICP) is an ion source, which generates mainly monoatomic positively charged ions of most metals, metalloids and some non-metals with an efficiency close to unity, and under similar operating conditions. The plasma operates at high temperatures (6000 – 10000 K) and at atmospheric pressure. The plasma is formed within a quartz torch, which consists of three concentric tubes, through which argon gas is introduced at different flow rates. Radiofrequency (RF) electromagnetic field (27.12 or 40.68 MHz) is generated at the top of the torch using a “load coil”. When argon stream is introduced through the outer tube of the torch (10 – 16 l min^{-1}), the plasma is ignited by a spark discharge and the charged particles are forced to flow in a closed annular path. These fast moving ions and electrons collide with other atoms of argon to produce further ionization. High thermal energy results from the resistance of charged particles to their flow and the plasma becomes self-sustained. The argon stream introduced by the external tube is used to cool the torch as well and also helps to center and stabilize the plasma because of the helical gas flow. The liquid or gaseous sample is then introduced to plasma through the inner torch tube and the thermal energy causes sample atomization, excitation and ionization [10,59]. Typically, the liquid samples are introduced to plasma via nebulization, using different pneumatic (concentric, V-groove, cross-flow, etc.) or ultrasonic nebulizers. A spray chamber is also necessary for elimination of large droplets and reduction of the solvent load to plasma [44,50,60].

The ICP has been used as ion source for optical emission spectrometry for about 30 years, but since the introduction of the first commercial inductively coupled plasma mass spectrometers (ICP-MS) in the mid-1980s, this technique has quickly become recognized as a powerful tool and primary detection technique for element determination and speciation

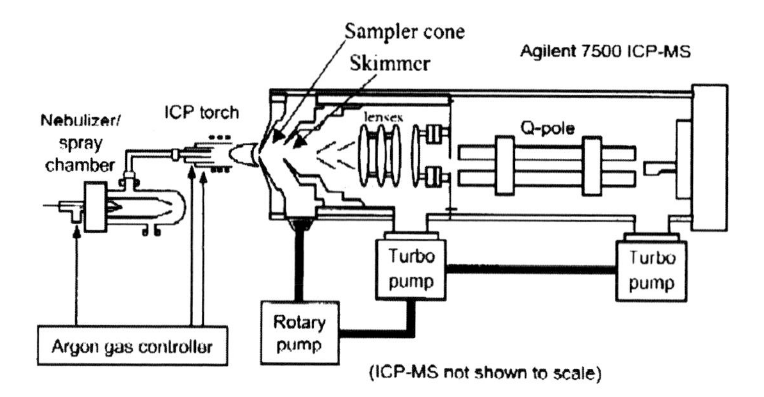

Fig. 7.1. Scheme of a quadrupole-based ICP-MS instrument. (Reproduced from Ref. [46] with the permission of Elsevier Science).

[3,6,49,61]. In ICP-MS, the ions are extracted from the atmospheric plasma into the low-pressure mass spectrometer through a cooled nickel or platinum sampling cone (1 mm diameter). Expansion of the gas then occurs in the low-pressure region behind the cone before a fraction of the ions pass through another cone (skimmer) and the majority of the argon is pumped away. Finally, the ions are focused into a more direct path to the mass analyzer using a series of ion "lenses" which are simply a series of electrodes, held at variable dc voltages. After separation of ions according to their mass to charge ratio in the mass analyzer, the detector (channel electron multiplier, Faraday cup, or discrete dynode electron multiplier) converts the ion stream into an electrical signal, which can be correlated through the calibration procedure to the concentration of targeted elements within the sample [62,63]. The most popular mass filter in the ICP-MS instruments is the quadrupole analyzer, which gives better than unit mass resolution over a mass range up to $m/z = 300$. The scheme of quadrupole-based ICP-MS instrument is shown in Fig. 7.1 [46]. On the other hand, the increasing applications of high resolution mass filters (sector-field or double-focusing analyzers) and simultaneous mass analyzers (quadrupole ion trap, time of flight and multichannel devices) should be noted [6,64,65].

The combination of the high ionization efficiency of ICP and excellent selectivity and sensitivity of MS drive the exceptional features of ICP-MS for element-specific detection. This technique offers: (i) low detection limits (routinely sub-nanogram per liter); (ii) high sensitivity; (iii) wide dynamic range (up to 9 orders of magnitude); (iv) excellent precision in counts rates near the signal baseline (0.2%–3% RSD); (v) less pronounced interferences as compared to other atomic spectrometry techniques and (vi) multielement and isotopic capabilities.

As a counter-argument, especially in the early applications, ICP-MS presented several shortcomings mainly due to troublesome polyatomic interferences, low ionization efficiency for biologically important elements (Se, S, P, halogens) and sequentially scanning nature of mass analyzer and detection system. The need for reliable results of total element determinations and, in particular, increasing research interest in speciation analysis have spurred the instrumental and methodological development, thus extending considerably the scope and capabilities of ICP-MS.

7.2.1 Polyatomic interferences

As already mentioned, the most important advantages of ICP as an MS ion source include high temperatures, relatively long analyte residence times, high electron density, efficient ionization (for most elements) and a relatively inert environment. In addition, singly charged positive element ions are generated (with exception of those with low second ionization energy, like barium), making the mass spectra relatively simple. It should be stressed however that, high solvent and/or salt loads alter considerably the stability of plasma and its ionization characteristics. Depending on the plasma operating conditions and also on the composition of sample, different charged species originated from plasma argon, air and/or sample may be generated with possible isobaric overlap between their *m/z* values and those of the target ions. In Table 7.2, common polyatomic interferences observed for arsenic, iron, phosphorus, sulfur and selenium are listed together with key strategies proposed for their elimination [6,10,18,23,25,26,37,43,65–103].

7.2.1.1 Optimization of the plasma operating conditions

For recognition of polyatomic interferences, the monitoring of few isotopes of one element can be helpful. For these poly-isotopic elements (Se, S), the isobaric interference can be avoided just switching to interference-free *m/z* of another isotope, which however involves the loss of sensitivity. Mathematical correction equations have also been developed [43,63,104]. On the other hand, the formation of charged species originated from plasma gas depends on the operating conditions applied. It has been shown that, using reduced RF forward power (500 – 800 W) and increased nebulizer Ar flow rates (1.5 – 1.8 l min^{-1}) plasma becomes “cooler”, hindering generation of Ar-based ions [43,63,79,05]. The weak point of the cold-plasma strategy is the increased contribution of polyatomic charged species originating from the sample matrix. Such interference can be avoided by separating the analyte from the potentially interfering sample components, prior to its introduction to ICP-MS [106,107]. Thus, different chromatographic, extraction and dialysis procedures have been reported for the elimination of chloride ions prior to As determination/speciation [73,108–111].

7.2.1.2 High resolution instruments

About 90% of all ICP-MS systems are equipped with quadrupole analyzer, the resolution of which (0.7–1.0 amu) is not adequate to differentiate many metalloids and non-metal elements from argon-, solvent- and matrix-based spectral interferences (especially in the mass range 40–80) [1,112]. Therefore, the commercial availability of high resolution ICP-MS (ICP–HRMS) instruments has enabled a significant advance in the analysis/speciation of non-metal elements. Mass analyzers typically used are composed of either magnetic sector or double-focusing sector device. In the magnetic sector, the field is perpendicular to the ion beam from the plasma. The trajectory of the ions in a circular path depends on mass to charge ratio, permitting the separation of different elements and isotopes [113,114]. In double-focusing sector mass analyzers, the ions are additionally subjected to an electrostatic field with the resultant trajectory being dependent on the

TABLE 7.2

POTENTIAL POLYATOMIC INTERFERENCES IN ICP-MS DETERMINATION OF AS, FE, P, S AND SE AND ACTUAL STRATEGIES USED FOR THEIR ELIMINATION

Element	Most abundant isotopes	Interfering species	Elimination strategy	Ref.
Arsenic	^{75}As (100%)	$^{40}Ar^{35}Cl^{+}$	High resolution ICP-MS instrumentation	[65–67]
			ICP operated under cool plasma conditions	[43,68]
			Collision/reaction cell technology	[18,69–71]
			Elimination of chloride ion (extraction, chromatographic separation, hydride generation)	[67,72–75]
Iron	^{56}Fe (91.75%)	$^{40}Ar^{16}O^{+}$	Collision/reaction cell technology	[71,76–79]
			High resolution ICP-MS instrumentation	[80–82]
			Optimization of plasma operating conditions	[6,79,83]
			Membrane desolvation	[79,84]
Phosphorus	^{31}P (100%)	$^{15}N^{16}O^{+}$, $^{14}N^{16}OH^{+}$, $^{12}CH_3{}^{16}O^{+}$, $^{62}Ni^{2+}$	Collision/reaction cell technology	[23,25,26, 85,89,90]
			High resolution ICP-MS instrumentation	[86–88]
			Desolvation to minimize plasma loading with $^{15}N^{16}O^{+}$ and $^{14}N^{16}OH^{+}$	[91]
			Optimization of plasma parameters	[92]
Sulfur	^{32}S (95.02%)	$^{16}O_2{}^{+}$, $^{14}N^{18}O^{+}$, $^{15}N^{16}OH^{+}$	Monitoring less abundant ^{34}Se (4.29%)	[2]
			High resolution ICP-MS instrumentation (R = 3000)	[93,94]
			Collision/reaction cell technology	[37,85,90, 95]
			Desolvation to minimize the formation of $^{16}O_2{}^{+}$	[96]
Selenium	^{78}Se (23.77%) ^{80}Se (49.61%)	$^{40}Ar\,^{38}Ar^{+}$, $^{31}P_2{}^{16}O^{+}$ $^{40}Ar_2{}^{+}$, $^{79}BrH^{+}$	High resolution ICP-MS instrumentation	[97]

(*Continued on next page*)

TABLE 7.2 (*contd.*)

Element	Most abundant isotopes	Interfering species	Elimination strategy	Ref.
	^{82}Se (8.73%)	$^{40}Ar_2H_2^+$, $^{12}C^{35}Cl_2^+$, $^{34}S^{16}O_3^+$, $^{82}Kr^+$, $^{81}BrH^+$,	Collision /reaction cell technology	[10, 98–102]
		$^{65}Cu^{17}O^+$, $^{64}Zn^{18}O^+$, $^{68}Zn^{14}N^+$	Deconvolution using the mass domain and signal intensity based domains	[103]

energy of the ion [113,114]. In spectrometers based on reversed Nier–Johnson geometry the electromagnet is placed before the electrostatic analyzer (ESA) [112]. When the energy dispersion of both the magnetic sector and ESA are equal in magnitude, but opposite in direction, both ion angles and ion energies are focused. Switching the direction of the electric field during the cycle time of the magnet, however, stops the ions of the particular mass passing through the exit slit, allowing for their measurement. Consequently, only the masses of interest can be monitored, rather than having to scan the entire mass range [115,116].

The resolving power ($R = m/\Delta m$) of ICP–HRMS can be changed (between 300 and 10,000), as compared to the range of 300–400 typically offered by traditional quadrupole mass analyzers [117]. The desired resolution is achieved by selecting different entrance and exit slit widths. However, the narrower slit widths used to yield higher resolution also limit the ions transmitted to the detector, subsequently reducing sensitivity. For that reason, it is necessary to use the minimal resolution required to resolve the analyte of interest from the interfering ion. The selection of R (high, medium or low) must be based on the difference between m/z values and also on the actual concentrations of the analyte and interfering species. For example, a theoretical resolution necessary to separate $^{32}S^+$ signal (m/z 31.97207) from $^{16}O_2^+$ (m/z 31.98982) is 1801. However, when the interfering species was much more abundant than $^{32}S^+$, a distinct low-mass tail was observed and a higher resolution (at least 3000) was required [93].

An important benefit from sector-field instruments is a high signal to noise ratio, as a result of very good ion transmission (high sensitivity at low R) and low background (<0.2 ions s^{-1}). As a consequence the detection limits in the range of pg l^{-1} are typically obtained [2,64]. Finally, excellent precision (0.1–0.5% RSD) of ICP–HRMS should be mentioned.

Despite the obvious advantages of ICP–HRMS instruments, they are significantly more expensive than the quadrupole mass analyzers commonly employed. They also require longer analysis times, which make them less than ideal for high-throughput applications. However, these instruments have dramatically enhanced the analytical capabilities of ICP-MS by increasing the number of elements that can effectively be measured [6, 2,64,118]. In particular, ICP–HRMS technology has been used to mitigate polyatomic interferences in the determination of biologically important As, Fe, P, S and Se (Table 7.2).

The important limitation of the quadrupole and sector-field analyzers for multielemental and LC–ICP-MS applications is their sequentially scanning character. When the instrument is operated in such non-continuous manner, duty cycle is reduced producing lower signal to noise ratio, poorer precision and also lower mass resolution for increasing number of elements/isotopes monitored [119]. Logically, longer observation time can be used to circumvent this problem. However, when ICP-MS is used as element-specific detector for LC, ions are generated from relatively fast transient events and the observation time is restricted by chromatographic peak width. Common error in such measurements (especially while observing multiple elements/isotopes) is spectral skew [6,48,120]. Recently available ICP–time of flight (TOF) mass spectrometers with scan rates up to 20,000 spectra/s can overcome this drawback, because rather than time-consuming scanning from one mass to the other, the mass filter distinguishes the different isotopes by their individual (*m/z*-dependent) time of flight to the detector [1,107,120,121]. For simultaneous detection of an entire atomic mass range, ICP has also been used with the quadrupole ion traps (QIT), Fourier transform ion cyclotron resonance mass spectrometers (FTICR) and multichannel mass spectrometers [6,28,107]. As referred to LC–ICP-MS coupling, the increasing applications of TOF and multichannel mass spectrometers should be noted [49,64,107,122,123].

7.2.1.3 Collision/reaction cell technology

Over the past decade, the development of quadrupole ICP-MS instruments equipped with collision/reaction cells inserted between the ion optics and the mass filter has offered an interesting alternative to a high resolution systems for the removal of polyatomic interferences [71,119,124–127]. Even though this technology was first introduced decades ago and intensively used as a tool enabling structural characterization of organic molecules by MS, it has more recently been employed in atomic mass spectrometry. Conceptually, a reaction cell for ICP-MS provides adequate conditions for controlled reaction or conversion of the interfering species or the analyte itself, which results in a change of their *m/z* and thereby enables safe determination of the element at quadrupole resolution [119]. A collision/reaction cell instrument essentially resembles the configuration of a traditional ICP-MS except that a collision/reaction cell with gas inlet is positioned just before the quadrupole mass analyzer. The cell consists of another quadrupole, hexapole or octopole to which only RF-voltages are applied. When subjected to this field, incoming ions are focused and collide or react with the collision/reaction gas that is admitted through the cell gas inlet. The collisions and reactions that occur between the analyte ions and the gas molecules can eliminate problematic interferences by (1) collisional deactivation where the energy of the collision gas exceeds the bond energy of the interfering polyatomic, (2) the reaction of the analyte with the cell gas shifting its *m/z* or (3) energy discrimination by preferentially reducing the polyatomic ion energy over the analyte ion energy (because the polyatomic has a larger collision cross-section), followed by carefully adjusting the quadrupole bias voltage to admit only the more energetic analyte ions.

Unfortunately, under these conditions rather complex secondary collisions and reactions can occur within the cell which can produce other interfering species, which must also be eliminated before reaching the detector. These secondary interferences are rejected in different ways, depending on the multipole of the cell. Quadrupoles act through mass discrimination, whereas higher order multipoles, such as hexapoles and octopoles can discriminate analytes from interferences based on kinetic energy [3,128]. In the view of these authors, the best operation of the cell is one that minimizes the chemical reactions taking place.

Quadrupoles, as a selective bandpass filters, can be optimized and easily changed for various interferences. Also, in a practice known as dynamic reaction cell (DRC) technology, quadrupoles permit the use of highly reactive gases, including ammonia and methane, as the collision/reaction gas [119,129]. Such gases can act as catalysts for ion molecule chemistry, converting any interfering species into neutral or otherwise uninhibiting species. Because they are highly reactive, the number of potential ion-molecule interactions is increased. Furthermore, by-products of these reactions can be prevented from forming other interfering species simply by scanning the bandpass of the cell's quadrupole. DRC technology has been successfully used for elimination of interferences on $^{52}Cr^+$ ($^{40}Ar^{12}C^+$) [130–132], $^{40}Ca^+$ ($^{40}Ar^+$) [133], $^{31}P^+$ ($^{15}N^{16}O^+$, $^{14}N^{16}OH^+$) [133], $^{51}V^+$ ($^{35}Cl^{16}O^+$) [132,134], $^{56}Fe^+$ ($^{40}Ar^{16}O^+$) [79], $^{80}Se^+$ and $^{82}Se^+$ (argon dimers) [135] and also in multielemental ICP-MS analysis [136,137]. The use of oxygen as the reaction gas to convert target phosphorus, sulfur, vanadium to their oxides should also be mentioned [85,107,134]. However, derivatization reactions, as these are, also lead to an inefficient use of analyte.

Hexapoles, on the other hand, do not permit the use of highly reactive gases due to their inability to act as mass filters and thus eliminate the undesired products of secondary reactions. Rather, these multipoles act as kinetic energy discriminators and require strictly low reactivity gases like helium, hydrogen or xenon. Analyte ions can be distinguished from other ions by allowing the cell bias to be slightly less positive than the mass filter bias. This causes ions with the same energy as the cell bias to be rejected, whereas the analyte ions, which have an energy higher than the cell bias, can be transmitted. Octopoles operate in a similar manner, but offer higher transmission characteristics, particularly at the lower end of the mass range. Several applications of multipole instruments have been reported for ICP-MS quantification of phosphorus [2,23–25,29], sulfur [2,15,37,138], selenium [10,98–101,119,139–143], arsenic [69,70,144,145], nickel [146–148] and also in the multielement and isotopic analyses [107,149–152].

7.2.2 Isotope analysis by ICP-MS

One of the main characteristics of mass spectrometry is the possibility of precise and accurate isotope ratio measurements. However, the analytical potential of such measurements is limited by several effects encompassing instrumental mass bias, spectral interferences, instrumental background, mass scale drift, plasma instabilities, or ion intensity drift [64,107,153,154]. Mass discrimination occurs because in mass analyzer, the light ions are deflected to a greater extent than the heavy ones, so the isotope ratio of lighter to heavier ions is biased (smaller than the true value). This effect is more pronounced for elements with low atomic mass and can be corrected using a suitable internal isotopic standard (e.g., Tl) [154, 255]. The dead time of the ion detector should also be mentioned as an uncertainty source. It is especially important for counting rates higher than 10^6 cps, since the lower number of counts are registered than actually occur. These effects depend on the ICP-MS instrumentation employed [64,107,153–155]. In particular, the quadrupole-based spectrometers with collision/reaction cells, high resolution instruments and those offering simultaneous monitoring of different ions are well suited for isotopic analysis. If the sector-field mass analyzer is operated at low resolution ($R \sim 300$), the isotope ratios can be measured with a precision as good as 0.02% (RSD) and the detection limits in pg l^{-1} region [107]. Through applying the multiple ion collectors to sector-field MS under the conditions of continous sample introduction, the precision has further been improved (0.001 – 0.002% RSD) [153, 154]. These instruments, as coupled to liquid chromatography for speciation analysis, enable reducing the external precision of transient signals by less than one order of magnitude [156,157].

When the natural isotope ratio is altered by spiking a sample with an analyte standard of a different isotopic abundance, the change in the measured isotopic ratio can be used to elucidate the original elemental composition of the sample. The only real requirements of the isotope dilution technique (ID) are that the analyte of interest has more than one stable isotope and that the spike standards are isotopically pure. Over the past 40 years, isotope dilution mass spectrometry (IDMS) has proven to be a very effective method for the measurement of total element concentration [158–161], and has routinely been used as part of EPA Method 6800 [162]. It should be stressed that ID not only is considered as a definitive technique mitigating the problems associated with instrumental drift and the poor ionization efficiency exhibited by a number of non-metal elements, but it also enables correcting for extraction of analyte from the sample and species inter-conversion during the entire analytical procedure [157,163]. The latter is of particular interest for elemental speciation studies. Heumann was the first to describe two different ID methodologies by which isotope dilution can be accomplished in such analytical applications [164,165] although a number of other papers have subsequently been published [163,166–174]. The "species-specific" mode entails spiking the sample with a specific elemental species that is isotopically labeled with an enriched isotope. This method is useful in the validation of speciation procedures, particularly in situations where certified reference materials (CRMs) are not available [57,175–177]. The "species-unspecific" mode, on the other hand, involves the addition of a simple enriched isotope of the desired element after physically separating the various species present in the sample, thereby allowing reliable quantification of unknown species [2,102,104,174,178–181].

Both of these methods offer clear benefits over other methods of species quantification, including shorter analysis times and less organic solvent consumption due to lower extraction efficiency requirements. However, they do not take into consideration the inter-conversion of species that may occur during sample preparation. For that reason much attention has been focused on speciated isotope dilution mass spectrometry (SIDMS) [162,173,182–186], a technique specifically designed to account for this potential change in native species distribution. SIDMS requires that each of the species spiked into the sample be labeled with a different isotope, thereby allowing any inter-conversions to be identified and mathematically calculated.

7.2.3 Progress in the analysis of "difficult" elements

Historically, ICP-MS has preferentially been used for the detection and quantification of metals and their species. Majority of these elements possess much lower first ionization potentials (e.g., 6.77, 7.73 and 7.58 eV for Cr, Cu and Ag, respectively) as compared to argon (15.8 eV) and therefore are efficiently ionized within the plasma. On the other hand, for halogens the first ionization energies range from 10.4 to 17.4 eV hindering their efficient ionization. Other "difficult" elements include phosphorus, sulfur and selenium, for which the values of first ionization energies are 10.49, 10.36 and 9.75 eV, respectively. In the case of selenium, theoretical calculations predicted that, under local thermodynamic equilibrium within a typical ICP, only 33% of Se atoms would be ionized in Ar-sustained plasma [187]. Thus, ICP-MS determination of the above mentioned elements has not been attractive due to the lower detection power, poorer analytical sensitivity and considerable imprecision. The use of isotope dilution method, mass analyzers with high ion transmission (sector-field instruments at low resolution) and instruments that measure different *m/z* values simultaneously (multiple ion collector spectrometers) enabled improving the analytical figures of merit and detection limits in the range of pg l^{-1} typically achieved today [2,107].

The second difficulty in the analysis of non-metals and some metalloids are spectral polyatomic interferences, already discussed in the previous sections (Table 7.2). The formation of charged species in the plasma may affect the analytical signal directly (isobaric overlaps) or may cause increased background [25,26,43,188]. Current strategies mitigating these troublesome interferences include collision/reaction cell technology, high resolution instrumentation, isotope dilution analysis and also different pre-ICP-MS procedures that conveniently decrease plasma loading with solvents and sample matrix [1,2,4,17,43,107,118,189].

The comprehensive reviews covering different aspects of atomic mass spectrometric detection, quantification and speciation of "difficult" elements are available [2,3,18,20,22, 43,189,190].

7.3 LIQUID CHROMATOGRAPHIC TECHNIQUES FOR SPECIES SEPARATION

As already mentioned, the use of LC separations is limited to non-volatile compounds. Typically, element species analyzed by LC can be classified according to the type of element bonding (chemical bonds versus weak interactions, covalent versus coordination bonds, etc.), electrical charge, molecular structure (inorganic salts, organic and bio-metals or biometalloid compounds) and also molecular size (in the case of biomolecules).

7.3.1 Basic separation modes

Depending on the relative polarity, solubility, electrical charge, molecular mass of the species of interest and possible presence of optical enantiomers, the LC modes used in speciation analysis encompass reversed phase (RP), reversed phase ion-pairing (IP-RP), micellar, ion exchange, size exclusion chromatograpy (SEC) and chiral LC [1,2,4,17,20,44,50,191]. A few applications of affinity chromatography have also been reported [104,192].

In the reversed phase technique, the mobile phase is more polar than the stationary phase and the solutes are separated according to their relative hydrophobicity. The resolution relies on the selection of a suitable column (type of the stationary phase, particle size, dimensions, etc.) and of mobile phase composition. Silica-bonded stationary phases, with different lengths of hydrocarbon chains are typically used (2, 8 or 18 carbon atoms in the chain known as C2, C8 and C18 phases). The mobile phase is an aqueous solution (pH buffers, inorganic salts) combined with organic modifier (acetonitrile, methanol, tetrahydrofuran or their mixtures). Species with the target element covalently bound are most suitable for analysis by RP. The separations of organotin compounds [193–197], organic species of arsenic [18,144], selenium [97,179,198,199], phosphorus [23,25,87,200], platinum [201–204] and iodine [31,32,34, 205,206] have been reported.

Ion-pairing chromatography is more popular in speciation analysis, since it enables the resolution of non-polar, polar and ionic species in one chromatographic run. The columns and mobile phases are essentially these same as those used in RP. However, a robust counter ion reagent is added to the mobile phase (typical concentrations 1–5 mmol l^{-1}) to form ion-pairs with target species, thus enhancing their hydrophobicity. The feasibility of IP-RP for separation of different charged and uncharged species makes this technique particularly attractive for analytical speciation of selenium [100,207–211] and arsenic [73,212–215]. In real-world samples, separation/quantification of As(III), As(V), MMA, DMA and AsB (pK_{a1}, respectively, 9.3, 2.3, 2.6, 6.2 and 2.2 [216]) is of primary importance and it has been achieved mainly using tetrabutylammonium ion and various alkylsulfonates as IP agents. In the case of selenium, IP-RP has been used for the analysis of plant extracts and food-related products [188,217]. Different perfluorinated carboxylic acids have been examined as the IP agents (pH 2.0–4.5), and the feasibility for separation of a large number of species was demonstrated in the enriched yeast and vegetable extracts [210,218]. Alkylsulfonic acids have been successfully used in speciation analyses carried out on nuts, onion leaves and yeast [208,209,219,220].

If the ion-pairing reagent is a surfactant and its concentration is above the critical micellar concentration, the technique is known as micellar chromatography. This liquid chromatographic mode is used to enhance solubility of hydrophobic compounds (organoarsenic [221,222] and organotin species [194,196]) in aqueous solutions. The sonication of the micellar medium causes re-arrangement of its structure and this modification of the mobile phase has also been used in speciation of small molecules (vesicle-mediated liquid chromatography) [18,43].

The separation by ion chromatography is based on the exchange equilibria between charged solute ions and the oppositely charged surface of the stationary phase. Solute ions and ions of equivalent charge in the mobile phase compete for the counterpart on the stationary phase. In principle, the relative retention of the ions is determined by the extent of this competition, which is dependent on three variables, namely pH, the ionic strength of the mobile phase and the nature of the ion exchanger [1,44,50]. Most of the ion exchange columns contain styrene-divinylbenzene or silica modified by sulfonate groups (cation exchange) or ammonium group (anion exchange). Common mobile phases are diluted aqueous solutions of inorganic salts and buffers. For key arsenic species in environmental and toxicological applications, anion exchange has provided good resolution while using phosphate-, carbonate- , hydroxide- or phthalate-based mobile phases (pH in the range 5–11) [43,223–232]. Ammonium salts are preferred in order to minimize the residue formed in the sampler/skimmer cones of ICP-MS. The most common elution order is AsB (minimum retention), As(III), DMA, MMA, As(V), but it is sometimes different, depending on the pH, gradient conditions, etc. The cation-exchange mode is often used to separate AsB, AsC,

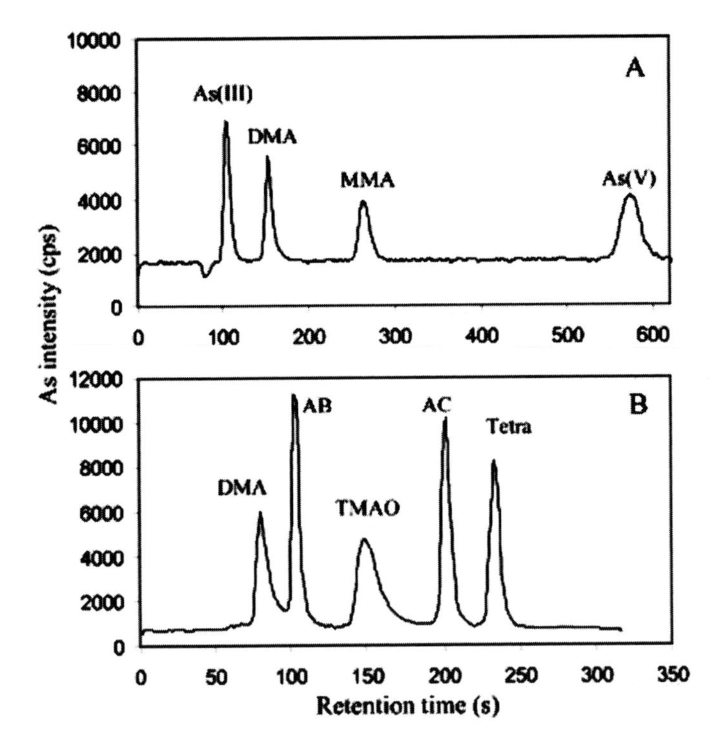

Fig. 7.2. Separation of arsenic compounds (0.5 μg l^{-1} As) by ion exchange LC with sector-field ICP-MS detection operated at low resolution ($m/\Delta m \sim 300$): (A) anion exchange chromatography (Hamilton PRP-X100, 20 mmol l^{-1} $NH_4H_2PO_4$, pH 5.6); (B) cation exchange chromatography (Zorbax 300-SCX, 20 mmol l^{-1} pyridine, pH 2.31). (Reproduced from Ref. [238] with the permission of The Royal Society of Chemistry).

TMAO and TMAs. These separations are achieved at generally lower pH values (pH < 4) with the mobile phases containing formate or pyridine/formate [233–237]. In Fig. 7.2, the anion- and cation-exchange ICP-MS chromatograms of arsenic species are shown [238]. A number of organo-Se compounds (selenoamino acids, peptides, trimethylselonium ion, etc.) are cations, or can be easily converted to cationic species by lowering the pH of the solution, thus enabling separation by cation exchange mechanisms. Up to 10 Se compounds (methyl-selenocysteine, allyl-selenocysteine, propyl-selenocysteine, selenomethionine, selenoethionine, selenocystine, dimethyl-seloniumpropionic acid, selenohomocystine, trimethylselonium ion, methyl-selenomethionine) have been resolved with different mobile phases at pH 2–5.7 (pyridine, ammonium formate, etc.) [44,127,239]. The representative ICP-MS chromatogram is presented in Fig. 7.3 [240]. Other applications of ion exchange LC in speciation analysis encompass separation of charged aluminum, chromium, phosphorus, antimony or platinum species [36,241–250].

In (SEC), the separation of solutes is based on their different molecular sizes. The calibration of column is achieved relating the retention time to the molecular mass of standards and, in speciation analysis this technique has been typically used to attain the distribution of

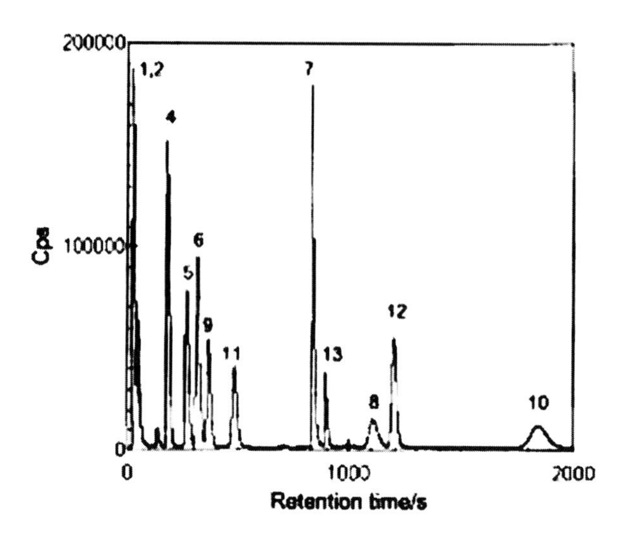

Fig. 7.3. Cation exchange LC separation of a mixture of 12 selenium standard compounds (100 μg l^{-1} Se): 1 – Se(IV); 2 – Se(VI); 4 – methyl-selenocysteine; 5 – allyl-selenocysteine; 6 – propyl-selenocysteine; 7 – selenocystine; 8 – selenohomocystine; 9 – selenomethionine; 10 – methyl-selenomethionine; 11 – selenoethionine; 12 – trimethylselonium ion; 13 – dimethylselonium ion. Separation was achieved on a silica-based strong cation exchange column, 100 (3 mm (id) (Ionosphere-C, Chrompack International) with an aqueous solution of pyridinium formate as mobile phase in a step elution programme (pyridinium formate from 0.75 mM, pH 3.0 to 8.0 mM, pH 3.2). (Reproduced from Ref. [240] with the permission of The Royal Society of Chemistry).

elements among different molecular mass fractions [2,3,10,22,143,251–255]. It should be stressed that SEC is a gentle separation method, so it usually does not cause any loss of element species, nor their on-column alterations. Consequently, SEC has been used to discriminate between chemical bonds versus weak interactions of element with biomolecules [50, 209]. As an example, in the analysis of protein extract from Brazil nut, Kannamkumarath *et al.* [209] observed different elution profiles of Se in the absence and in the presence of reducing agent (Fig. 7.4). When β-mercaptoethanol was added to the extract, the contribution of Se eluting in the region of molecular mass lower than 10 kDa increased. The semi-quantitative evaluation of the results revealed that about 12% of total Se in nuts was weakly bound to proteins.

7.3.2 Orthogonal chromatographic approach

In addition to the chromatographically important properties of solutes, the composition of sample and the requirement to preserve native element speciation are also important factors to be considered while choosing a suitable separation mechanism. In spite of unquestionable versatility of LC, chemical lability of element species, their interactions with the stationary phase and/or the components of mobile phase (buffers, organic modifiers) may cause unwanted changes in the original element distribution among different species [4,44,50,256]. That is why, in many applications that compromise conditions have to be used, sacrificing chromatographic resolution, but assuring the integrity of species during separation. The optimum

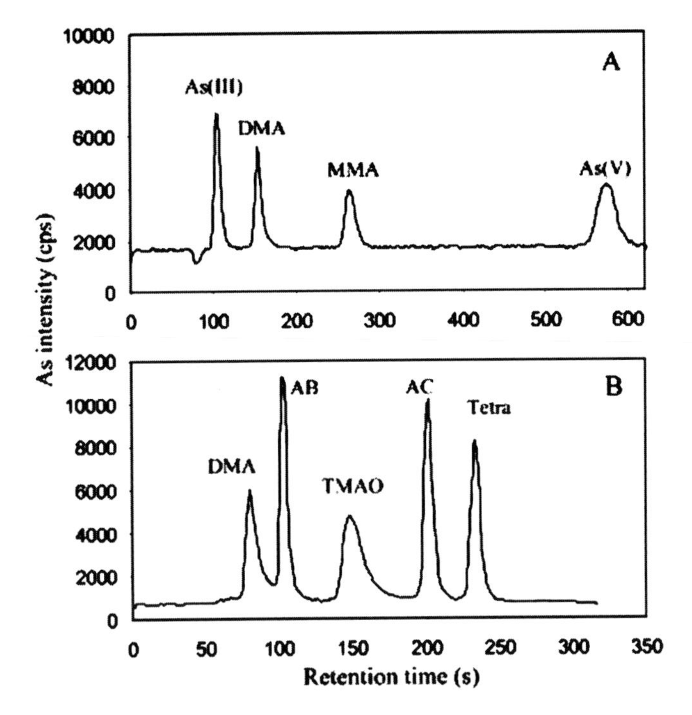

Fig. 7.4. Typical chromatograms of protein extracts from Brazil nuts obtained by SEC–ICP-MS (^{78}Se): (A) in the absence and (B) in the presence of reducing agent (β-mercaptoethanol, 100 mM) (separation achieved on Superdex Peptide HR 10/30 column with acetate buffer (50 mM, pH 4.5) + 0.02% SDS mobile phase, at a flow rate 0.6 ml min^{-1}). (Reproduced from Ref. [209] with the permission of Springer).

separation conditions may not always be used in speciation studies. Compromises are sometimes necessary to avoid possible changes in natural species distribution due to their interaction with stationary and mobile phases (pH conditions, complexing agents, organic modifiers). To improve the separation, especially in the analysis of complex matrices containing unknown species, the multidimensional chromatographic approach has often been explored [52]. In such an approach, few (two or three) orthogonal (complementary) separation techniques are applied in series, thus assuring progressive purification of chromatographic fractions. The most common combinations incorporate SEC followed by ion exchange or reversed-phase HPLC. The initial SEC separation facilitates removal of large biological polymers, partial desalting of analytes and a rough fractionation based upon molecular mass. Worth mentioning is that the multidimensional separation provides a more simplified composition of the solution to be introduced to ICP-MS with lower risk of plasma instability and/or spectral interferences. Multidimensional LC separations in biochemical speciation analysis had been reviewed [52] and more recent applications can be found in the literature [2,34,147,251,257–261]. Since excellent peak purity is usually achieved in the orthogonal separations, this approach has been extensively explored for structural characterization of unknown species by mass spectrometry [2,9,47].

7.3.3 Miniaturization of LC in analytical speciation

Since the early applications of liquid chromatography, continuous technological and methodological progress has enabled substantial improvement of the analytical performance. The main trends in technology encompass development of new packing materials and packing methods, production of small particle stationary phases and, in consequence, the use of columns with small internal diameter.

In the literature concerning element speciation, micro columns are typically classified as microbore (0.5–2.0 mm i.d.) and nanobore or capillary (less than 0.5 mm i.d.) [2,99,199, 262–264]. Since these small columns offer increased mass sensitivity, better resolving power, smaller sample volumes, no sample dilution and less solvent consumption as compared to conventional (4.6 mm i.d.) columns, their use in speciation analysis is particularly attractive. The lower peak volume and higher peak concentration of the eluted species (by a factor of 10–100 times) generally improve the absolute detection limits with respect to normal bore columns. However, due to small injection volume, the sensitivity of detection system is a critical factor, for which in elemental speciation ICP-MS has been the method of choice. The analytical applications have been focused on selenium in yeast [199,264–266] and urine [211], separation of phosphopeptides [25,87,91,267] and oxidation states of chromium [268,269]. In the recent study, Stefanka *et al.* [270] applied three reversed phase columns (2.1, 1 and 0.32 mm i.d.) for separation of carboplatin and oxaliplatin. Using the optimized separation and interface conditions, the analytical performance for carboplatin was evaluated in each case. Thus, for microbore columns (2.1 and 1 mm i.d.) at eluent flow rates 250 and 60 $\mu l\ min^{-1}$ and for injection volumes 12.5 and 3 μl (150 and 36 pg of carboplatin), the absolute detection limits were 1.13 and 0.81 pg respectively, which corresponded to relative detection limits of 0.09 and 0.27 $\mu g\ l^{-1}$. For capillary column (0.32 mm i.d.) at a flow rate 4 $\mu l\ min^{-1}$, a substantially lower absolute detection limit was achieved (0.09 pg); however, taking into account that only 0.2 μl were injected (2.4 pg of carboplatin), the concentration detection limit was poorer (0.45 $\mu g\ l^{-1}$).

It should be noted that, the effluent from the micro column is introduced to ICP-MS at the low volumetric flow, thus its composition has negligible effect on plasma stability. Consequently, organic solvents and/or high salt concentrations can be effectively used with capillary-based LC–ICP-MS to provide enhanced separation performance. On the other hand, such coupling requires specialized interfacing devices, which are discussed in the following section.

7.4 COUPLING ICP-MS TO LIQUID CHROMATOGRAPHY

Technically speaking, interfacing LC separations to ICP-MS detection does not present major difficulties. The column effluent is introduced to the ICP through a nebulization system. Typical flow rates for standard-sized analytical columns (0.5–1.2 $ml\ min^{-1}$) are compatible with the aspiration rate of traditional pneumatic nebulizers (concentric Meinhard, Babington or cross-flow), so the column outlet can be connected directly with the liquid inlet of the nebulizer using a polymeric or stainless steel tubing. Obviously, while using as short and narrow connection as possible, peak broadening resulting from the transfer line can be minimized. However, low transport efficiency of traditional pneumatic nebulizers (1–5%) is an important limiting factor. Furthermore, in such coupling the composition of column effluent can affect the performance of plasma. In particular, ICP-MS does not tolerate large amounts of organic solvents that cause lower plasma stability and the formation of carbon deposits on the torch and the sampling cone. The high salt concentration is also problematic, because of nebulizer and cone clogging, plasma destabilization and a potential increase of polyatomic interferences. In many applications of LC–ICP-MS, the chromatographic conditions have been compromised for plasma/detector compatibility rather than optimized for separation. The capability of element-specific detection by ICP-MS to compensate for incomplete chromatographic resolution is though, an important advantage of this hyphenated technique [4,20,44].

Current strategies developed to circumvent typical problems in LC–ICP-MS are focused on (1) high transport efficiency, (2) minimum plasma loading with solvents, salts and buffers, (3) elimination of polyatomic interferences, (4) sensitive and accurate detection of fast transient peaks.

The increased efficiency of sample introduction to ICP can be achieved by using ultrasonic (USN), micro concentric (MCN), hydraulic high-pressure (HHPN), direct injection (DIN), high efficiency (HEN) or direct injection high efficiency (DIHEN) [46,217,241, 271,272] nebulizers. In USN, a thin film of the solution flows over the surface of a piezoelectric transducer and is converted to aerosol by the high frequency vibrations from the transducer. About 10–30% of the aspirated solution reaches the plasma, however a desolvation unit is necessary to remove most of the organic solvent, and the chromatographic conditions should involve low salt/volatile buffer concentrations [46,196,250,273,274]. Other nebulizers listed above are low-flow devices and are particularly well suited for microbore and capillary LC. The commercially available micro concentric nebulizers operate at the flow rates 10–100 μl min^{-1} yielding a transport efficiency close to 50% [23,233,275, 276]. The DIN was developed by Shum *et al.* [277] and subsequently used in a number of speciation studies [241,278–280]. This nebulizer is positioned inside the ICP torch and its nozzle is placed only a few millimeters from the base of the plasma. The sample uptake requires a flow rate in the range 30–150 μl min^{-1} at relatively high gas flow rate (1 l min^{-1}). The DIN does not use a spray chamber and thus offers 100% analyte transport efficiency, low dead volume (<2 μl), rapid wash-in and wash-out times, low memory effects, enhanced sensitivity and improved precision as compared to nebulizer–spray chamber arrangements. However, the easy clogging of DIN capillary hinders the introduction of elevated salt/buffer concentrations. Plasma cooling has also been observed [50,262]. An alternative low-flow design is the high efficiency nebulizer (HEN), which is composed of a small capillary and a spray chamber. The efficient formation of uniform fine droplets is achieved at 40–100 μl min^{-1} sample flow rate (gas flow rate 1 l min^{-1}, as in DIN). The designs of HEN and DIN have been combined and modified to produce a simple and low cost direct injection high efficiency nebulizer (DIHEN) [281]. The DIHEN offers optimal sensitivity at 1–100 μl min^{-1} sample flow rate and significantly lower injector gas flow rates (0.25 l min^{-1}). Few modifications of the original design have been proposed that enabled reduced dead volume for introduction of effluents from capillary LC columns [267,282]. The detection limits in the range of pg l^{-1} have been reported, similarly to those obtained with normal-size columns and traditional pneumatic nebulizers [270,281,283]. Recently, a sheathless interface between capillary HPLC and ICP-MS via a micro flow total consumption nebulizer has been developed [199].

The design of spray chamber, which is present in most actual introduction systems also affect the transport efficiency and may contribute to the peak broadening. The double pass and cyclone spray chambers are most commonly used in LC–ICP-MS (see Table 7.3), yet some recent laboratory-made designs should also be mentioned [2,25]. Finally, post-column hydride generation, converting column effluent into gaseous phase mainly composed of analyte hydrides is an excellent approach toward efficient transport to ICP-MS [284–289].

The RP and IP-RP chromatographic modes inevitably involve the use of organic modifiers. Due to their incompatibility with ICP, several possibilities have been explored to reduce the introduction of organic solvents to plasma. Membrane desolvatation systems and cooling the spray chamber are common approaches [4,23,50,256]. The post-column dilution with aqueous solution (diluted nitric acid) has also been proposed [31,32]. Furthermore, the stability of plasma can be improved by applying high forward power and/or by the addition of oxygen to the nebulizer argon flow, thus promoting the combustion of organic solvents [4,17,196]. It should be stressed that, while using microbore and capillary columns, the minimum volume of solution is introduced to ICP with much lower risk of its harmful effect on the plasma performance. Consequently, the use of organic modifiers has become less critical for LC–ICP-MS coupling [199,283].

Another problem in interfacing originates in relatively high concentration of salts and buffers in LC mobile phases (ion exchange, IP-RP, micellar LC, etc.) and also in the sample matrix. One possible solution is to adjust the chromatographic conditions to the requirements of ICP-MS; however, it usually involves poorer separation performance. The use of mobile phases based on relatively volatile compounds (such as ammonium salts and/or organic acids) has also been helpful [43,50]. The increasing applications of micro columns in speciation analysis is partly due to lower plasma loading with undesired compounds potentially affecting the analytical performance of the entire procedure.

On the other hand, progress in high resolution instrumentation and collision/reaction cell technology has substantially improved the ICP-MS capabilities as LC detector, since the modern instruments conveniently eliminate troublesome polyatomic interferences [3,71,126, 128]. The development of novel ion detectors that use single ion and multiple ion collectors has enhanced detection capabilities for fast transient signals, both in multielemental and isotope ratio detection [64,107].

Even though, the important improvements in LC–ICP-MS coupling can be noted, factors such as the separation mechanism, composition and flow of the mobile phase and type of nebulizer and sample introduction device still must be carefully considered and have subsequently been topics of discussion in many papers and reviews [44,50,290–293].

7.5 ELEMENT SPECIATION BY LC–ICP-MS IN THE REAL-WORLD SAMPLES

The environmental, biological and clinical relevance of element speciation is well demonstrated. Regarding the applications of LC–ICP-MS, this hyphenated technique has been most often used for speciation of arsenic, selenium, iodine, and to a less extent, mercury. The

TABLE 7.3

THE LITERATURE SURVEY OVER RECENT LC–ICP-MS APPLICATIONS IN THE SPECIATION ANALYSIS OF ENVIRONMENTAL AND BIOLOGICAL RELEVANCE

Sample	Species of interest	LC separation	ICP-MS detection	Comments	Ref.
River water	Phosphorus herbicides: glufosinate, glyphosate and aminomethyl-phosphonic acid	Ion-pairing HPLC (Zorbax SB-C8 column) with a mobile phase composed of tetrabutyl ammonium hydroxide + acetate buffer at pH 4.7	Column effluent introduced through Meinhard nebulizer with double-pass Scott Spray chamber to ICP-MS with octopole reaction cell (^{31}P)	Detection limits were found to be in the low ppt range (25–32 ng l^{-1} as P). The developed method was applied to the analysis of water samples collected from the Ohio River and spiked with a standard compound at a level of 20 μg l^{-1}	[26]
Natural water	Glyphosate and phosphate	Anion exchange LC (Dionex IonPac AS16 column) with 20 mM citric acid as a mobile phase, flow rate 0.5 ml min^{-1}	Meinhard concentric nebulizer with double-pass, cooled Scott spray chamber, ICP-MS (^{31}P)	The chromatographic and ICP-MS (quadrupole) operating parameters were carefully selected, which minimized typical polyatomic interferences on ^{31}P, the DL was 0.7 μg l^{-1} as P	[27]
Natural waters	Heavy metals with biogenic chelators	Narrow bore anion exchange column, using the synthetic hydrophilic quaternary ammonium anion exchanger and nitrate-based mobile phase	Micro concentric nebulizer and ICP-MS. In river water of different pollution levels beside CuEDTA other anionic Cu-complexes were found at nanomolar concentrations	Low DL (in the range of nM) enabled for detection of different copper complexes in river water. In addition to CuEDTA, other anionic Cu complexes were found at nM concentrations, depending on the pollution level	[275]

(Continued on next page)

TABLE 7.3 (*contd.*)

Sample	Species of interest	LC separation	ICP-MS detection	Comments	Ref
Rice flour	As(III), AB, As(V), MMA, DMA	The methanol–water extracts analyzed by anion exchange LC (Hamilton PRPX-100 column with 10 mM phosphate mobile phase at pH 6) and cation exchange LC (Hamilton PRP-X200 column with 4 mM pyridine formiate mobile phase an pH 2.8)	Meinhard-type concentric nebulizer with a double-pass, cooled Scott spray chamber, ICP-MS (^{75}As)	The DLs for dry flour rice were 2 and 3 ng g^{-1} for As(III) and AsB on the cation column (separation time 6 min) and 3, 6 and 5 ng g^{-1} for As(V), MMA and DMA, respectively, on the anion column (16 min). At least 1 month stability of As species in the extracts was demonstrated	[233]
Beverages	As(III), DMA, MMA, As(V)	The samples were cleaned-up by SPE (C18 cartridge) and analyzed by anion exchange LC (Hamilton PRX-100 column) with a phosphate buffer based mobile phase (5 mM, pH 8.5) at a flow rate 1 ml min^{-1}.	Cross-flow nebulizer with a Scott-type spray chamber, ICP-MS (^{75}As)	As species separated in 13 min, the DLs for were in the range 0.2–0.5 μg l^{-1} (as As)	[303]
Antractic krill	MW distribution of element-binding molecules (Ag, As, Cd, Co,	Water extracts fractionated by preparative SEC and analyzed on three SEC columns (300 × 10 mm): Superdex	PFA MicroFlow 100 nebulizer, ICP-MS, The interference from chloride was traced by monitoring	Two-dimensional chromatographic approach: the distribution of target elements in nine molecular size ranges was obtained by using	[304]

	Cr, Cu, Fe, Mn, Ni, Pb, Se) and As species	200 10/300 GL; Superdex 75 HR 10/30 and Superdex peptide HR 10/30 with ammonium acetate mobile phase; As speciation on anion exchange column (IonPak AS14)	$m/z = 37$ ($^{37}Cl^+$) and $m/z = 51$ ($^{35}Cl^{16}O^+$) simultaneously	SEC columns and further separation/identification of As species carried out by reversed phase and ion exchange LC	
Marine organisms	As(III), MMA, DMA, As(V), AB, TMAO, AC, TETRA	LC separation by anion exchange (IonPac AG4 guard column and IonPac AS4A analytical column in a gradient of nitric acid (from 0.4 to 50 mM, pH 3.3–1.3)	Micro concentric nebulizer (cyclonic spray chamber) and ICP-MS (^{75}As)	The separation was accomplished in 12 min with the DLs for the eight As species in the range 0.03–1.6 $\mu g\ l^{-1}$ (as As)	[276]
Trout liver microsomes	Fe associated with P-450 enzymes	Anion exchange FPLC (Mono Q HR 5/5 column), elution with 20 mM Tris/acetate buffer containing 15% glycerol (pH 7.5) in a gradient of ammonium acetate, flow rate 0.6 ml min^{-1}	FPLC effluent introduced through Meinhard nebulizer to ICP-MS with octopole reaction system (^{54}Fe ^{56}Fe ^{57}Fe)	Chromatographic conditions adopted to ICP-MS detection; the separation of different P-450 isoenzymes was achieved. The Fe elution profile obtained matched that observed for isoenzymes	[305]
Herring gull eggs	SeCys, SeCa, SeMet, SeEt	The methanol–water extracts were separated on reversed phase column (Eurospher 100 Å or Nucleosil 120 Å, C18, 5 μm) with a mobile phase composed of 30 mM $HCOONH_4$ (pH 3.0) + 5% (v/v) methanol, a flow rate 1.2 ml min^{-1}	Hydraulic high pressure nebulizar, ICP-sector-field MS at resolution 400 (for separation of ^{34}S signal from interfering $^{16}O^{18}O^+$, a resolution 3000 was used). Isotopes monitored: ^{82}Se, ^{77}Se, ^{34}S, ^{88}Sr, ^{85}Rb, ^{75}As, ^{63}Cu and ^{65}Cu	The SeCa was detected in biological samples and, in this column fraction, Se correlated with Cu, As and S	[306]

(Continued on next page)

TABLE 7.3 (*contd.*)

Sample	Species of interest	LC separation	ICP-MS detection	Comments	Ref.
Salmon egg cell cytoplasm	Hg binding to proteins	Surfactant mediated LC separation on CHAPS-coated ODS column with mobile phase containing 0.1 M Tris/HNO_3 (pH 7.4) and 0.2 mM CHAPS; flow rate 0.7 ml min^{-1}. The protein fraction further separated on SEC column	On-line UV(254 nm) and ICP-MS (^{202}Hg, ^{32}S and ^{78}Se) detection	The results indicate that Hg in salmon egg cell cytoplasm binds with proteins containing selenocysteine and/or cysteine residues in proteins	[8]
Rat tissue after oral intake of $HgCl_2$	Hg complexes with MTs	MTs were purified from liver and kidney homogenate extracts by preparative SEC (Sephadex G-75 column, isocratic elution with 10 mM Tris/HCl buffer at pH 8.0) and, after desalting, separated by reversed phase LC (narrow-bore Vydac C8 column) using methanol gradient in 5 mM CH_3COONH_4	The column effluent was mixed with 20% HNO_3 by a T-joint prior to its introduction to ICP-MS (Hg, Cu, Zn)	Large amounts of Hg and Cu (but not Zn) were found in kidney MTs, which were structurally characterized by ESI-MS, however the distinction between copper and zinc was hindered by the limited instrument resolution	[307]

Human milk (fractionated by centrifugation)	Mn binding to different MW fractions and inorganic Mn species	Two-dimensional LC separation: (1) SEC (10–150 kDa on Toyo Pearl TSK HW 55 F and 100–2000 Da on Toyo Pearl TSK HW 40 S stationary phase) with 10 mM acetate buffer mobile phase (pH 6.3) and (2) anion exchange (Dionex AS 11 column) elution with 10 mM acetate buffer (pH 6.3) in NaOH gradient	Column effluent introduced through Meinhard nebulizer (cyclon spray chamber) to ICP-MS (^{55}Mn)	Mn concentrations were approximately 3 μg l^{-1} in total human milk, 2.85 μg l^{-1} in the defatted fraction, 0.25 μg l^{-1} in the pellet fraction and 2.6 μg l^{-1} in the low molecular weight supernatant fraction. Inorganic manganese species and Mn–citrate complex were identified	[308]
Human milk	SeMet enantiomers	Elimination of fat and proteins by centrifugation and ultrafiltration, LC on a glycopeptide teicoplanin-based chiral stationary phase (Chirobiotic T column) with deionized water as a mobile phase, the flow rate 1 ml min^{-1}	Meinhard nebulizer with a cooled double-pass spray chamber, ICP-MS (^{77}Se, ^{82}Se)	The analytical performance of ICP-MS detection (DL 0.9 μg l^{-1} as Se) was compared with that obtained for on-line post-column microwave-assisted digestion-hydride generation AFS (about 3 μg l^{-1} as Se). The authors reported poorer signal-to-noise ratio for ICP-MS detection	[309]
Human plasma	Selenoprotein P	LC separation on Shodex AFpak AHR-394 Heparin Affinity column in line with a Shodex Asahipak GS-520HQ size exclusion column, elution with phosphate buffer in a gradient of heparin	Magnetic sector ICP-MS at low-resolution mode (^{77}Se, ^{82}Se, ^{35}Cl, ^{79}Br and ^{81}Br)	Column fractions analyzed by SDS-PAGE to determine the relative effectiveness of chromatographic separation	[192]
Human plasma	Al and Fe binding to isoforms of hTf	LC separation on anion exchange column (Mono-Q; Amershem), gradient elution	ICP–HRMS (^{56}Fe, ^{27}Al, ^{32}S)	The binding of Fe was similar between asialo-hTf and native-hTf, while that of Al for asialo-hTf was larger than that for native-hTf. Possible toxicological relevance was discussed	[310]

(Continued on next page)

TABLE 7.3 (*contd.*)

Sample	Species of interest	LC separation	ICP-MS detection	Comments	Ref.
Human serum proteins	V binding to Tf	LC separation on anion exchange MonoQ (HR5/5), elution in gradient of CH_3OONH_4 (0–0.5 M)	Meinhard nebulizer with cooled Scott double-pass spray chamber, ICP-MS with octopole reaction cell (3 ml min^{-1} He), molecular mass analysis of different transferrin forms by MALDI-TOF	The distribution profile of V was similar to that previously reported for Fe(III) and Al(III) in human serum. The analysis by MALDI-TOF confirmed the molecular structure of transferrin as the protein associated to the V fraction	[311]
Serum of cystic fibrosis patients	SeU, SeMet, SeCys, SeEt, Se(IV), Se(VI)	LC separation on strong anion exchange column (Dionex AS 11)] with 0.6 mM NaOH in gradient of tetramethylammonium hydroxide, flow rate 1 ml min^{-1}	Meinhard nebulizer and a cyclone spray chamber, ICP-MS (^{77}Se, ^{78}Se, ^{82}Se)	The DLs for six species separated were in the range 0.09–0.22 μg l^{-1} (as Se), the values of SeCys and those of cationic/neutral Se compounds (probably Se enzymes) were lower in pathological sera as compared to healthy subjects	[312, 313]
Serum and prostate tissue from prostate cancer patients	SeMet, MeSeCys	Protein hydrolysis (Protease), ion-pairing HPLC (Symmetry Shield Guard-Column; SymmetryShield RP8 analytical column with a mobile phase composed of water–methanol (97:3, v/v) with 0.1% HFBA	ICP-MS detection (^{82}Se) and ESI–MS for the confirmation of species	The analytical methodology applied would help to measure the amount of Se delivered to a target tissue of interest and further investigate whether tissue Se levels are dependent on the form of Se administered	[314]
Calf thymus DNA digest	Phosphorylated deoxyribo-	Reversed phase LC (Discovery RP16 Amide column) with	PFA microflow 20 nebulizer with Peltier-	LC and CE separation with ICP-MS detection were compared,	[24]

	nucleotides	15 mM CH_3COONH_4 (pH 5.8) + 2.5% methanol as a mobile phase	cooled quartz spray chamber, ICP-MS with octopole reaction system (He, 2.21 min^{-1})	showing the baseline resolution in both techniques, but with different felution order. The advantage of LC was in better detection limit (as P). It was shown that ESI–MS could be operated under the same HPLC conditions that had been used for the ICP-MS experiments	
Cardiovascular tissues	As binding to proteins	Two-dimensional LC separation: (1) SEC (Superdex Peptide, 0.1–7 kDa and Phenomenex, 1–300 kDa columns), elution with 10 mM Tris/HCl (pH 7.4) + 0.1 mM PMSF and 25 mM NaCl at 0.5 ml min^{-1} and (2) anion exchange FPLC (MonoQ HR 5/5 column), elution with 10 mM Tris/HCl (pH 7.4) in a gradient of CH_3COONH_4, 0.8 ml min^{-1}	Babington glass nebulizer and a Scott double-pass spray chamber cooled by a Peltier system, ICP-MS (^{75}As, ^{32}S, ^{65}Cu and ^{31}P), chloride elution was investigated	The results obtained from SEC separation suggested As binding to MTs, glutathione (As(SG)3), cysteine (As(Cys)3), transferrin and hemoglobin. The simultaneous monitoring of As, S, Cu and P in FPLC effluent showed that As and Cu are usually bound to the same type of proteins, mainly through vicinal sulfur groups	[260]
Blood, urine, fingernails, hair	AC, AB, DMA(III), DMA, MMA(III), MMA, As(III), As(V)	The sample extracts were analyzed by anion exchange exchange LC (Shodex Asahipak ES-502N 7C column) with 15 mM citric acid (pH 2) as a mobile	Pneumatic nebulizer, ICP-MS (^{75}As and m/z = 77 to check for possible chlorine interferences)	Feasibility of LC–ICP-MS procedure for the determination of eight As species in different biological samples was demonstrated. The DLs were in the range 0.14–0.33 μg l^{-} (as As)	[315, 316]

(*Continued on next page*)

TABLE 7.3 (*contd.*)

Sample	Species of interest	LC separation	ICP-MS detection	Comments	Ref.
		phase, flow rate 1 ml min^{-1}. Since the mobile phase had had no buffering capacity, the equilibration of column was necessary			
Urine	As(III), As(V), AB, AC, MMA, DMA, DMAA, DMAE, TETRA, TMAO, TMAP, As-sugars	(1) Anion exchange LC (ICSep ION-120, 10 μm or Hamilton PRP X-100, 3 μm columns), elution with aqueous mobile phase in gradient of $NH_4(CO_3)_2$ (5–50 mM) (2) cation exchange LC (silica-based ChromPack Ionospher 5C, 5 μm column), elution in a gradient of pyridinium ion	Meinhard nebulizer with double pass, cooled spray chamber, ICP-MS with octopole reaction system	Several species eluting in the void volume in anion-exchange mode (AB, AC, TETRA, TMAO, TMAP) were separated by cation exchange mechanism. The DLs were in the range 0.04–0.2 μg l^{-1} (as As), nine species identified in urine samples and in urine reference materials	[235]
Urine, fish tissues	As(III), As(V), AB, MMA, DMA	Sample acidification with H_3PO_4 (10 mM), ion-pairing LC (C19 Altima, 5 μm), mobile phase composed 5 mM hexanesulfonic acid + 5 mM citric acid/NaOH (pH 4.5) at a flow rate 0.9 ml min^{-1}	Meinhard nebulizer with double pass, cooled spray chamber, ICP-MS	Base line separation of five species in 4 min with quantification limits in the range 44–94 ng l^{-1} (as As)	[73]
Urine	TMSe, MeSeMet, MeSeCys, SeCa,	(1) Reversed phase LC (two Luna C18 (2), 2 mm i.d. micro bore	MicroMist AR30-1-F02 glass concentric nebulizer	The DLs for eight Se species were evaluated for different	[282]

	SeMet, SeGal, SeGlu, SeEt	columns in series), elution with 200 mM CH_3COONH_4 + 5% methanol, 200 μl min^{-1} and (2) ion-pairing LC (Luna C8(2), 2 or 1 mm i.d. micro bore column), elution 0.2% HFBA) + 20% or 30% methanol, 200 or 50 μl min^{-1}	and a cyclonic spray chamber was used with columns of i.d. = 2 mm and modified fied direct injection nebulizer (MDIN) with column of i.d. = 1 mm (or CE), ICP-MS (^{77}Se, ^{78}Se, ^{82}Se). The	separation and introduction modes. Experiments with ^{77}Se-labeled yeast suggested that after Se ingestion, metabolites in urine are formed very quickly, being SeGal the main metabolite	
Standards	Cobalamin species	Reversed phase LC on C18 column (2 or 3 μm) and normal phase LC on Cyano column (3 μm), both columns of 75 μm i.d. and the total length 37 cm for ICP-MS (30 cm for UV), injection volume 20 nL, gradient elution and methanol make-up flow	Non-pneumatic Burgener Mira Mist CE nebulizer (low-flow, parallel-path, sample flow rates 1–2500 μl min^{-1})with cooled cyclonic spray chamber, ICP-MS detection (^{59}Co)	Baseline separation of five cobalt species achieved using reversed-phase conditions with a C18 packed capillary column and elution in a gradient of methanol.	[317, 318]
Cl-containing pharmaceutics	Diclofenac chlorpromazine	Reversed phase separation (Apex C2 column) with mobile phase composed of 0.05% (v/v) HCOOH + 35% methanol. Gradient elution also tested	Pneumatic nebulizer, ICP-MS (^{35}Cl, ^{37}Cl)	The feasibility of gradient elution HPLC–ICP-MS for the determination of clorine-containing xenobiotis was demonstrated	[319]

(Continued on next page)

TABLE 7.3 (*contd.*)

Sample	Species of interest	LC separation	ICP-MS detection	Comments	Ref.
Rat urine and bile	2-, 3- and 4-Iodobenzoic acids	The samples after base hydrolysis were analyzed by reversed phase LC (Polaris C18-A, 3 μm column) with 0.1% (v/v) HCOOH + 30% (v/v) acetonitrile mobile phase at a flow rate 1 ml min^{-1}	The column effluent was split allowing 50 μl min^{-1} into the ICP-MS (^{127}I) and the remainder into UV detector (254 nm). For ICP-MS, O_2 was added to the nebulizer gas (5% v/v), concentric nebulizer was used with a cooled double pass spray chamber	Metabolite profiles for urine and bile (separation time 20 min) showed extensive metabolism with unchanged iodobenzoic acids forming a minor part of the total	[320]
Yeast lipid extracts	Phospholipids	Reversed phase LC separation (YMC-Pack Diol-120 column, 5 μm), gradient elution with mobile phases acetone–hexane–acetic acid–triethlyamine mobile and methanol–hexane–acetic acid–triethylamine, flow rate 0.6 ml min^{-1}, column temperature 40°C	The column effluent was split and 130 μl min^{-1} introduced to ICP-MS through micro concentric nebulizer and a and a Scott double pass spray chamber. An optional gas (O_2 20%, v/v) was applied through a T-piece. The octopole reaction cell (He, 4 ml min^{-1}) enabled to control polyatopmic interferences on ^{31}P	Six phospholipid standards were baseline Separated in 50 min The achieved absolute detection limits were between 0.21 and 1.2 ng (as P). The analysis of yeast extracts revealed eight P-containing species: four of them were identified using the standards	[23]

Se-enriched Onion	Se species: inorganic Se, SeMet, MeSeCys	Onion and bulb extracts analyzed by SEC (Superdex HR 10/30 column) with a mobile phase 10 mM CAPS buffer (pH 10.0) and by ion pairing HPLC (Altima C8 column) with a mobile phase 0.1% v/v HFBA + 5% (v/v) methanol (pH 2.5) or 5 mM citric acid + 5 mM hexane-sulfonic acid + 5% (v/v) methanol (pH 4.5)	Column effluent introduced through concentric or cross-flow nebulizer (Scott-type double-pass spray chamber) to ICP–MS (^{77}Se, ^{78}Se, ^{82}Se); ESI–MS for species identification/confirmation.	Different growth/enrichment and sample pretreatment procedures tested; similar results obtained while using different LC conditions	[321, 322]
Se-enriched *Brassica juncea*	Se binding to proteins	Plant extracts were analyzed by two-dimensionas LC: (1) SEC on Superdex 75 10/30 column with 30 mM Tris/ HCl (pH 7.5) mobile phase; (2) FPLC on anion exchange Mono Q 5/50 GL column, elution with 30 mM Tris/HCl (pH 7) in salt gradient (NH_4Cl, CH_3COONH_4)	Meinhard nebulizer with Peltier-cooled spray chamber, shielded torch ICP-MS with octopole reaction cell (H_2 3 ml min^{-1})	Different pretreatment procedures were tested (based on SDS-mediated protein solubilization, acetone pre-cipitation and re-solubilization) and careful optimization of the reaction cell conditions was carried out	[323]
Se-enriched *Brassica juncea*	Cationic Se species: SeMet, MeSeMet, MeSeCys, DMSeP	Pyridinium formate plant extracts analyzed by strong cation exchange column (Phenosphere SCX, 5 μm), with step elution program of pyridinium formate buffer (0.75–8.0 mM, pH 3.0–3.2)	Meinhard concentric nebu-lizer, a Peltier-cooled spray chamber and a shielded torch ICP-MS with octo-pole reaction cell (4 ml min^{-1} H_2), isotopes moni-tored: ^{77}Se, ^{78}Se, ^{80}Se	The results obtained suggest that MeSeMet (primary species found) is the likely precursor of volatile selenium species when plants are supplemented with Se(IV)	[101]

(Continued on next page)

TABLE 7.3 (*contd.*)

Sample	Species of interest	LC separation	ICP-MS detection	Comments	Ref.
Tl hyper-accumulator plant (*Iberis intermedia)*	Tl(I)/Tl(III)	After stabilization of Tl(III) (Tl(III)-DTPA), plant extracts were analyzed by SEC (Superdex Peptide HR 10/30 column with 100 mM CH_3COONH_4 at pH 6.2 as the mobile phase), cation exchange (Supelguard LC–SCX column with 10 mM nitric acid as a mobile phase) and anion exchange (Hamilton PRPX100 column with 100 mM CH_3COONH_4 at pH 6.2) LC	Column effluent introduced through a cross-flow nebulizer (Rayton spray chamber) to ICP-MS (^{205}Tl), ESI-MS analysis for the identification of species	Better separation was achieved by anion exchange mechanism as referred to cation exchange. The results obtained enabled to identify Tl(I) as a predominant element from in plant. The DLs for Tl(I) and Tl(III) as low as 0.3 $\mu g\ l^{-1}$ (as Tl)	[324]
Arabidopsis Thaliana	Cd bound to PCs and related peptides	Sequential extraction (Tris/HCl, driselase, SDS in Tris/HCl and CH_3COONH_4) and two-dimensional LC separation: SEC (Superdex Peptide HR 10/30 or Superdex 75 HR 10/30 column with 30 mM Tris buffer + 10 mM of NaCl eluent, pH 7.4) and reversed phase LC (Zorbax 300 SB–C18, 3.5 μm column, gradient elution with acetonitrile in 0.1% TFA)	Babington nebulizer with a Scott spray chamber, ICP-MS (^{111}Cd, ^{112}Cd, ^{114}Cd) and ESI–MS for species identification	The application of reversed phase HPLC to the fractions cut from SEC enabled to link the ligands identified by ESI–MS with the metal monitored by ICP-MS	[325]

Pharmaceutically important standards of organic acids	Organic acids (maleic, sorbic and fumaric) after tagging with TMPP	The TMPP derivatives of organic acids were separated by reversed phase LC (Phenomenex Luna C18(2), 3 μm), elution with 0.5% (v/v) formic acid in a gradient of acetonitrile, 1 ml min^{-1}	Micromist nebulizer with cyclonic, cooled spray chamber, ICP- sector-field MS at resolution 3000 (^{31}P)	The efficiency of the derivatization reaction was estimated to be between 10% and 20% and the values of DLs were 0.046 nM for TMPP and 0.25 nM for derivatized maleic acid (5 μl injection)	[88]
Cosmetic products	Bromine-containing preservatives	Ultra performance liquid chromatography (UPLC), 1.7 μm column packing material	the sensitivity of ICP-MS detection decreased when the linear flow velocity was incr-eased from 0.5 to 1.9 mm s^{-1}	The precision was better than 2.2% (RSD) and regression analysis showed a linear response at both flow rates ($R^2 > 0.9993$, $n = 36$). The analysis time was 2.7 min at a flow rate 90 μl min^{-1} with the limits of detection and quantifica-cation 20 and 65 $μl^{-1}$ (as Br)	[40]

Notes: AB – arsenobetaine; AC – arsenocholine; AFS – atomic fluorescence spectrometry; DL – detection limit; DMA – dimethylarsinic acid; DMA(III) – dimethylarsinic acid; DMAA – dimethylarsinoylacetic acid; DMAE – dimethylarsinoylethanol; DMSeP – dimethylselenoniumpropionate; CAPS – 3(cyclohexylamino)-1-propanesulfonic acid; CHAPS – (3-[3-cholamidopropyl)dimethylammonio]-1-propane sulfonate); ESI–MS – electrospray ionization–mass spectrometry; FPLC – fast protein liquid chromatography; HFBA – heptafluorobutyric acid; hTf – human transferring; MALDI-TOF – matrix assisted laser desorption ionization – time of flight mass analyzer; MeSeCys – methylselenocysteine; MeSeMet – methylselenomethionine; MMA – monomethylarsonic acid; MMA(III) – monomethylarsonic acid; MTs – metallothioneins; MW – molecular mass; PCs – phytochelatins; SDS-PAGE – sodium dodecylsulfate–polyacrylamide gel electrophoresis; SEC – size exclusion chromatography; SeCa – selenocystamine; SeEt – selenoethionine; SeGal – Se-methyl-*N*-acetylgalactosamine; SeGlu – Se-methyl-*N*-acetylglucosamine; SeMet – selenomethionine; SeU – selenourea; SFMS – sector-field MS; TFA – trifluoroacetic acid; TETRA – tetramethylarsonium ion; TMAO – trimethylarsine oxide; TMAP – trimethylarsoniopropionate; TMPP – tris(2,4,6-trimethoxyphenyl)phosphonium propylamine bromide; TMSe – trimethylselonium ion; TPA – diethylenetriaminepentaacetic acid.

primary species of these elements include different oxidation states, alkylated metal and/or metalloid compounds [1,5,17,20,44]. More recently, the research interest has expanded on the analysis of biologically important non-metals, including phosphorus, sulfur, iodine and other halogens [2,22,38,294]. In protein research, analytical targets involve different types of metal-binding, phosphorylated proteins and/or those incorporating the elements amenable for modern ICP-MS detection [2,3,91]. As mentioned in the Introduction, LC–ICP-MS tool has also been useful in studies on the pharmacokinetics of metal-based drugs (Pt, V, Au), metalloporphyrins (Ni, V, Fe, etc.), heavy metals in phytochelatins (Cd, Cu, Zn, Hg, etc.) and in humic substances [3,21].

The variety of real-world samples explored by speciation analysis involve natural waters, sediments, soils, air, plants, microorganisms (fungi, yeast, bacteria) and different samples from higher animals. Owing to the complex chemical matrix and often low concentrations of analyte species, suitable pretreatment of the sample becomes a critical step in any speciation procedure. In principle, such procedure is necessary to convert the original sample to a form that can be introduced and analyzed by LC–ICP-MS without altering the native species distribution. If the goal of speciation analysis is the characterization of unknown species and/or elucidation of elemental pathways and specific functions in a living system, the pretreatment and/or fractionation procedure has to be even more carefully selected. The macroscopic separation of organs and tissues, isolation of certain types of cells, cell disruption and separation of sub-cellular fractions, as well as isolation of specific biomolecules become important. Comprehensive reviews on this important topic are available [43,189,295–297]. It should be stressed that, ultimate elucidation of molecular mechanisms relies on the identification of species involved. Since ICP-MS detection provides only elemental information, the complementary structural characterization by mass spectrometry becomes inevitable.

In Table 7.3, the survey over recent LC–ICP-MS applications in the analysis of real-world samples is presented. The sample type together with species of interest, details of analytical procedure and some relevant comments is listed in each case.

7.6 APPROACHING A HIGHER RELIABILITY OF SPECIATION RESULTS

Without exception, every analytical procedure has different source of errors related to sampling, sample pretreatment, instrumental calibration and measurement. Consequently, the results obtained can only be considered as an approach to the actual analyte content in the sample. Owing to the specific difficulties in element speciation analysis, quality control and assurance strategies are emerging in this area. Recent developments presented in this chapter concerning LC, ICP-MS and their coupling, seem to meet the general requirement of high precision in speciation analysis. In particular, the enhanced LC resolution through miniaturization, minimum peak broadening in the interface and precise isotopic measurements by modern ICP-MS instruments should be noted. Importantly for accuracy, troublesome polyatomic interferences can be now better controlled and the use of stable or radioactive artificial isotopes enhances substantially the quantification capabilities of LC–ICP-MS coupling. Nevertheless, the best accuracy test is through the analysis of appropriate certified reference material, which should be introduced as early as possible

TABLE 7.4

METHODOLOGICAL APPROACHES FOR IMPROVED RELIABILITY OF LC–ICP-MS RESULTS

Analytical task	Sample pretreatment	LC–ICP-MS procedure reliability	Approach to enhanced	Comments	Ref.
Sb(III)/Sb(V) Speciation in soils	Extraction with citric acid (100 mM, pH 2.1) in ultrasonic bath (45 min)	Anion exchange column (PRP-X100) and 10 mM EDTA + 1 mM phthalic acid at pH 4.5 as a mobile phase, conical nebulizer and ICP-MS	Post-column, on-line ID: the column effluent mixed with an enriched (94.2%) ^{123}Sb spike solution	The DLs were 20 and 65 ng l^{-1} for Sb(V) and Sb(III) respectively. The precision (RSD at 100 ng l^{-1} was 2.7% and 3.2% ($n = 6$) for Sb(V) and Sb(III)	[326, 327]
Se species in mushrooms	Hot water extraction	(1) SEC (Superdex peptide 300 column), 25 m MCH_3COONH_4 (pH 5.6) at a flow rate 0.7 ml min^{-1}; (2) RP (Waters Spherisorb ODS2), 0.1 M CH_3COONH_4 (pH 4.5) with 2% methanol at a flow rate of 0.9 ml min^{-1}; (3) anion exchange (Hamilton PRP-X100), 5 mM ammonium citrate (pH 5) with 2% methanol, 0.9 ml min^{-1}	Different LC techniques and post-column IDA, ICP-MS with octopole reaction cell (H_2, 4 ml min^{-1})	Complementary chromato graphic techniques, based on different separation mechanisms were used, enabling the identification of not protein bound selenomethionine in mushroom extracts. Quantitative results of speciation analysis obtained by IDA	[180]
Metal-binding to dissolved organic matter-(DOM)	Reverse osmosis	SEC (YMC-Pack Diol-300 column with a Drop-in Guard column YMC Diol S-5 μm), 10 mM KClO4	ICP–HRMS detection (resolution 4000 for ^{66}Zn, ^{63}Cu, ^{59}Co, ^{56}Fe, ^{52}Cr, ^{51}V)	Possible polyatomic interferences that can be avoided by HRMS are listed. The quantitative information regarding the distribution	[328]

(Continued on next page)

TABLE 7.4 (*contd.*)

Analytical task	Sample pretreatment	LC–ICP-MS procedure reliability	Approach to enhanced	Comments	Ref.
		(pH 7.0) mobile phase at a flow rate 0.5 ml min^{-1}		of both DOM and its metal-bound complexes was obtained, interesting for characterizing metal–DOM interactions and metal speciation in natural waters	
SeMet quantification in Se-enriched yeast and commercial supplements	Different enzymatic extractions (protease XIV, driselase, in vitro gastrointestinal enzymolysis)	Anion exchange column (Hamilton PRP-X100), gradient elution with phosphate buffer (pH 7) + 2% (v/v) methanol	The isotopically enriched SeMet was used for isotope dilution analysis with ICP-MS detection using octopole reaction system (H_2, 4 ml min^{-1})	The isotopically enriched SeMet (^{77}Se) was biosynthesized in the cultures of *Saccharomyces cerevisiae*. The quantification results obtained were affected by the extraction procedure applied and also by the chemical composition of the sample	[329, 330]
Quantification of TBT in sediments	Accelerated solvent extraction with acetate buffer	Reversed phase C-18 ACE micro bore column (3 μm) with a mobile phase of 65:23:12:0.05% (v/v) acetonitrile–water–acetic acid–TEA, flow rate 0.2 ml min^{-1}, micro concentric nebulizer with Peltier-cooled spray-chamber, 0.1 l min^{-1} O_2 added post-nebulization, shield torch ICP-MS	Species-specific ID (^{117}Sn TBTCl)	The analytical performances of LC and GC as coupled to ICP-MS were compared. The uncertainty of the results was larger for LC–ICP-MS (10.5–12.8%) as compared to GC – ICP-MS (4.6–5.1%) for analysis of the same extracts	[331]
Hg inorganic and methylated species in the environmental samples	Cold acid extraction and sonication	Reversed phase separation on Supelcosil LC-18 (5 μm) column with 30% (v/v) methanol + 0.005% 2-mercaptoethanol + 0.06 M	Species-specific ID using ^{201}Hg-enriched methylmercury, synthesized in the laboratory	The synthesis yielded more than 90% as ^{201}Hg in CH_3Hg, was stable over 6 months (in dark) and could be used in for isotope dilution analysis as applied to	[301]

		CH_3COONH_4 mobile phase, sample introduction to ICP-MS via concentric nebulizer with quartz spray chamber		environmental samples	
Bromide and iodine species in bottled and tap water	Spiking experiments	Anion exchange columns: (1) 100 × 4 mm, filled with PS/DVB functionalized with diethanol-methyl-amine and elution with 25 mM HNO_3 (pH 4–6); (2) 250 × 4 mm, filled with PS/DVB-functionalized with dimethyl-ethanol-amine and elution with 120 mM HNO_3 + 1 mM $HClO_4$. In both cases, internal standard was added to the mobile phase (15 μg l^{-1} Ge). ICP-MS detection for ^{52}Cr, ^{127}I, ^{79}Br and ^{74}Ge	The use of GeO_2 as internal standard: the analyte to GeO_2 ratio of four bromine and iodine species was nearly constant over 4 months and almost independent from the ICP-MS instrumental settings	The use of GeO_2 as internal standard improved external calibration and allowed semi-quantitative determination of bromate, bromide, iodate and iodide without any calibration procedure	[332]
Fe-binding to transferring isoforms in human serum	Dilution	Mono-Q HR 5/5 anion-exchange column, elution in gradient of ammonium acetate (from 0 to 250 mM)	ICP-MS detection (^{56}Fe); octopole reaction system to minimize polyatomic interferences	DLs in the range of 0.02–0.04 mM for Tf isoforms with 2, 3, 4, 5 and 6 sialic acid residues (S(2), S(3),	[78]

(Continued on next page)

TABLE 7.4 (*contd.*)

Analytical task	Sample pretreatment	LC–ICP-MS procedure reliability	Approach to enhanced	Comments	Ref.
		buffered at pH 6 (Tris/HCl)		S(4), S(5) and S(6)) in real serum samples	
(1) Silicones in blood, (2) organo-phosphates in plasma and (3) inorganic phosphates in foods	Dilution or extraction	(1) SEC on Phenogel 5 M3 column; isocratic elution with xylene at a flow rate 1 ml min^{-1}	Sector-field high resolution inductively coupled plasma mass spectrometry (ICP-HRMS) at resolution 3000, to resolve spectral interferences caused by N_2^+ and CO^+ on ^{28}Si , and NOH^+ on $^{31}P^+$, thereby facilitating the the speciation of these elements	The feasibility of LC–ICP-HRMS was demonstrated to determine organosilicon polymers in biological samples of organophosphorus pesticides in food products	[333]
		(2) RP on Watares Nova Pak C18 column, water – methanol (1:1) mobile phase and flow rate 1 ml min^{-1}; (3) ion exchange on Phenomenre PRP–X100 column with 0.15 mM 1, 3–6-naphthalene sodium trisulfonate + 5% methanol at flow rate 1 ml min^{-1}			
As(III), As(V), MMA, DMA is soft drinks	Sample dilution	Hamilton PRX-100 anionic-exchange column, elution with phosphate buffer (5 mM, pH 8.5)	Analysis of certified reference materials	Method validation	[303]
As species in urine	Sample dilution (1 + 3)	Cation exchange column with gradient elution and ICP-MS detection	Analysis of certified reference materials	Method validation	[235]

into the chart of the analytical procedure [11,98]. Unfortunately, the list of CRMs for speciation analysis is still limited. Several materials are commercially available for methylmercury and arsenic species in marine matrices, arsenic in urine, organotin compounds in solutions and sediments, as well as inorganic selenium and chromium oxidation states in solutions [1,20,299]. This area is under dynamic development [226,233,300] Recent interest in species-specific ID has encouraged the synthesis of isotope enriched selenium and mercury species [13,175,185,301,302]. As already stressed in this text, species-specific ID is useful for validation of speciation procedure, while speciated isotope dilution additionally enables for tracing possible species interconversion.

Some representative examples from the literature, illustrating LC–ICP-MS approaches toward better reliability of speciation results are extensively presented in Table 7.4.

7.7 CONCLUSIONS

Completely reliable analytical tools to deal with element speciation analysis are needed, owing to the increasing importance of speciation results in all research areas related to human health. Hyphenation of liquid chromatography to ICP-MS is well established and considered as a primary speciation tool for non-volatile compounds. Such on-line coupling helps to minimize possible analyte loss or sample contamination, to avoid problems related with the storage of fractions, provides higher reproducibility and enables automation. The importance of compatibility between column effluent and plasma as well as the importance of state-of-the-art interface designs should be emphasized. Within this context, several shortcomings of LC–ICP-MS described in the early applications have been mitigated as a result of both technological and methodological development. Miniaturization in LC not only has improved separation performance, but also enabled enhanced compatibility of this technique with ICP-MS detection. On the other hand, the use of high-resolution ICP-MS instruments and collision/reaction cell technology eliminates troublesome polyatomic interferences and offers better capabilities for non-metal determinations. Several elements can be monitored in one chromatographic run, providing information on their possible binding in the sample. Relatively poor precision, sensitivity and mass bias typical for monitoring different ions entering plasma in form of fast transient signals is now less problematic while using multiple collector ICP-MS instruments. Finally, the environmental, toxicological and clinical importance of speciation results call for more reliable analytical results, which has been reflected in the increasing applications of isotope dilution and, in particular, species-specific and speciated isotope dilution analysis.

REFERENCES

1 B. Michalke, *Ecotoxicol. Environ. Saf.*, 56 (2003) 122–139.
2 J. Szpunar, *Analyst.*, 130 (2005) 442–465.
3 K. Wrobel, K. DeNicola, K. Wrobel and J.A. Caruso, *Advances in Mass Spectrometry*, Elsevier Science, Amsterdam, 2004, Chapter 4.
4 J.A. Caruso and M. Montes-Bayon, *Ecotoxicol. Environ. Saf.*, 56 (2003) 148–163.

5 J.A. Caruso, R.G. Wuilloud, J.C. Altamirano and W.R. Harris, *J. Toxicol. Environ. Health B Crit. Rev.*, 9 (2006) 41–61.
6 S.J. Ray, F. Andrade, G. Gamez, D. McClenathan and D. Rogers, *J. Chromatogr. A*, 1050 (2004) 3–34.
7 J.L. Gómez-Ariza, T. García-Barrera, F. Lorenzo, V. Bernal and M.J. Villegas, *Anal. Chim. Acta.*, 524 (2005) 15–22.
8 T. Hasegawa, M. Asano, K. Takatani, H. Matsuura and T. Umemura, *Talanta*, 68 (2005) 465–469.
9 N. Jakubowski, R. Lobinski and L. Moens, *J. Anal. Atom. Spectrom.*, 19 (2004) 1–4.
10 A.H. Serafin Muñoz, K. Kubachka, K. Wrobel, S.K.V. Yathavakilla and F. Gutierrez Corona, *J. Agric. Food Chem.*, 53 (2005) 5138–5143.
11 P. Quevauviller, Certified Reference Materials: A tool for quality control of elemental speciation analysis. *Elemental Speciation. New Approaches for Trace Element Analysis*; Elsevier Science, Amsterdam, 2000, pp. 531–569.
12 R. Cornelis, H. Crews, O.F. Donard, L. Ebdon and P. Quevauviller, *Fresenius J. Anal. Chem.*, 370 (2001) 120–125.
13 M. Leermakers, W. Baeyens, P. Quevauviller and M. Horvat, *Trends Anal. Chem.*, 24 (2005) 383–393.
14 A.R. Timerbaev, O. Semenova, C.G. Hartinger, M. Galanski and B.K. Keppler, *Electrophoresis*, 25 (2004) 1988–1995.
15 B.P. Jensen, C. Smith, I.D. Wilson and L. Weidolf, *Rapid Commun. Mass Spectrom.*, 18 (2004) 181–183.
16 D. Hagrman, A.-K. Souid and J. Goodisman, *J. Pharmacol. Exp. Ther.*, 308 (2004) 658–666.
17 J.A. Caruso, B. Klaue, B. Michalke and D.M. Rocke, *Ecotoxicol. Environ. Saf.*, 56 (2003) 32–44.
18 K.A. Francesconi and D. Kuehnelt, *Analyst.*, 129 (2004) 373–395.
19 P.D. Whanger, *Br. J. Nutr.*, 91 (2004) 11–28.
20 B. Michalke, *Trends Anal. Chem.*, 21 (2002) 154–165.
21 A. Sanz-Medel, M. Montes-Bayon and M. Luisa Fernandez Sanchez, *Anal. Bioanal. Chem.*, 377 (2003) 236–247.
22 M. Shah and J.A. Caruso, *J. Sep. Sci.*, 28 (2005) 1969–1984.
23 M. Kovacevic, L. Leber, S.D. Kohlwein and W. Goessler, *J. Anal. Atom. Spectrom.*, 19 (2004) 80–84.
24 D. Profrock, P. Leonhard and A. Prange, *J. Anal. Atom. Spectrom.*, 18 (2003) 708–713.
25 D. Profrock, P. Leonhard, W. Ruck and A. Prange, *Anal. Bioanal. Chem.*, 381 (2005) 194–204.
26 B.B. Sadi, A.P. Vonderheide and J.A. Caruso, *J. Chromatogr. A*, 1050 (2004) 95–101.
27 Z.X. Guo, Q. Cai and Z.G. Yang, *J. Chromatogr. A*, 1100 (2005) 160–167.
28 W.T. Cooper, J.M. Llewelyn, G.L. Bennet and V.J.M. Salters, *Talanta*, 66 (2005) 348–358.
29 A.P. Vonderheide, J. Meija, M. Montes-Bayon and J.A. Caruso, *J. Anal. Atom. Spectrom.*, 18 (2003), 1097–1102.
30 B. Michalke, P. Schramel and H. Witte, *Biol. Trace Elem. Res.*, 78 (2000) 81–91.
31 S.S. Kannamkumarath, R.G. Wuilloud, A. Stalcup, J.A. Caruso and H. Patel, *J. Anal. Atom. Spectrom.*, 19 (2004) 107–113.
32 B. Michalke, P. Schramel and H. Witte, *Biol. Trace Elem. Res.*, 78 (2001) 67–79.
33 R. Simon, J.E. Tietge, B. Michalke, S. Degitz and K.W. Schramm, *Anal. Bioanal. Chem.*, 372 (2002) 481–485.
34 M. Shah, S.S. Kannamkumarath, J.A. Caruso and R.G. Wuilloud, *J. Anal. Atom. Spectrom.*, 20 (2005) 176–182.
35 Z. Huang, K. Ito, A.R. Timerbaev and T. Hirokawa, *Anal. Bioanal. Chem.*, 378, 1836–1841.

36 K.A. Schwehr and P.H. Santschi, *Anal. Chim. Acta.*, 482 (2003) 59–71.
37 D. Profrock, P. Leonhard and A. Prange, *Anal. Bioanal. Chem.*, 377 (2003) 132–139.
38 Y. Ogra, K. Ishiwata, Y. Iwashita and K.T. Suzuki, *J. Chromatogr. A*, 1093 (2005) 118–125.
39 J. Meija, M. Montes-Bayon, J.A. Caruso, D.L. Leduc and N. Terry, *Se. Pu.*, 22 (2004) 16–19.
40 L. Bendahl, S.H. Hansen, B. Gammelgaard, S. Sturup and C. Nielsen, *J. Pharm. Biomed. Anal.*, 40 (2006) 648–652.
41 J.T. Creed and C.A. Brockhoff, *Anal. Chem.*, 71 (1999) 722–726.
42 A. Seuber, G. Schminke, M. Nowak, W. Ahrer and W. Buchberger, *J. Chromatogr. A*, 884 (2000) 191–199.
43 C. B'Hymer and J.A. Caruso, *J. Chromatogr. A*, 1045 (2004) 1–13.
44 M. Montes-Bayon, K. DeNicola and J.A. Caruso, *J. Chromatogr. A*, 1000 (2003) 457–476.
45 A. Prange and D. Profrock, *Anal. Bioanal. Chem.*, 383 (2005) 372–389.
46 S.S. Kannamkumarath, K. Wrobel, C. B'Hymer and J.A. Caruso, *J. Chromatogr. A*, 975 (2002) 245–266.
47 J. Szpunar, *Analyst*, 125 (2000) 963–988.
48 B. Bouyssiere, J. Szpunar and R. Lobinski, *Spectrochim. Acta Part B*, 57 (2002) 805–828.
49 A.L. Rosen and G.M. Hieftje, *Spectrochim. Acta Part B*, 59 (2004) 135–146.
50 B. Michalke, *Trends Anal. Chem.*, 21 (2002) 142–153.
51 C.A. Ponce de Leon, M. Montes-Bayon and J.A. Caruso, *J. Chromatogr. A*, 974 (2002) 1–21.
52 J. Szpunar and R. Lobinski, *Anal. Bioanal. Chem.*, 373 (2002) 404–411.
53 G. Centineo, E.B. Gonzalez and A. Sanz-Medel, *J. Chromatogr. A*, 1034 (2004) 191–197.
54 J.M. Liu and J.K. Cheng, *Electrophoresis*, 24 (2003) 1993–2012.
55 B. Michalke, *Electrophoresis*, 26 (2005) 1584–1597.
56 F. Cubadda, *J. AOAC Int.*, 87 (2004) 173–204.
57 K.G. Heumann, *Anal. Bioanal. Chem.*, 378 (2004) 318–329.
58 R.N. Collins, *J. Chromatogr. A*, 1059 (2004) 1–12.
59 G.J. Shugar and J.A. Dean, *The chemist's ready reference handbook*; McGraw-Hill, New York, 1990.
60 K.L. Sutton, J.A. Caruso, *J. Chromatogr. A*, 856 (1999) 243–258.
61 R.S. Houk, *Anal. Chem.*, 97A (1986) 58.
62 R. Thomas, *Spectroscopy*, 17 (2002) 42–48.
63 A. Taylor, S. Branch, A. Fisher, D. Halls and M. White, *J. Anal. Atom. Spectrom.*, 16 (2000), 421–446.
64 M. Moldovan, E.M. Krupp, A.E. Holliday and O.F.X. Donard, *J. Anal. Atom. Spectrom.*, 19 (2004) 815–822.
65 U. Kohlmeyer, E. Jantzen, J. Kuballa and S. Jakubik, *Anal. Bioanal. Chem.*, 377 (2003) 6–13.
66 J. Zheng, H. Hintelmann, B. Dimock and M.S. Dzurko, *Anal. Bioanal. Chem.*, 377 (2003) 14–24.
67 B. Klaue and J.D. Blum, *Anal. Chem.*, 71 (1999) 1408–1414.
68 K. Sakata and K. Kawabata, *Spectrochim. Acta. Part B*, 49B (1994) 1027.
69 D.A. Polya, P.R. Lythgoe, A.G. Gault, J.R. Brydie and F. Abou-Shakra, *Mineralogical Magazine*, 67 (2003) 247–261.
70 A. Shinohara, M. Chiba, Y. Inaba, M. Kondo and F.R. Abou-Ahakra, *Bunseki Kagaku*, 53 (2004) 589–593.
71 D.R. Bandura, V.I. Baranov and S.D. Tanner, *Fresenius J. Anal. Chem.*, 370 (2001) 454–470.
72 M. Okina, K. Yoshida, K. Kuroda, H. Wanibuchi and S. Fukushima, *J. Chromatogr. B Analyt. Technol. Biomed. Life Sci.*, 799 (2004) 209–215.
73 K. Wrobel, K. Wrobel, B. Parker, S.S. Kannamkumarath and J.A. Caruso, *Talanta*, 58 (2002) 899–907.

74 Y.-L. Feng, *Bunseki Kagaku*, 51 (2002) 331–332.
75 R. Ritsema, L. Dukan, W. Van Leeuwen, N. Oliveira and P. Wolfs, *Appl. Organomet. Chem.*, 12 (1998) 591–599.
76 H.M. Zhang, D.Y. Chen, M. Jing and X.L. Liu, *Wei Sheng Yan Jiu*, 34 (2005) 603–606.
77 C.F. Yeh and S.J. Jiang, *J. Chromatogr. A*, 1029 (2004) 255–261.
78 S. Arizaga Rodriguez, E. Blanco Gonzalez, G. Alvarez Llamas, M. Montes-Bayon and A. Sanz-Medel, *Anal. Bioanal. Chem.*, 383 (2005) 390–397.
79 F. Vanhaecke, L. Balcaen, G. De Wannemacker and L. Moens, *J. Anal. Atom. Spectrom.*, 17 (2002) 933–943.
80 C.F. Harrington, S. Elahi, S.A. Merson and P. Ponnampalavanar, *J. AOAC Int.*, 87 (2004) 253–258.
81 M.H. Nagaoka and T. Maitani, *Analyst.*, 125 (2000) 1962–1965.
82 C. Sariego Muñiz, J.M. Marchante Gayón, J.I. García Alonso and A. Sanz-Medel, *J. Anal. Atom. Spectrom.*, 14 (1999) 1505–1510.
83 C.J. Park and J.K. Suh, *J. Anal. Atom. Spectrom.*, 12 (1997) 573.
84 F. Vanhaecke, *Anal. Bioanal. Chem.*, 372 (2002) 20–21.
85 D.R. Bandura, V.I. Baranov and S.D. Tanner, *Anal. Chem.*, 74 (2002) 1497–1502.
86 J.S. Becker, S.F. Boulyga, C. Pickhardt, J. Becker and S. Buddrus, *Anal. Bioanal. Chem.*, 375 (2003) 561–566.
87 M. Wind, M. Edler, N. Jakubowski, M. Linscheid and H. Wesch, *Anal. Chem.*, 73 (2001) 29–35.
88 A.J. Cartwright, J. Jones, J.C. Wolff and E.H. Evans, *J. Anal. Atom. Spectrom.*, 20 (2005) 75–80.
89 C.F. Yeh and S.J. Jiang, *Analyst.*, 127 (2002) 1324.
90 Smith, C.J. Wilson, I.D. Weidolf, L. Abou-Shakra, F. Thomsen, M. *Chromatographia*, 59 Suppl 2 (2004) S165.
91 M. Wind, H. Wesch and W.D. Lehmann, *Anal. Chem.*, 73 (2001) 3006–3010.
92 M. Kovacevic, A. Gartner and M. Novic, *J. Chromatogr. A*, 1039 (2004) 77–82.
93 E.H. Evans, J.C. Wolff and C. Eckers, *Anal. Chem.*, 73 (2001) 4722–4728.
94 P. Evans, C. Wolff-Briche and B. Fairman, *J. Anal. Atom. Spectrom.*, 16 (2001) 964–969.
95 P.R.D. Mason, K. Kaspers and M.J. van Bergen, *J. Anal. Atom. Spectrom.*, 14 (1999) 1067–1074.
96 T. Prohaska, C. Latkoczy and G. Stingeder, *J. Anal. Atom. Spectrom.*, 14 (1999) 1501.
97 N. Jakubowski, D. Stuewer, D. Klockow, C. Thomas and H. Emons, *J. Anal. Atom. Spectrom.*, 16 (2001) 135–139.
98 J. Darrouzès, M. Bueno, G. Lespès and M. Potin-Gautier, *J. Anal. Atom. Spectrom.*, 20 (2005) 88–94.
99 J.R. Encinar, D. Schaumlöffel, R. Lobinski and Y. Ogra, *Anal. Chem.*, 76 (2004) 6635–6642.
100 A.P. Vonderheide, S. Mounicou, J. Meija, Henry and J.A. Caruso, *Analyst.*, 131 (2006) 33–40.
101 S.K.V. Yathavakilla, M. Shah, S. Mounicou and J.A. Caruso, *J. Chromatogr. A*, 1100 (2005) 153–159.
102 V. Díaz Huerta, L. Hinojosa Reyes, J.M. Marchante-Gayón, M.L. Fernández Sánchez and A. Sanz-Medel, *J. Anal. Atom. Spectrom.*, 18 (2003) 1243–1247.
103 J. Meija and J.A. Caruso, *J. Am. Soc. Mass Spectrom.*, 15 (2004) 654–658.
104 L. Hinojosa Reyes, J.M. Marchante-Gayón, J.I. García Alonso and A. Sanz-Medel, *J. Anal. Atom. Spectrom.*, 18 (2003) 1210–1216.
105 B. Fairman, M.W. Hinds, S.M. Nelms, D.M. Pennyd and P. Goodal, *J. Anal. Atom. Spectrom.*, 15 (2000) 1606–1631.
106 Y. Gao, K. Oshita, K.H. Lee, M. Oshima and S. Motomizu, *Analyst.*, 127 (2002) 1713–1719.
107 J.S. Becker, *J. Anal. Atom. Spectrom.*, 20 (2005) 1173–1184.

108 Y. Martinez-Bravo, A.F. Roig-Navarro, F.J. Lopez and F. Hernandez, *J. Chromatogr. A*, 926 (2001) 265–274.
109 J. Zheng, W. Kosmus, F. Pichler-Semmelrock and M. Kock, *J. Trace Elem. Med. Biol.*, 13 (1999) 150–156.
110 A.G. Gault, J. Jana, S. Chakraborty, P. Mukherjee and M. Sarkar, *Anal. Bioanal. Chem.*, 381 (2005) 347–353.
111 M.F. Giné, A.C.S. Bellato and A.A. Menegário, *J. Anal. Atom. Spectrom.*, 19 (2004) 1252–1256.
112 R. Thomas, *Spectroscopy*, 16 (2001) 22–27.
113 K.L. Ackley, K.L. Sutton and J.A. Caruso, The use of ICP-MS as a detector for elemental speciation studies. *Comprehensive Analytical Chemistry*, Elsevier, Amsterdam, 2000, pp. 249–276.
114 G. O'Connor and E.H. Evans, *Inductively Coupled Plasma Spectrometry and its Applications*, Sheffield Academic Press, Sheffield, 1999, pp. 370.
115 R. Hutton, A. Walsh, D. Milton and J. Cantle, *Chem. SA.*, 17 (1991) 213–215.
116 U. Geismann and U. Greb, *Fresenius J. Anal. Chem.*, 350 (1994) 186–193.
117 F. Adams, R. Gijbels and R. Van Grieken, *Inorganic Mass Spectrometry*, Wiley, New York, 1988.
118 A.M. Featherstone, A.T. Townsend, G.A. Jacobson and G.M. Peterson, *Anal. Chim. Acta.*, 512 (2004) 319–327.
119 S. Mazan, N. Gilon, G. Crétier, J.L. Rocca and J.M. Mermet, *J. Anal. Atom. Spectrom.*, 17 (2002) 366–370.
120 C.N. Ferrarello, M.R. Fernández de la Campa and A. Sanz-Medel, *Anal. Bioanal. Chem.*, 373 (2002) 412–421.
121 M. Balcerzak, *Anal. Sci.*, 19 (2003) 979–989.
122 J.H.T. Barnes, G.D. Schilling, S.F. Stone, R.P. Sperline and M.B. Denton *et al., Anal. Bioanal. Chem.*, 380 (2004) 227–234.
123 K. Dash, K. Chandrasekaran, S. Thangavel, S.M. Dhaville and J. Arunachalam, *J. Chromatogr. A*, 1022 (2004) 25–31.
124 S.D. Tanner and V.I. Baranov, *J. Am. Soc. Mass Spectrom.*, 10 (1999) 1083.
125 V.I. Baranov and S.D. Tanner, *J. Anal. Atom. Spectrom.*, 14 (1999) 1133.
126 I. Feldmann, *Trends Anal. Chem.*, 24 (2005) 228–242.
127 J.J. Sloth and E.H. Larsen, *J. Anal. Atom. Spectrom.*, 15 (2000) 669–672.
128 R.A. Thomas, *Spectroscopy*, 17 (2002) 42–48.
129 S.D. Tanner and V.I. Baranov, *Atom. Spectrosc.*, 20 (1999) 45–52.
130 C. Bonnefoy, A. Menudier, C. Moesch, G. Lachatre and J.M. Mermet, *Anal. Bioanal. Chem.*, 383 (2005) 167–173.
131 Y.-L. Chang and S.-J. Jiang, *J. Anal. Atom. Spectrom.*, 16 (2001) 858–862.
132 D.E. Nixon, K.R. Neubauer, S.J. Eckdahl, J.A. Butz and M.F. Burritt, *Spectrochimica Acta – Part B Atom. Spectrosc.*, 57 (2002) 951–966.
133 M.C. Wu, S.J. Jiang and T.S. Hsi, *Anal. Bioanal. Chem.*, 377 (2003) 154–158.
134 C.C. Chéry, K. De Cremer, R. Cornelis, F. Vanhaecke and L. Moens, *J. Anal. Atom. Spectrom.*, 18 (2003) 1113–1118.
135 D.E. Nixon, S.J. Eckdahl, J.A. Butz, M.F. Burritt and K.R. Neubauer, *Spectrochimica Acta – Part B Atom. Spectrosc.*, 58 (2003) 97–110.
136 F. Cubadda, A. Raggi and E. Coni, *Anal. Bioanal. Chem.*, 384 (2006) 887–896.
137 S. Ciardullo, G. Taviani, R. Mattei and S. Caroli, *J. Environ. Monit.*, 7 (2005) 1332–1334.
138 B. Bouyssiere, P. Leonhard, D. Profrock, A. Prange and F. Baco *et al., J. Anal. Atom. Spectrom.*, 19 (2004) 700–702.
139 J.R. Encinar, D. Schaumlöffel, Y. Ogra and R. Lobinski, *Anal. Chem.*, 76 (2004) 6635–6642.

140 O. Palacios, J.R. Encinar, G. Bertin and R. Lobinski, *Anal. Bioanal. Chem.*, 383 (2005) 516–522.
141 T. Lindemann and H. Hintelmann, *Anal. Chem.*, 74 (2002) 4602–4610.
142 J.M. Marchante-Gayón, I. Feldmann, C. Thomas and N. Jakubowski, *J. Anal. Atom. Spectrom.*, 16 (2001) 457–463.
143 A.H. Serafin Muñoz, K. Kubachka, K. Wrobek, F.J. Gutierrez Corona, S.K.V. Yathavakilla, *et al.*, *J. Agric. Food. Chem.*, 54 (2006) 3440–3444.
144 M. Miguens-Rodriguez, R. Pickford, J.E. Thomas-Oates and S.A. Pergantis, *Rapid Commun. Mass Spectrom.*, 16 (2002) 323–331.
145 X. Qianli, R. Kerrich, E. Irving, K. Liber and F. Abou-Shakra, *J. Anal. Atom. Spectrom.*, 17 (2002) 1037–1041.
146 V. Vacchina, D. Schaumlöffel, R. Lobinski, S. Mari and P. Czernic *et al.*, *Anal. Chem.*, 75 (2003) 2740–2745.
147 V. Vacchina, S. Mari, P. Czernic, L. Marques and K. Pianelli *et al.*, *Anal. Chem.*, 75 (2003) 2740–2745.
148 D. Schaumlöffel, L. Ouerdane, B. Bouyssiere and R. Lobinski, *J. Anal. Atom. Spectrom.*, 18 (2003) 120–127.
149 P. Heitland and H.D. Koster, *Clin. Chim. Acta*, 365 (2006) 310–318.
150 S.F. Boulyga and J.S. Becker, *Fresenius J. Anal. Chem.*, 370 (2001) 618–623.
151 M. Iglesias, N. Gilon, E. Poussel and J.M. Mermet, *J. Anal. Atom. Spectrom.*, 17 (2002) 1240–1247.
152 M. Niemela, H. Kola, K. Eilola and P. Peramaki, *J. Pharm. Biomed. Anal.*, 35 (2004) 433–439.
153 J.S. Becker and H.J. Dietze, *Fresenius J. Anal. Chem.*, 368 (2000) 23–30.
154 J. Ruiz Ençinar, J.I. Garcia Alonso, A. Sanz-Medel, S. Main and T.J. Turner, *J. Anal. Atom. Spectrom.*, 16 (2001) 315–321.
155 J. Ruiz Encinar, J.I. García Alonso, A. Sanz-Medel, S. Main and P.J. Turner, *J. Anal. Atom. Spectrom.*, 16 (2001) 322–326.
156 I. Gunther-Leopold, B. Wernli, Z. Kopajtic and D. Gunther, *Anal. Bioanal. Chem.*, 378 (2004) 241–249.
157 S.J. Hill, L.J. Pitts and A.S. Fisher, *Trends Anal. Chem.*, 19 (2000) 120–126.
158 G.N. Bowers Jr., J.D. Fassett and E. White, *Anal. Chem.*, 65 (1993) 475r.
159 J.D. Fassett and P.J. Paulsen, *Anal. hem.*, 61 (1989) 643a.
160 H.G. Heumann, *Inorganic Mass Spectrometry*, John Wiley, New York, 1988, p. 301.
161 H.P. Longerich, *Atom. Spectrosc.*, 10 (1989) 112.
162 SW-846 EPA, Method 6800: Elemental and speciated isotope dilution mass spectrometry. *Test Methods for Evaluating Solid Waste*, Environmental Protection Agency, Washington, DC, 1998.
163 S.J. Hill, L.J. Pitts and A.S. Fisher, *Trends Anal. Chem.*, 19 (2000) 120–126.
164 K.G. Heumann, *Metal Speciation in the Environment*, Springer, Berlin, Heidleberg, New York, 1990.
165 K.G. Heumann, *Int. J. Mass Spectrom. Ion Processes*, 118/119 (1992) 575.
166 D. Tanzer and K.G. Heumann, *Anal. Chem.*, 63 (1991) 1984.
167 L. Rottmann and K.G. Heumann, *Fresenius J. Anal. Chem.*, 350 (1994) 221.
168 R. Nusko and K.G. Heumann, *Fresenius J. Anal. Chem.*, 357 (1997) 1050.
169 R. Nusko and K.G. Heumann, *Anal. Chim. Acta*, 286 (1994) 283.
170 K.G. Heumann, S.M. Gallus, G. Radlinger and J. Vogl, *Spectrochim. Acta, Part B*, 53 (1998) 273.
171 K.G. Heumann, L. Rottman and J. Vogl, *J. Anal. Atom. Spectrom.*, 9 (1994) 1351.
172 A.A. Brown, L. Ebdon and S.J. Hill, *Anal. Chim. Acta*, 286 (1994) 391.
173 D. Huo, H.M. Kingston and B. Larget, *Comprehensive Analytical Chemistry*, Elsevier, Amsterdam, 2000, pp. 277–313.

174 J.R. Encinar, *Anal. Bioanal. Chem.*, 375 (2003) 41–43.
175 L. Lambertsson and E. Bjorn, *Anal. Bioanal. Chem.*, 380 (2004) 871–875.
176 M. Monperrus, R.C. Rodriguez Martin-Doimeadios, J. Scancar, D. Amouroux and O.F. Donard, *Anal. Chem.*, 75 (2003) 4095–4102.
177 L. Yang, Z. Mester and R.E. Sturgeon, *Anal. Chem.*, 74 (2002) 2968–2976.
178 C.N. Ferrarello, J. Ruiz Encinar, G. Centineo, J.I. García Alonso, M.R. Fernández de la Campa *et al.*, *J. Anal. Atom. Spectrom.*, 17 (2002) 1024–1029.
179 V.D. Huerta, M.L.F. Sánchez and A. Sanz-Medel, *J. Anal. Atom. Spectrom.*, 19 (2004) 644–648.
180 V. Diaz Huerta, M.F. Fernandez Sanchez and A. Sanz-Medel, *Anal. Chim. Acta*, 538 (2005) 99–105.
181 A. Rodríguez-Cea, M.D.R. Fernández de la Campa, E.B. González, A. Sanz-Medel and B.A. Fernández, *J. Anal. Atom. Spectrom.*, 18 (2003) 1357–1364.
182 H.M. Kingston, D. Huo, Y. Lu and S. Chalk, *Spectrochim. Acta, Part B*, 53 (1998) 299.
183 H.M. Kingston, R. Cain, D. Huo and G.M. Rahman, *J. Environ. Monit.*, 7 (2005) 899–905.
184 D. Huo and H.M. Kingston, *Anal. Chem.*, 72 (2000) 5047–5054.
185 G.M. Rahman and H.M. Kingston, *Anal. Chem.*, 76 (2004) 3548–3555.
186 P. Rodriguez-Gonzalez, J.R. Encinar, J.I. Garcia Alonso and A. Sanz-Medel, *Analyst.*, 128 (2003) 447–452.
187 R.S. Houk, *Anal. Chem.*, 58 (1986) 97A.
188 A.P. Vonderheide, K. Wrobel, S.S. Kannamkumarath, C. B'Hymer, M. Montes-Bayon *et al.*, *J. Agric. Food Chem.*, 50 (2002) 5722–5728.
189 K. Wrobel, K. Wrobel and J.A. Caruso, *Anal. Bioanal. Chem.*, 381 (2005) 317–331.
190 J. Huang, X. Hu, J. Zhang, K. Li, Y. Yan *et al.*, *J. Pharm. Biomed. Anal.*, 40 (2006) 227–234.
191 S.P. Mendez, E.B. Gonzalez and A. Sanz-Medel, *Biomed. Chromatogr.*, 15 (2001) 181–188.
192 G.A. Jacobson, A.M. Featherstone, A.T. Townsend, R. Lord and G.M. Peterson, *Biol. Trace. Elem. Res.*, 107 (2005) 213–220.
193 X. Dauchy, R. Cottier, A. Batel, M. Borsier, A. Astruc *et al.*, *Environ. Technol.*, 15 (1994) 569–576.
194 L. Ebdon, S.J. Hill and C. Rivas, *Trends Anal. Chem.*, 17 (1998) 277–288.
195 B. Fairman and R. Wahlen, *Spectrosc. Eur.*, 13 (2001) 16–22.
196 E. González-Toledo, R. Compaño, M. Granados and M.D. Prat, *Trends Anal. Chem.*, 22 (2003) 26–33.
197 R. Wahlen and T. Catterick, *J. Chromatogr. B Analyt. Technol. Biomed. Life Sci.*, 783 (2003) 221–229.
198 J.L. Gomez-Ariza, D. Sanchez-Rodas, M.A. Caro de la Torre, I. Giraldez and E. Morales, *J. Chromatogr. A*, 889 (2000) 33–39.
199 D. Schaumlöffel, J. Ruiz Encinar and R. Lobinski, *Anal. Chem.*, 75 (2003) 6837–6842.
200 C. Siethoff, I. Feldmann, N. Jakubowski and M. Linscheid, *J. Mass Spectrom.*, 34 (1999) 421–426.
201 W.R. Cairns, L. Ebdon and S.J. Hill, *Anal. Bioanal. Chem.*, 355 (1996) 202–208.
202 P. Galettis, J.L. Carr, J.W. Paxton and M.J. McKeage, *J. Anal. Atom. Spectrom.*, 14 (1999) 953–956.
203 M.H. Hanigan, D.M. Townsend, M. Deng, J.A. Marto and T.J. MacDonald, *Drug Metabolism and Disposition*, 31 (2003) 705–713.
204 T. Oe, Y. Tian, P.J. O'Dwyer, I.A. Blair, D.W. Roberts *et al.*, *J. Chromatogr. B: Anal. Technol. Biomed. Life Sci.*, 792 (2003) 217–227.
205 B.O. Axelsson, M. Jornten-Karlsson, P. Michelsen and F. Abou-Shakra, *Rapid Commun. Mass Spectrom.*, 15 (2001) 375–385.
206 B. Michalke, P. Schramel and H. Witte, *Biol. Trace Elem. Res.*, 78 (2001) 81–91.

207 M. Kotrebai, M. Birringer, J.F. Tyson, E. Block and P.C. Uden, *Analyst.*, 125 (2000) 71–78.
208 K. Wrobel, S.S. Kannamkumarath and J.A. Caruso, *Anal. Bioanal. Chem.*, 375 (2003) 133–138.
209 S.S. Kannamkumarath, K. Wrobel, A.Vonderheide and J.A. Caruso, *Anal. Bioanal. Chem.*, 373 (2002) 454–460.
210 M. Kotrebai, J.F. Tyson, E. Block and P.C. Uden, *J. Chromatogr. A*, 866 (2000) 51–63.
211 B. Gammelgaard, L. Bendahl, U. Sidenius and O. Jøns, *J. Anal. Atom. Spectrom.*, 17 (2002) 570–575.
212 T. Guerin, A. Astruc and M. Astruc, *Talanta*, 50 (1999) 1–24.
213 S. Londesborough, J. Mattusch and R. Wennrich, *Fresenius J. Anal. Chem.*, 363 (1999) 577–581.
214 X.C. Le, M. Ma, W.R. Cullen, H.V. Aposhian, X. Lu *et al.*, *Environ. Health Perspect.*, 108 (2000) 1015–1018.
215 U. Kohlmeyer, J. Kuballa and E. Jantzen, *Rapid Commun. Mass Spectrom.*, 16 (2002) 965–974.
216 E. Vassileva, A. Becker and J.A. Broekaert, *Anal. Chim. Acta*, 441 (2001) 135.
217 J.M. Marchante Gayón, C. Thomas, I. Feldmann and N. Jakubowski, *J. Anal. Atom. Spectrom.*, 15 (2000) 1093–1102.
218 S.M. Bird, H. Ge, P.C. Uden, J.F. Tyson, E. Block *et al.*, *J. Chromatogr. A*, 789 (1997) 349–359.
219 K. Wrobel, K. Wrobel and J.A. Caruso, *J. Anal. Atom. Spectrom.*, 17 (2002) 1048–1054.
220 K. Wrobel, K. Wrobel, S.S. Kannamkumarath, J.A. Caruso, I.A. Wysocka *et al.*, *Food Chem.*, 86 (2004) 617–623.
221 H. Ding, J. Wang, J.G. Dorsey and J.A. Caruso, *J. Chromatogr. A*, 694 (1995) 425–431.
222 L. Benramdane, F. Bressolle and J.J. Vallon, *J. Chromatogr. Sci.*, 37 (1999) 330–344.
223 J.A. Day, M. Montes-Bayón, A.P. Vonderheide and J.A. Caruso, *Anal. Bioanal. Chem.*, 373 (2002) 664–668.
224 M. Bissen and F.H. Frimmel, *Fresenius J. Anal. Chem.*, 367 (2000) 51–55.
225 B.P. Jackson and P.M. Bertsch, *Environ. Sci. Technol.*, 35 (2001) 4868–4873.
226 A. Polatajko and J. Szpunar, *J. AOAC Int.*, 87 (2004) 233–237.
227 D.T. Heitkemper, N.P. Vela, K.R. Stewart and C.S. Westphal, *J. Anal. Atom. Spectrom.*, 16 (2001) 299–306.
228 H. Helgesen, E.H. Larsen, *Analyst.*, 123 (1998) 791–796.
229 J.A. Caruso, D.T. Heitkemper and C. B'Hymer, *Analyst.*, 126 (2001) 136–140.
230 S.S. Kannamkumarath, K. Wrobel and J.A. Caruso, *J. Agric. Food Chem.*, 52 (2004) 1458–1463.
231 N.P. Vela and D.T. Heitkemper, *J. AOAC Int.*, 87 (2004) 244–252.
232 X. Wei, C.A. Brockhoff-Schwegel and J.T. Creed, *J. Anal. Atom. Spectrom.*, 16 (2001) 12–19.
233 I. Pizarro, M. Gomez, M.A. Palacios and C. Camara, *Anal. Bioanal. Chem.*, 376 (2003) 102–109.
234 K.T. Suzuki, T. Tomita, Y. Ogra and M. Ohmichi, *Chem. Res. Toxicol.*, 14 (2001) 1604–1611.
235 J.J. Sloth, K. Julshamn and E.H. Larsen, *J. Anal. Atom. Spectrom.*, 19 (2004) 973–978.
236 L.S. Milstein, A. Essader, E.D. Pellizzari, R.A. Fernando, J.H. Raymer *et al.*, *Environ. Health Perspect.*, 111 (2003) 293–296.
237 K.T. Suzuki, B.K. Mandal and Y. Ogra, *Talanta*, 58 (2002) 111–119.
238 B. Zheng and H. Hintelmann, *J. Anal. Atom. Spectrom.*, 19 (2004) 191–195.
239 W. Goessler, D. Kuehnelt, C. Schlagenhaufen, Z. Slejkovec and K.J. Irgolic, *J. Anal. Atom. Spectrom.*, 13 (1998) 183.
240 E.H. Larsen, M. Hansen, T. Fan and M. Vahl, *J. Anal. Atom. Spectrom.*, 16 (2001) 1403–1408.
241 R. Garcia-Sanchez, R. Feldhaus, J. Bettmer and L. Ebdon, *J. Anal. Atom. Spectrom.*, 16 (2001) 1028–1034.
242 F.A. Byrdy, L.K. Olson, N.P. Vela and J.A. Caruso, *J. Chromatogr. A*, 712 (1995) 311–320.

243 A. Sanz-Medel, A.B. Soldado Cabezuelo, R. Milacic and T.B. Polak, *Coordination Chem. Rev.*, 228 (2002) 373–383.
244 M. Krachler and H. Emons, *J. Anal. Atom. Spectrom.*, 16 (2001) 20–25.
245 M.T. Siles Cordero, E.I. Vereda Alonso, A. García de Torres and J.M. Cano Pavón, *J. Anal. Atom. Spectrom.*, 19 (2004) 398–403.
246 K.-I. Tsunoda, T. Umemura, K. Ohshima, S.-I. Aizawa, E. Yoshimura *et al.*, *Water Air Soil Pollut.*, 130 (2001) 1589–1594.
247 N. Ulrich, P. Shaked and D. Zilberstein, *Fresenius J. Anal. Chem.*, 368 (2000) 62–66.
248 K. Wrobel, E.B. Gonzalez and A. Sanz-Medel, *Analyst.*, 120 (1995) 809–815.
249 P. Zöllner, A. Zenker, M. Galanski, B.K. Keppler and W. Lindner, *J. Mass Spectrom.*, 36 (2001) 742–753.
250 R. Falter and R.D. Wilken, *Sci. Total Environ.*, 225 (1999) 167–176.
251 A.P. Navaza, M. Montes-Bayon, D.L. Leduc, N. Terry A. Sanz-Medel, *J. Mass Spectrom.*, 41 (2006) 323–331.
252 B. Li, J. Bergmann, S. Lassen, P. Leonhard, A. Prange, *J. Chromatogr. B Analyt Technol. Biomed. Life Sci.*, 814 (2005) 83–91.
253 B. Kralj, I. Krizaj, P. Bukovec, S. Slejko and R. Milacic, *Anal. Bioanal. Chem.*, 383 (2005) 467–475.
254 R.G. Wuilloud, S.S. Kannamkumarath and J.A. Caruso, *Appl. Organometal. Chem.*, 18 (2004) 156–165.
255 K. Wrobel, B.B. Sadi, J.R. Castillo and J.A. Caruso, *Anal. Chem.*, 75 (2003) 761–767.
256 J.E. Carr, K. Kwok, G.K. Webster and J.W. Carnahan, *J. Pharm. Biomed. Anal.*, 40 (2006) 42–50.
257 V. Gergely, K.M. Kubachka, S. Mounicou, P. Fodor and J.A. Caruso, *J. Chromatogr. A*, 1101 (2006) 94–102.
258 J.-C. Shen, Z.-X. Zhuang, X.-R. Wang, F.S.C. Lee and Z.-Y. Huang, *Appl. Organometal. Chem.*, 19 (2005) 140–146.
259 R.J. Chalkley, P.R. Baker, K.C. Hansen, K.F. Medzihradszky, N.P. Allen *et al.*, *Mol. Cell. Proteomics*, 4 (2005) 1189–1193.
260 I. Pizarro, M. Gomez, C. Camara, M.A. Palacios and D.A. Roman-Silva, *J. Anal. Atom. Spectrom.*, 19 (2004) 292–296.
261 J.R. Encinar, L. Ouerdane, W. Buchmann, J. Tortajada, R. Lobinski *et al.*, *Anal. Chem.*, 75 (2003) 3765–3774.
262 H. Garraud, A. Woller, P. Fodor and O.F.X. Donard, *Analysis*, 25 (1997) 25–31.
263 X. Huang, Q. Wang and B. Huang, *Fenxi Huaxue*, 33 (2005) 467–470.
264 Y. Ogra and K.T. Suzuki, *J. Anal. Atom. Spectrom.*, 20 (2005) 35–39.
265 P. Giusti, D. Schaumlöffel, F. Preudhomme, J. Szpunar and R. Lobinski, *J. Anal. Atom. Spectrom.*, 21 (2006) 26–32.
266 S. McSheehy, L. Yang, R. Sturgeon and Z. Mester, *Anal. Chem.*, 77 (2005) 344–349.
267 M. Wind, A. Eisenmenger and W.D. Lehmann, *J. Anal. Atom. Spectrom.*, 17 (2001) 21–26.
268 F. Vanhaecke, S. Saverwyns, G. De Wannemacker, L. Moens and R. Dams, *Anal. Chim. Acta*, 419 (2000) 55–64.
269 S.Saverwyns, K. Van Hecke, F. Vanhaecke, L. Moens and R. Dams, *Fresenius J. Anal. Chem.*, 363 (1999) 490–494.
270 Z. Stefanka, G. Koellensperger, G. Stingeder and S. Hann, *J. Anal. Atom. Spectrom.*, 21 (2006) 86–89.
271 C. B'Hymer and J.A. Caruso, Nebulizer sample introduction for elemental speciation. *Comprehensive Analytical Chemistry*, Elsevier, Amsterdam, 2000, pp. 213–226.
272 Y.C. Sun, Y.S. Lee, T.L. Shiah, P.L. Lee, W.C. Tseng *et al. J. Chromatogr. A*, 1005 (2003) 207–213.

273 A. Chatterjee, Y. Shibata, H. Tao, A. Tanaka and M. Morita, *J. Chromatogr. A*, 1042 (2004) 99–106.
274 H.G. Infante, G. O'Connor, R. Wahlen, J. Entwisle, P. Norris *et al.*, *J. Anal. Atom. Spectrom.*, 19 (2004) 1529–1538.
275 A.A. Ammann, *Anal. Bioanal. Chem.*, 372 (2002) 448–452.
276 S. Karthikeyan and S. Hirata, *Appl. Organometal. Chem.*, 18 (2004) 3232–3330.
277 S.C. Shum, R. Neddersen and R.S. Houk, *Analyst.*, 117 (1992) 577–582.
278 H. Emteborg, G. Bordin and A.R. Rodriquez, *Analyst.*, 123 (1998) 245–253.
279 G. Zoorob, M. Tomlinson, J. Wang and J.A. Caruso, *J. Anal. Atom. Spectrom.*, 10 (1995) 853–858.
280 M.J. Powell, D.W. Boomer and D.R. Wiederin, *Anal. Chem.*, 67 (1995) 2474–2478.
281 J.A. McLean, H. Zhang and A. Montaser, *Anal. Chem.*, 70 (1998) 1012–1020.
282 B. Gammelgaard, L. Bendahl, *J. Anal. Atom. Spectrom.*, 19 (2004) 135–142.
283 E. Bjorn and W. Frech, *Anal. Bioanal. Chem.*, 376 (2003) 274–278.
284 D. Sanchez-Rodas, J. Luis Gomez-Ariza, I. Giraldez, A. Velasco and E. Morales, *Sci. Total. Environ.*, 345 (2005) 207–217.
285 H.G. Infante, M.L. Fernandez Sanchez and A. Sanz-Medel, *J. Anal. Atom. Spectrom.*, 15 (2000) 519–524.
286 A. Chatterjee, Y. Shibata and M. Morita, *J. Anal. Atom. Spectrom.*, 15 (2000) 913–919.
287 S.-J. Chorng-jev Hwang, *Anal. Chim. Acta*, 289 (1994) 205–213.
288 B. Do, P. Alet, D. Pradeau, J. Poupon, M. Guilley-Gaillot *et al., J. Chromatogr. B Biomed. Sci. Appl.*, 740 (2000) 179–186.
289 E. Schmeisser, W. Goessler, N. Kienzl and K.A. Francesconi, *Anal. Chem.*, 76 (2004) 418–423.
290 S.J. Hill, M.J. Bloxham and M.J. Worsfold, *J. Anal. Atom. Spectrom.*, 8 (1993) 499.
291 F.A. Byrdy and J.A. Caruso, *Environ. Sci. Technol.*, 28 (1994) 528A.
292 K. Sutton, R.M. Sutton and J.A. Caruso, *J. Chromatogr. A*, 789 (1997) 85–126.
293 C.A. Ponce de Leon, M. Montes-Bayon and J.A. Caruso, *J. Chromatogr. A*, 974 (2002) 1.
294 O. Corcoran, J.K. Nicholson, E.M. Lenz, F. Abou-Shakra, J. Castro-Perez *et al.*, *Rapid Commun. Mass Spectrom.*, 14 (2000) 2377–2384.
295 M.J. Marques, A. Salvador, A. Morales-Rubio and M. De la Guardia, *Fresenius J. Anal. Chem.*, 367 (2000) 601–613.
296 K. Wrobel, S.S. Kannamkumarath, K. Wrobel and J.A. Caruso, *Green Chem.*, 5 (2003) 250–259.
297 J.L. Gomez-Ariza, E. Morales, I. Giraldez, D. Sanchez-Rodas and A. Velasco, *J. Chromatogr. A*, 938 (2001) 211–224.
298 P. Quevauviller, *Fresenius J. Anal. Chem.*, 345 (1996) 515–520.
299 F.C. Adams and S.Slaets, *Trends Anal. Chem.*, 19 (2000) 80–85.
300 H. Chassaigne, C.C. Chery, G. Bordin and A.R. Rodriguez, *J. Chromatogr. A*, 976 (2002) 409–422.
301 G.M.M. Rahman, H.M.S. Kingston and S. Bhandari, *Appl. Organometal. Chem.*, 17 (2003) 913–920.
302 M. Monperrus, E. Tessier, S. Veschambre, D. Amouroux and O. Donard, *Anal. Bioanal. Chem.*, 381 (2005) 854–862.
303 M.M.M. Coelho, M.L. Coelho, E.S. De Lima, A. Pastord and M. De la Guardia, *Talanta*, 66 (2005) 818–822.
304 B. Li, J. Bergmann, S. Lassen, P. Leonhard and A. Prange, *J. Chromatogr. B*, 814 (2005) 83–91.
305 A. Rodriguez-Cea, M.R. De la Campa and A. Sanz-Medel, *Anal. Bioanal. Chem.*, 381 (2005) 388–393.
306 N. Jakubowski, D. Stuewer, D. Klockow, C. Thomas and H. Emons, *J. Anal. Atom. Spectrom.*, 16 (2001) 135–139.

307 J.C. Shen, Z.Y. Huang, Z.X. Zhuang, X.R. Wang and F.S.C. Lee, *Appl. Organometal. Chem.*, 19 (2005) 140–146.
308 B. Michalke and P. Schramel, *J. Anal. Atom. Spectrom.*, 19 (2004) 121–128.
309 J.L. Gomez-Ariza, F. Lorenzo and T. Garcia-Barrera, *Anal. Bioanal. Chem.*, 382 (2005) 485–492.
310 M.H. Nagaoka and T. Maitani, *J. Inorg. Biochem.*, 99 (2005) 1887–1894.
311 K.G. Fernandes, M. Montes-Bayon, E. Blanco Gonzalez, M. Del Castillo-Busto, J.A. Nobrega *et al.*, *J. Anal. Atom. Spectrom.*, 20 (2005) 210–215.
312 B. Michalke, *J. Chromatogr. A*, 1058 (2004) 203–208.
313 B. Michalke, P. Schramel and H. Witte, *J. Anal. Atom. Spectrom.*, 16 (2001) 593.
314 D.W. Nyman, M. Suzanne Stratton, M.J. Kopplin, B.L. Dalkin, R.B. Nagle *et al., Cancer Detect. Prev.*, 28 (2004) 8–16.
315 B.K. Mandal, Y. Ogra and Suzuki, *Toxicol. Appl. Pharmacol.*, 189 (2003) 73–83.
316 B.K. Mandal, Y. Ogra, K. Anzai and K.T. Suzuki, *Toxicol. Appl. Pharmacol.*, 198 (2004) 307–318.
317 E.G. Yanes and N.J. Miller-Ihli, *Spectrochim. Acta Part. B*, 59 (2004) 188–196.
318 E.G. Yanes and N.J. Miller-Ihli, *Spectrochim. Acta Part. B*, 59 (2004) 883–890.
319 C.J. Duckett, N.J.C. Bailey, H. Walker, F. Abou-Shakra, I.D. Wilson *et al., Rapid Commun. Mass Spectrom.*, 16 (2002) 245–247.
320 B. Packert Jensen, C.J. Smith, C. Bailey, C. Rodgers and I.D. Wilson *et al., J. Chromatogr. B.*, 809 (2004) 279–285.
321 M. Shah, S.S. Kannamkumarath, J.C.A. Wuilloud, R.G. Wuilloud, J.A. Caruso, *J. Anal. Atom. Spectrom.*, 19 (2004) 381–386.
322 K. Wrobel, K. Wrobel, S.S. Kannamkumarath, J.A. Caruso, A. Wysocka, *et al.*, *Food Chem.*, 86 (2004) 617–623.
323 S. Mounicou, J. Meija and J. Caruso, *Analyst.*, 129 (2004) 116–123.
324 A. Nolan, D. Schaumlöffel, E. Lombi, L. Ouerdane, R. Lobinski *et al. J. Anal. Atom. Spectrom.*, 19 (2004) 757–761.
325 K. Polec-Pawlak, R. Ruzika, K. Abramski, M. Ciurzynska and H. Gawronska, *Anal. Chim. Acta*, 540 (2005) 61–70.
326 S. Amereih, T. Meisel, R. Scholger and W. Wegscheider, *J. Environ. Monit.*, 7 (2005) 1200–1206.
327 S. Amereih, T. Meisel, E. Kahr and W. Wegscheider, *Anal. Bioanal. Chem.*, 383 (2005) 1052–1059.
328 F. Wu, D. Evans, P. Dillon and S. Schiffd, *J. Anal. Atom. Spectrom.*, 19 (2004) 979–983.
329 L. Hinojosa Reyes, F. Moreno Sanz, P. Herrero Espilez, J.M. Marchante Gayón, J.I. Garcia Alonso *et al., J. Anal. Atom. Spectrom.*, 19 (2004) 1230–1235.
330 L. Hinojosa Reyes, J.M. Marchante-Gayon, J.I. Garcia Alonso and A. Sanz-Medel, *J. Agric. Food Chem.*, 54 (2006) 1557–1563.
331 R. Wahlen and C. Wolff-Briche, *Anal. Bioanal. Chem.*, 377 (2003) 140–148.
332 T. Eickhorst and A. Seubert, *J. Chromatogr. A*, 1050 (2004) 103–109.
333 J. Carter, L. Ebdon and E.H. Evans, *Microchem. J.*, 76 (2004) 35–41.

Achille Cappiello (Editor)
Advances in LC–MS Instrumentation
Journal of Chromatography Library, Vol. 72

Chapter 8

HPLC-chip/MS: a new approach to nano-LC/MS

TOM A. VAN DE GOOR

8.1 INTRODUCTION

8.1.1 The need for improvements in nano-LC/MS

Mass spectrometry offers the ability to identify unknown substances based on either accurate mass or based on MS/MS fragmentation. The last decade has shown tremendous improvements in sensitivity in mass spectrometry allowing attomole-level identification of analytes in complex samples. This capability has been driving new levels of understanding of biological processes [1]. Coupled with high-resolution separation methods to reduce sample complexity and powerful informatics tools, proteomics has become the next frontier in life science to understand biology at the protein level [2]. Working with small quantities of sample however has been challenging. In order to get maximum sensitivity, the sample has to be handled in the smallest possible volume in order to keep concentration levels within the detection limit of the instrumentation. This has driven the development of nanoflow liquid chromatography using columns with diameters below 100 μm. At these low flow rates leaks in the system are difficult to detect and connecting small diameter tubing can be challenging as well. Electrospray interfaces that are compatible with these low flow rates consist of very small pulled capillaries with a tip below 10 μm [3]. Adjustment for optimal stability and performance requires high user skill level, typically present in core labs. In order to fully exploit the power of mass spectrometry in the biology field, simplification of instrumentation is a must. This chapter describes a new approach to nano-LC/MS based on microfluidic technology [4]. Integration of the column components as well as the electrospray emitter in a single device, combined with automated positioning and interfacing between the chromatography system and mass spectrometer, allows ease of operation similar to standard flow rate systems. Reduction of connections further enhances reliability as well as improved performance due to minimized dead volumes. Performance factors and applications of the HPLC-chip/MS technology are described both for proteomics as well as for small molecule applications.

8.1.2 High sensitivity requirements for biological applications

After the wave of genomics, mapping the genes that encode proteins, scientists have moved to linking this information to the actual protein population in biological fluids, cells and tissues to better understand disease. The search for markers that are characteristic for a specific disease before it can be observed from clinical symptoms is the next challenge for early diagnosis as well as for targeted drug development [5].

However, unlike the Genomics world where amplification techniques such as PCR can be utilized to get to low-level sensitivity, no such methods exist for proteins. In addition, fluorescent labeling techniques that are used in genomics based on hybridization are difficult to implement for proteins due to the far more complex nature of the analytes and the diversity at the protein level. Antibody-based techniques can be used when the target is known but for discovery work this would become too expensive to be practical [6].

Mass spectrometry has emerged as the dominant detection technique for the identification of proteins. Tandem mass spectral fragmentation information can determine the amino acid sequence of proteins and peptides and based on the existence of protein databases from genomics information, identification of proteins is possible [7].

Identifying proteins in complex biological samples however is still a very challenging task. The concentration range over which they are present can range from 10^5 in tissue samples to 10^{10} or more in serum [8]. Proteins of interest are typically present in low abundance, so in order to be able to measure them, one has to start with as large a sample amount as available. Through selective removal of abundant proteins and prefractionation and concentration techniques, the goal is to get the complexity to a manageable level and concentrations within reach of detection. Still the most sensitive LC/MS approach needs to be used as the final analysis step. This can be achieved by using nano-LC/MS (Fig. 8.1).

8.1.3 Advantages and challenges of nano-LC/MS

The main reason for moving to low-flow LC/MC for increased sensitivity for sample limited applications is to handle the analytes in the smallest possible volume. Using

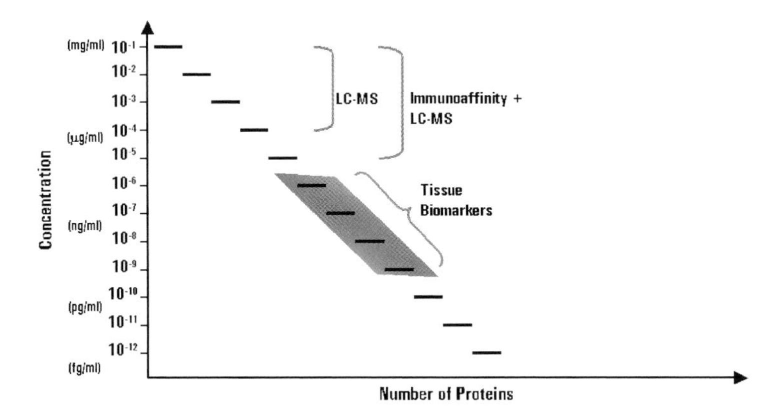

Fig. 8.1. Dynamic range of proteins in serum.

enrichment columns, the full sample can be trapped on column and then eluted in the smallest possible volume to reach maximum concentration levels. In addition, dispersion in small diameter columns is reduced and electrospary efficiency improved due to the small droplet formation in nanospray [9].

Nanopump technology has significantly improved, allowing reproducible delivery of solvent to the nanocolumns. Moving away from passive splitting of higher flow rates to active splitting using on-column flow sensing and regulated feedback as well as direct pumping in the nano flow range has resulted in reproducibility of nano-LC similar to that of higher flow systems [9].

Robustness issues have however been poorly addressed. Connecting nano flow components such as the enrichment columns to the analytical column as well as interfacing the nanospray emitter to the analytical column are still prone to failure. Small leaks go largely undetected due to the low volume, dead volumes between the different components can lead to dispersion, sample losses on exposed surfaces can skew results and positioning of the nanospray emitter for optimal results has shown to be dependent on user skill. Typical uptime of a nano-LC/MS system has therefore been limited and as such throughput has been troublesome (Fig. 8.2).

The need for robust operation is however significant. For a protein identification experiment in a complex sample, prefractionation is often required to reduce sample complexity. Several prefractionation techniques are used such as immunodepletion, ion exchange chromatography, isoelectric point (pI)-based fractionation, etc. The goal of these steps is to spread the diverse protein population over many fractions in order to have a better chance to sample the true population in the data-dependent MS/MS process. Proteins are digested down to peptides in this experimental approach further increasing the complexity. Although MS/MS scan speeds have improved, undersampling is still a serious issue, which drives the level of fractionation required. Long LC/MS runs (typically 2 hours) are then used to again spread the peptide population and allow as much time as possible for the mass spectrometer to sample

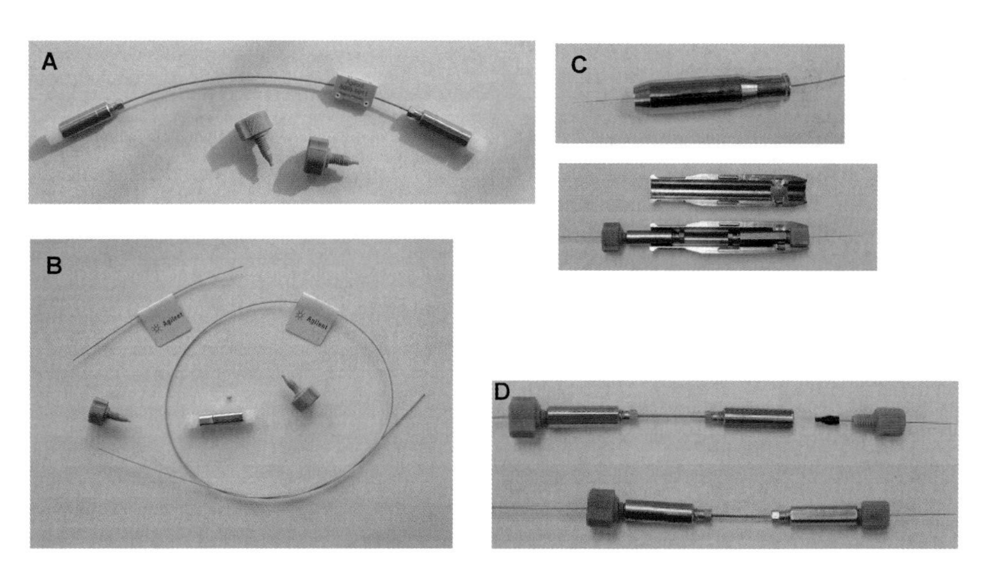

Fig. 8.2. Typical flow components of a nano-LC/MS system. (A) Nano column with connections; (B) enrichment column with connections; (C) nanospray emitter with connections and (D) additional fittings to complete the system.

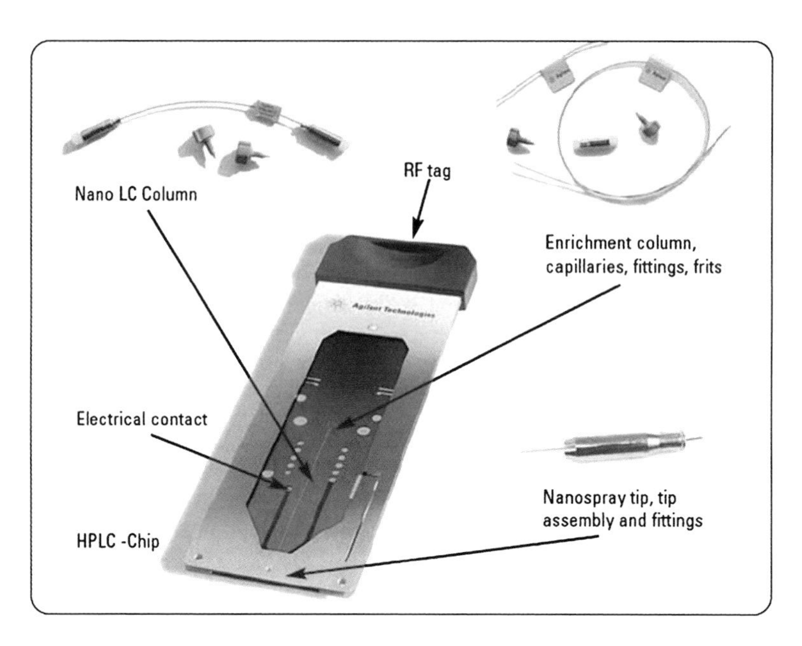

Fig. 8.3. HPLC-chip integrating components for nano-LC/MS.

the peptides that are eluting. This strategy leads to often a multiday experiment for a single sample and often weeks of work for a full biological experiment. Unattended operation in the nano-LC/MS mode is key in order to work efficiently and this has often not been possible using conventional nano-LC/MS systems.

8.1.4 HPLC-chip/MS, nano-LC/MS made easy

The HPLC-chip/MS implementation removes many of the integration issues by removing most of the fluid connections and providing robust interfacing with low dead volume. The main components are integrated on a microfluidic chip device. This includes the enrichment column, the separation column as well as the nanospray emitter. Standard nano LC components are still used to provide solvents to the chip device but the valving interface does not require connecting and disconnecting fittings when chips are replaced (Fig. 8.3). Furthermore, the positioning of the emitter is done through automation and does not require user adjustment for optimal performance. This allows users to simply insert the chip in the chip handler and run nano-LC/MS with the same level of robustness as they would for standard LC systems [10].

8.2 MICRO FABRICATION OF HPLC-CHIP DEVICES

8.2.1 Laser ablation technology

Laser ablation technology was applied in the 1990s as a low-cost means to deliver ink for inkjet printing applications. Multilayer polyimide structures were created having both

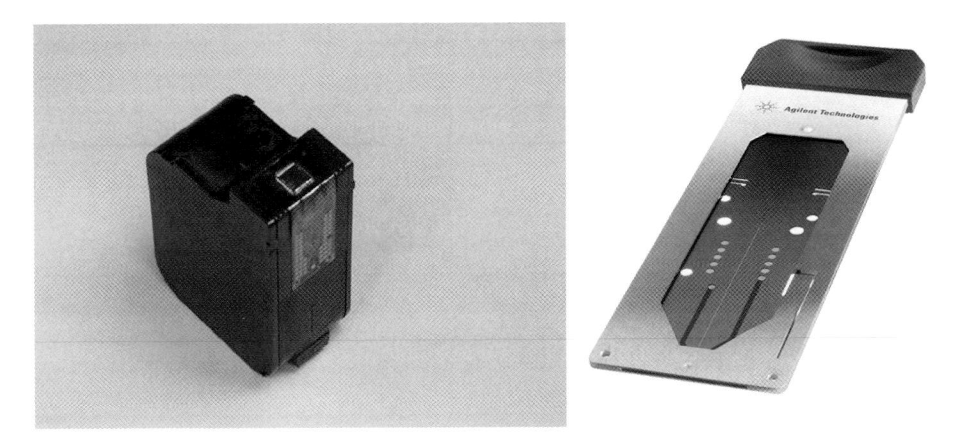

Fig. 8.4. Polyimide technology similarity between inkjet printing and HPLC-chip devices.

fluid channels for ink delivery from the reservoir to the print nozzles, as well as electrical components required for thermal ejection of the ink from the nozzle to the paper. This technology used processes from the flex circuit electronics industry to integrate electrical components in a circuit board with the flexibility needed for compact footprint (Fig. 8.4). Hewlett-Packard pioneered this technology and perfected its manufacturing to be a high volume, low-cost process needed for commercial success in the printer business. Researchers at their central research lab saw the opportunity for application in the analytical measurement area and applied similar techniques to prototype devices for chemical analysis [11].

A UV laser system is used to ablate features in plastic films that define the fluid paths needed to create a chip-based LC/MS device. The advantage of laser ablation is its flexibility to create structures in films based on direct writing of films. Other micro fabrication technologies often require creation of masks that are used in subsequent steps for etching of the structure in, e.g., glass. Prototyping new structures therefore often required a relatively long turn-around cycle, limiting the speed with which designs could be tested. Using the ablation approach, designs could be created using a CAD tool and designs could be transferred directly into instructions to ablate the features that were needed.

Plastics have additional advantages in terms of cost, flexibility and the ability to easily integrate electrical features. Certain plastics such as polyimide also offer the advantage of being more bio fouling resistant in that there is less non-specific adsorption of biomolecules such as proteins to the surface, helping in preserving the sample during analysis.

8.2.2 Multilayer devices and their fabrication

The fabrication process involves multiple steps. After structuring of the different films that make up the microfluidic structure, parts are cleaned to remove debris from the ablation process. The films used to create a full chip are alternating layers of fully imidized and not fully imidized polyimide. This is needed to laminate the layers together during a high temperature and pressure lamination process. Under these conditions the not fully imidized polyimide imidizes and the bond is formed between the layers, resulting in a uniform material for the integrated device.

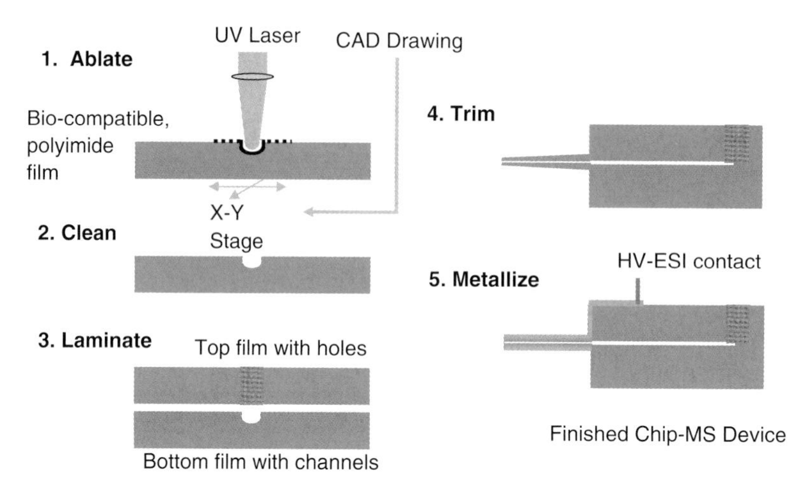

Fig. 8.5. Process steps to create microfluidic devices in polyimide.

Additional ablation steps are used to create the electrospray interface as well as metallization steps to bring electrical connections for the electrospray to the right locations on the chip [12]. After fabrication chips are packaged in a holder to make handling in the instrumentation possible as well as to protect the electrospray tip from being damaged (Fig. 8.5).

8.2.3 Use of stationary phases

In order to ensure consistent chromatography results, standard bead-based packing material is used in the HPLC-chip. This allows users to migrate their methods from standard or capillary HPLC to the chip device and obtain comparable results without renewed optimization. The packing material is packed into the chip devices using slurry-based packing procedures commonly available for packing nano columns [13]. Micro fabricated filters and frits are used to keep the packed bed in place under chromatographic conditions and high pressure for both the enrichment bed as well as the analytical column (Fig. 8.6).

8.2.4 Valve interfacing

In order to interface the fluids from the pumps to the chip, a new valve design was developed where the valve can open in a motorized way to allow loading of the chip. After insertion the rotor and stator are clamped back together making a high-pressure connection to the surface of the chip. Fluid lines connect from the stator side which includes both the sample delivery as well as delivery of the gradient to the column. The rotor is positioned on the other side of the chip and can communicate with the stator through holes in the chip. The rotor rotates directly on the surface of the chip without any leaks and enables high-pressure flow switching between the loading and analysis mode as shown in Fig. 8.7 (see also Fig. 8.8).

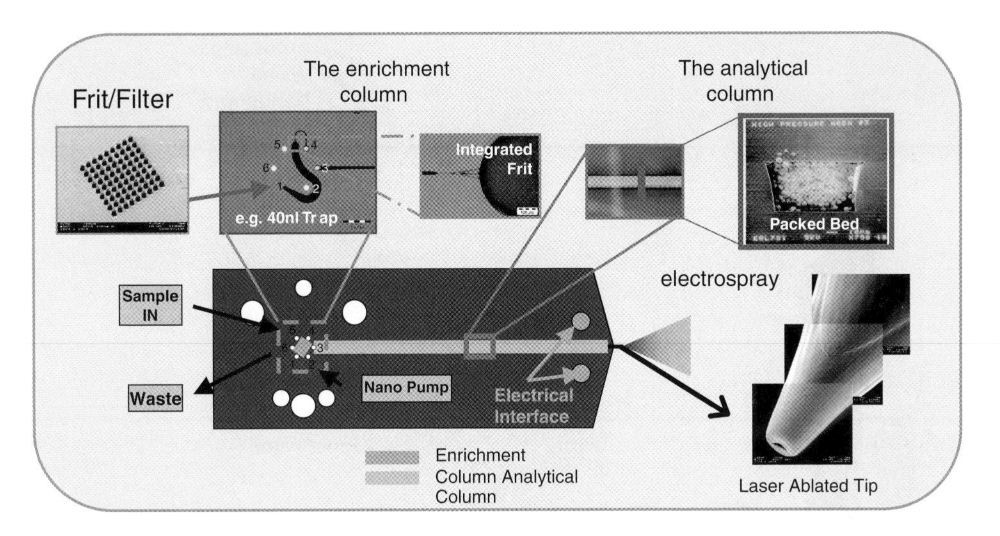

Fig. 8.6. Fabrication details of the HPLC-chip/MS device.

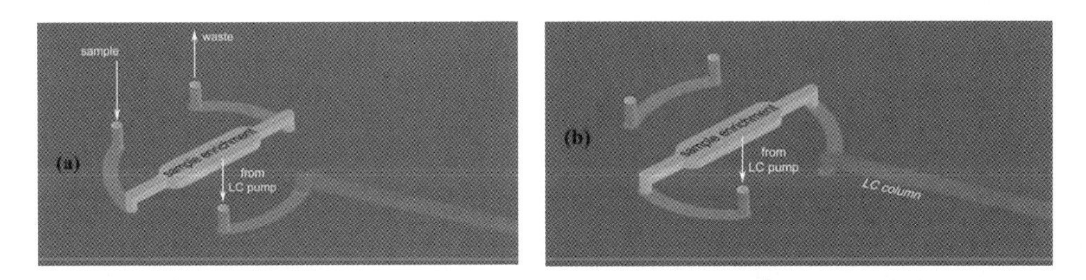

Fig. 8.7. Loading (a) and analysis (b) flow paths in the HPLC-chip [10]. In the design of the valve further flexibility was built in using a rotor in rotor approach. Two concentric random position rotors can interface with the chip, allowing maximum flexibility for future more integrated microfluidic designs. The compact nature of the rotor in rotor design ensures tight high-pressure interfacing for LC conditions.

8.2.5 MS interfacing

When inserted into the chip handler, the chip is automatically positioned in front of the orifice of the mass spectrometer. Due to the mechanical precision of the chip handler as well as the very tight tolerances achieved in the micro fabrication process of the chip, reproducible and optimal positioning ensures maximum sensitivity without the need for user intervention. When moving into position, the chip is moved out of its encapsulation, exposing the tip.

Single settings for electrospray voltage and position are ensured over the full range of organic/water solvent composition used in the gradient elution. Using the orthogonal electrospray interface, analyte ions are preferentially sampled, while solvent ions and droplets are reduced going into the mass spectrometer (Fig. 8.9).

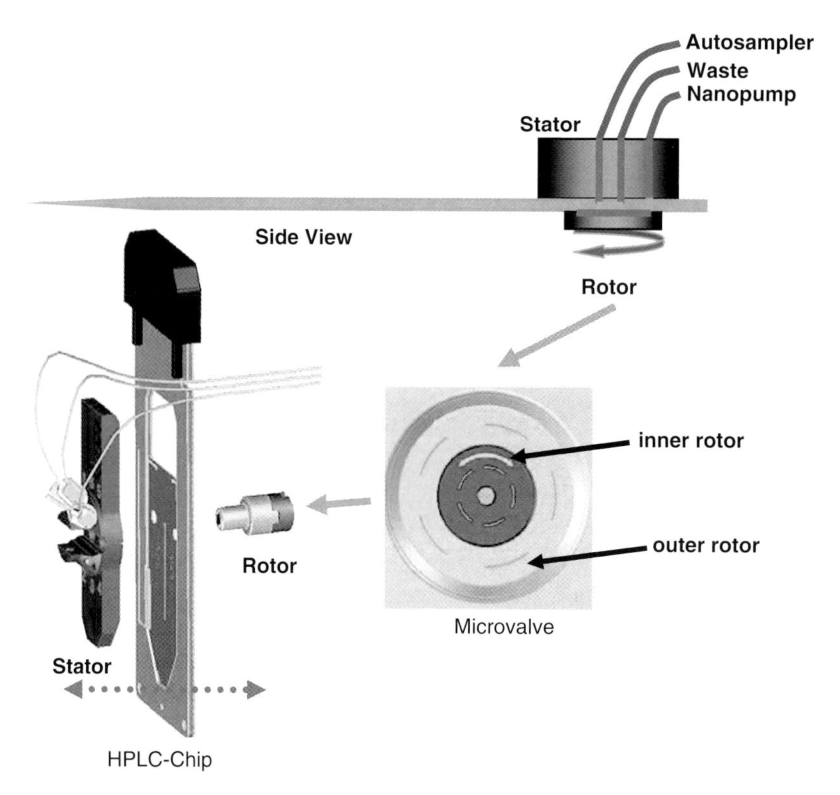

Fig. 8.8. Schematic representation of the chip to valve interface using rotor in rotor valve design.

8.2.6 Full system integration

The full HPLC-chip/MS system includes an autosampler and capillary LC pump for delivery of the sample to the enrichment column. This allows sample loading of larger volume samples in a short period of time using a higher loading flow rate (e.g., 4 μL/min). The analytical column is driven by a nano LC pump operating at typically 100–600 nL/min. This allows also for overlapped operation where the next sample can be loaded while the previous sample is being analyzed. Single point instrument control operates the full HPLC-chip/MS system (Fig. 8.10).

8.3 ANALYTICAL PERFORMANCE OF HPLC-CHIP/MS VERSUS NANO-LC/MS [10]

8.3.1 Reduced dispersion

After sample loading onto the chip, the analysis is started by rotation of the valve. This puts the enrichment column directly into the analytical flow path and the gradient elution is started. Due to the absence of any fittings and connections, dispersion is significantly

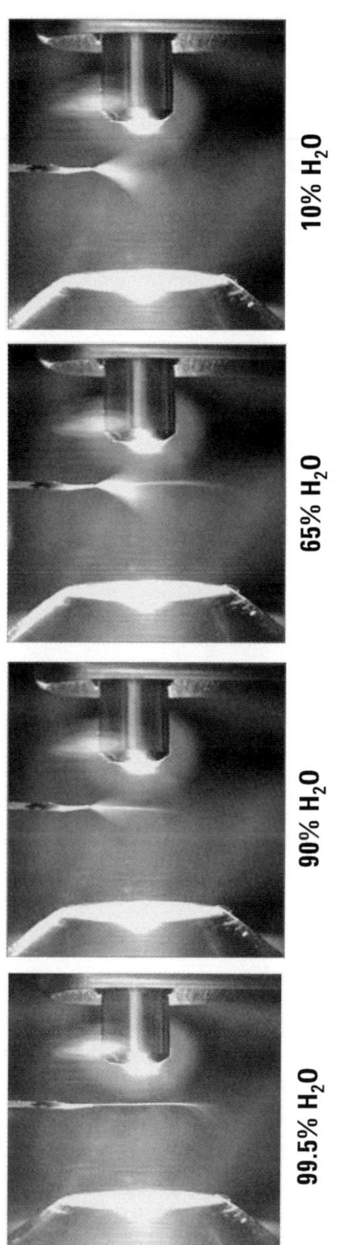

Fig. 8.9. Reliable electrospray over a wide range of organic/water solvent conditions. On the left is the orifice of the mass spectrometer, while on the right is a counter electrode to shape the electric field for efficient ion sampling.

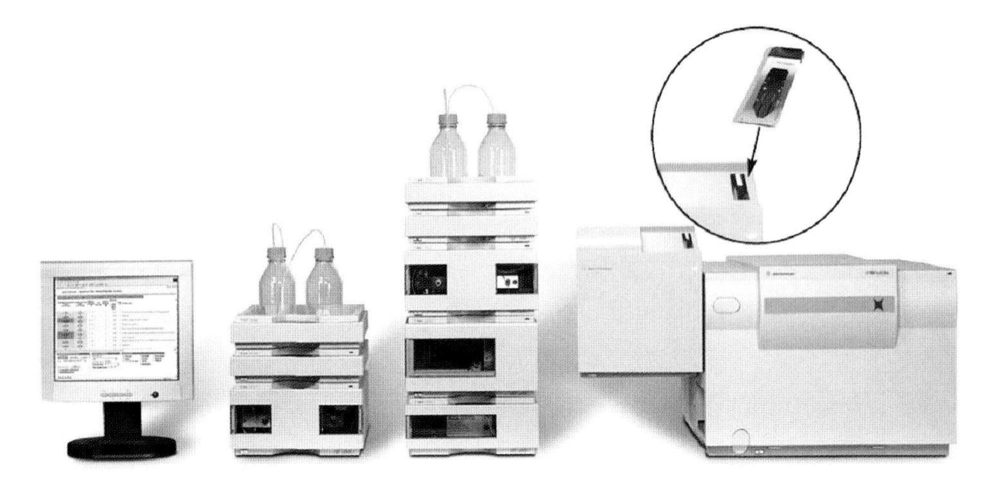

Fig. 8.10. Example of an HPLC-chip/MS system using an ion-trap mass spectrometer. The chip handler is mounted directly onto the front end of the mass spectrometer and connected to standard LC components.

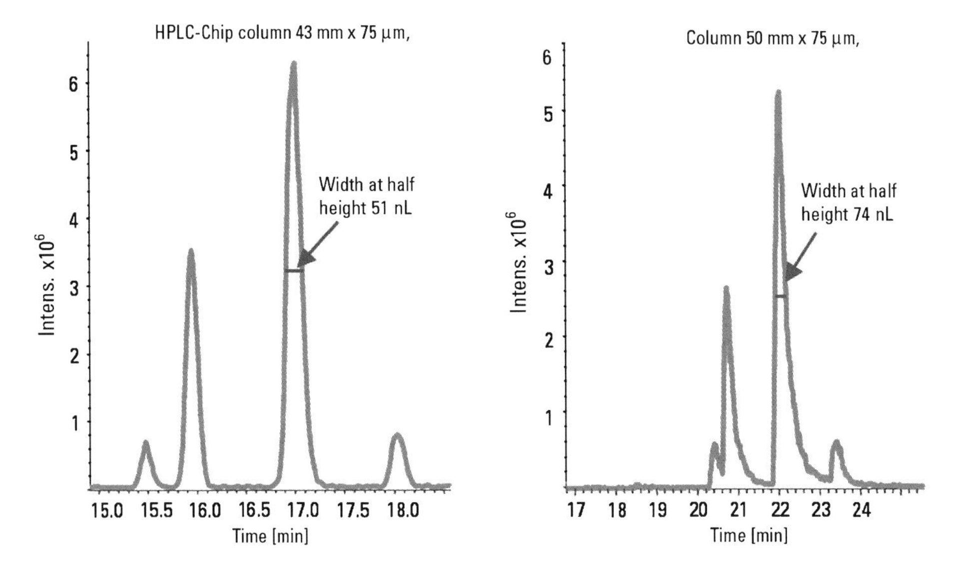

Fig. 8.11. Improved resolution using HPLC-chip.

reduced for the analysis. A significant contribution is also the direct integration of the spray tip at the end of the column, which allows the packing material to extend all the way to the spray tip. Fig. 8.11 shows an example of direct comparison between the HPLC-chip and conventional nano LC analysis using standard optimized conditions. As can be seen both dispersion and tailing is reduced allowing higher peak capacity on the column. This also reduces the chance of ion suppression in the electrospray process since fewer analytes are co-eluting at the same time and are competing for charge.

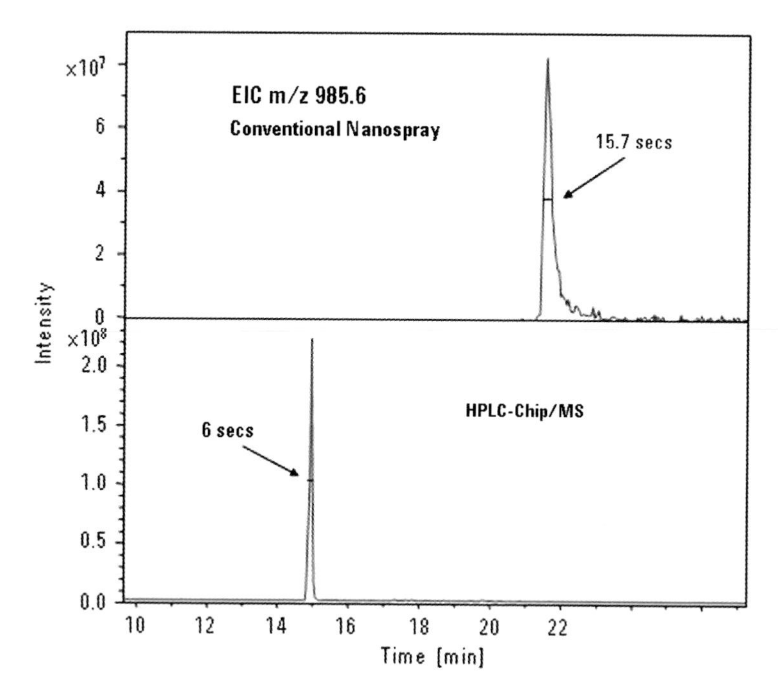

Fig. 8.12. Higher sensitivity as observed in ion abundance for HPLC-chip.

8.3.2 Increased sensitivity

Due to the fact that dispersion is reduced, signal intensity is increased. This translates directly into higher sensitivity. As can be seen in Fig. 8.12, about a 2-fold increase in ion abundance count is observed using the same settings on the mass spectrometer. The integrated signal count (abundance × time) remains of course constant.

The high sensitivity achievable is also demonstrated in the analysis of human serum albumin tryptic digest at very low levels. Fig. 8.13 shows the database search results for 100 amole HAS on column as analyzed by HPLC-chip-ion trap in data-dependent MS/MS mode. Six distinct peptides were found at this very low level with MS and MS/MS spectra for one of the peptides shown on the right. As can be clearly seen, the quality of the spectra at this low level is still very good and more than sufficient for database searching for positive identification.

8.3.3 Increased peak capacity

The increased peak capacity is shown in a simple analysis of the digest of bovine serum albumin. More individual peaks are observed in the total ion chromatogram. Peaks that are not fully resolved under conventional conditions are resolved using the HPLC-chip approach (Fig. 8.14).

Early eluting peaks also seem much better resolved and have a significant increase in response. This can be explained by more efficient sample trapping and elution, where less

Group (#)	Spectra (#)	Distinct Peptides (#)	Distinct Summed MS/MS Search Score	% AA Coverage	Mean Peptide Spectral Intensity	Database Accession #	Protein Name
1	6	6	83.98	12	1.40e+006	IPI00022434	Serum albumin

#	Sequence	m/z Measured (Da)	MH⁺ Matched (Da)	A100-1 Intensity
1	(K)FQNALLVR(Y)	480.97	960.5631	1.80e+006
2	(K)KVPQVSTPTLVEVSR(N)	547.47	1639.9383	1.50e+006
3	(K)LVNEVTEFAK(T)	575.46	1149.6156	1.93e+006
4	(K)QNCELFEQLGEYK(F)	829.49	1657.7532	1.04e+006
5	(R)RPCFSALEVDETYVPK(E)	637.96	1910.9322	1.02e+006
6	(K)YICENQDSISSK(L)	722.43	1443.6426	1.13e+006

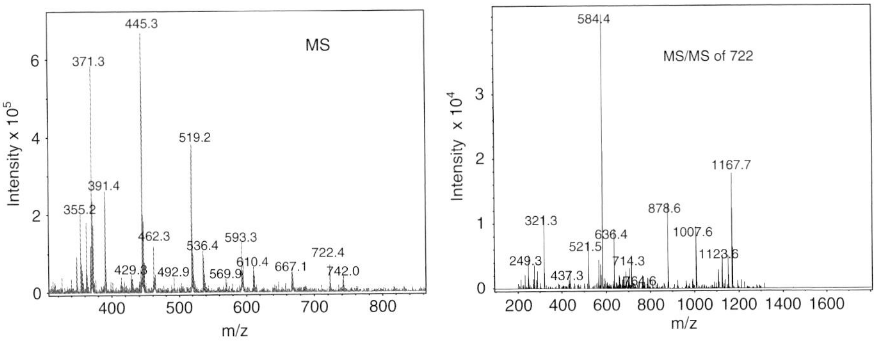

Fig. 8.13. Database search results for human serum albumin for 100 amole on-column.

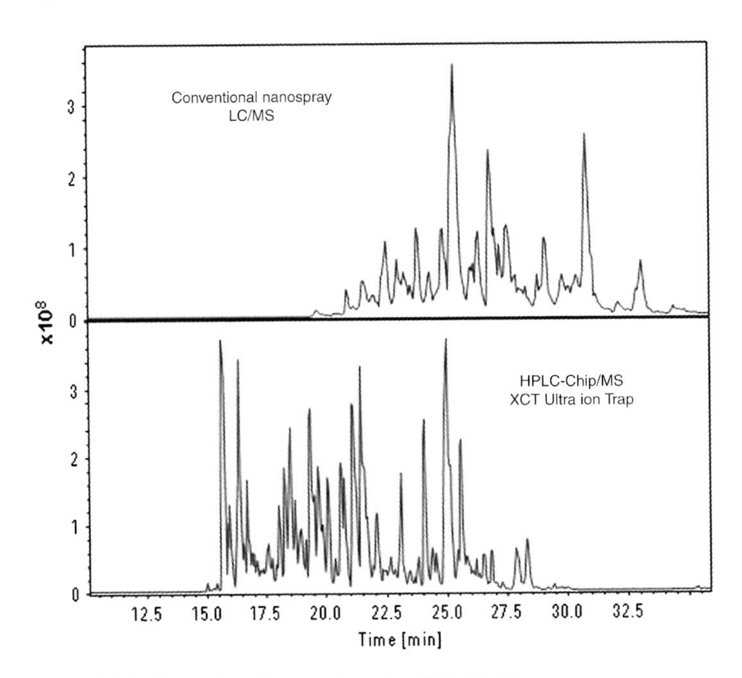

Fig. 8.14. Enhanced peak capacity using HPLC-chip.

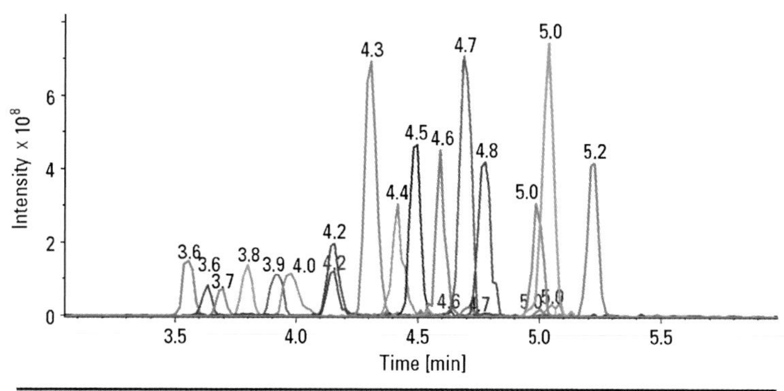

	Average RT	SD	%RSD
EIC 487.8 ±All MS	3.618	0.014	0.40
EIC 752	3.788	0.011	0.29
EIC 740.6 ±All MS	5.018	0.010	0.20
EIC 874.4	3.968	0.012	0.31
EIC 653.6	4.289	0.012	0.28
EIC 511.7	3.681	0.012	0.31
EIC 722.7	3.547	0.012	0.35
EIC 778	4.143	0.010	0.23
EIC 526.3	4.399	0.015	0.34
EIC 547.5	4.472	0.011	0.25
EIC 746.7	5.196	0.011	0.20
EIC 519.1	4.142	0.011	0.26
EIC 508.2	4.972	0.011	0.23
EIC 582.4	4.679	0.011	0.23
EIC 461.9	3.905	0.012	0.30
EIC 474	4.759	0.011	0.22
EIC 628	4.584	0.010	0.22

Fig. 8.15. Reproducibility of peptide mapping using HPLC-chip/MS.

sample losses are observed. Possibly material differences between the glass surfaces in the capillary approach as compared to the polymeric surfaces of the chip can contribute to the increased recovery for polar analytes.

8.3.4 Increased reproducibility

Reproducibility of the HPLC-chip system was tested using repeat injections of the tryptic digest of bovine serum albumin using a short gradient run. Extracted Ion Chromatogram (EIC) signals for the different peptides were extracted and retention times for each of the peptides recorded for 70 consecutive runs. From this information standard deviation and percent relative standard deviation were calculated. Better than 0.4% relative standard deviation (RSD) was observed for all peptides, showing the excellent reproducibility from run to run. Chip to chip reproducibility was also excellent and found to be around 1% (Fig. 8.15).

In addition, measured ion abundance reproducibility was also explored using a mixture of almost 5000 glycopeptides enriched from a cell lysate measured using a time-of-flight mass spectrometer, coupled to the HPLC-chip system. Fig. 8.16 shows the ion intensity for all 5000 ions as measured in 10 repetitive experiments. Run 1 is plotted versus run 10 and

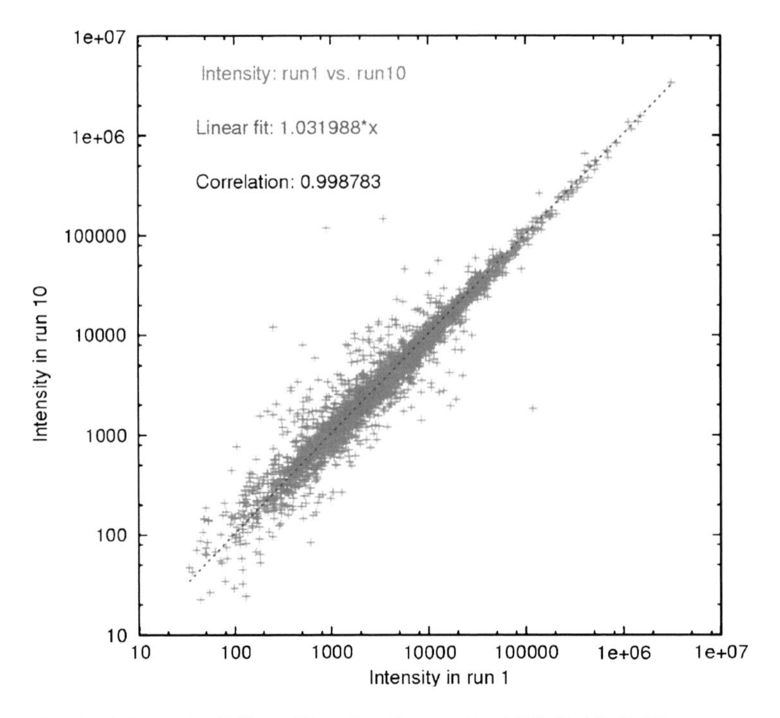

Fig. 8.16. Reproducibility of ion abundance using HPLC-chip/TOF.

shows a correlation coefficient of better than 0.998. This indicates that in addition to retention time, the ionization efficiency of the nanospray process is excellent and that ion abundance can be used for relative quantitation of analytes.

These experiments all indicate the robust and high performance of HPLC-chip/MS. These benefits help in routine analysis using nano-LC/MS and fulfill the requirements for unattended, reproducible operation in proteomics applications.

8.4 APPLICATIONS OF HPLC-CHIP/MS

8.4.1 Protein identification [7]

The most commonly used Protein Identification Strategy using LC/MS is based on a data-dependent MS/MS strategy using sensitive, high scan rate mass spectrometers. Complex protein mixtures are digested into peptides prior to analysis. The scan rate is important in order to capture as many possible MS/MS spectra for the peptides present in order to identify as many proteins as possible through database searching. However, when samples are very complex, a limited number of proteins can be identified. Extending the LC gradient run can improve protein identification but often additional fractionation techniques are required to reduce complexity in order to boost the number of lower abundance proteins found. This results in many more LC/MS runs needed and robust, high sensitivity methods are a must. The HPLC-chip technology combined with a high scan speed ion trap mass spectrometer was used in this example of protein identification.

Membrane proteins play pivotal roles in various physiological processes such as signal transduction, molecular transport, and cell–cell interactions and a comprehensive analysis of these proteins is essential to uncovering diagnostic disease biomarkers, therapeutic agents and drug receptor candidates [14]. However, profiling membrane proteins has proven to be particularly challenging because of their hydrophobic nature and low abundance. These obstacles pose major limitations for proteomic techniques such as gel electrophoresis or chromatography-based separation methods. For gel electrophorectic analyses, many hydrophobic proteins are not readily soluble causing poor gel performance and recoveries, while liquid chromatography separation techniques may suffer from poor separation characteristics, non-reproducibility and low protein recoveries. To overcome these limitations, a novel and highly robust method for the separation and identification of HeLa cell membrane proteins was used based on an LC only separation strategy. Employing optimized reversed-phase (RP) conditions and a macroporous reversed-phase column intact membrane proteins were prefractionated into 47 fractions. These fractions were then digested and the resulting peptides analyzed using 2D LC/MS/MS using strong cation exchange followed by reversed-phase chromatography. The strong cation exchange column was directly in-line with the HPLC-chip reversed-phase column and fractions were eluted onto the chip using 11 increasing salt steps. Each combined 2D LC run took 6 hours of run time over which MS/MS spectra were collected for the eluting peptides. In total almost 300 hours of MS/MS spectra were collected for all the 47 intact protein fractions and data submitted for database searching. Clearly robust and reliable operation is an absolute must for this experiment, which was enabled by using the HPLC-chip approach [15].

Fig. 8.17 shows the workflow for the experiment. After intact protein fractionation, the 2D LC analysis of the peptides was compared to further fractionation at the protein level used a 1D SDS-PAGE gel followed by in-gel digestion of over 200 bands resolved on the gel and 1D LC/MS of the peptides using the reversed-phase HPLC-chip method.

Fig. 8.18 shows the intact protein separation of the HeLa membrane proteins using the reversed-phase macroporous column. High-resolution separation was obtained for the intact proteins which resulted in clean fractionation into 47 fractions. Due to the nature of the column and the chromatographic conditions, very high recovery of the proteins was obtained.

Using Spectrum Mill database search software 954 proteins (470 membrane and 337 integral membrane proteins) were identified by HPLC-chip 2D LC/MS/MS. The reversed-phase separation and fractionation protocol for intact proteins, combined with in-solution digestion, represents a fast, reliable and reproducible tool for the proteomic characterization of complex hydrophobic proteins samples. This methodology is a robust alternative to traditional 1D SDS-PAGE after RP fractionation, which requires much more time and yields less protein identifications, in this case 688 proteins.

8.4.2 Protein profiling

Protein profiling is used to characterize known proteins, e.g., in qualification of recombinant protein production batches. Consistency from lot to lot can be assessed by comparing the profile of the digested protein if the resolving power in both the separation dimension as well as in the mass dimension is sufficiently high. Abundance measurement for the individual

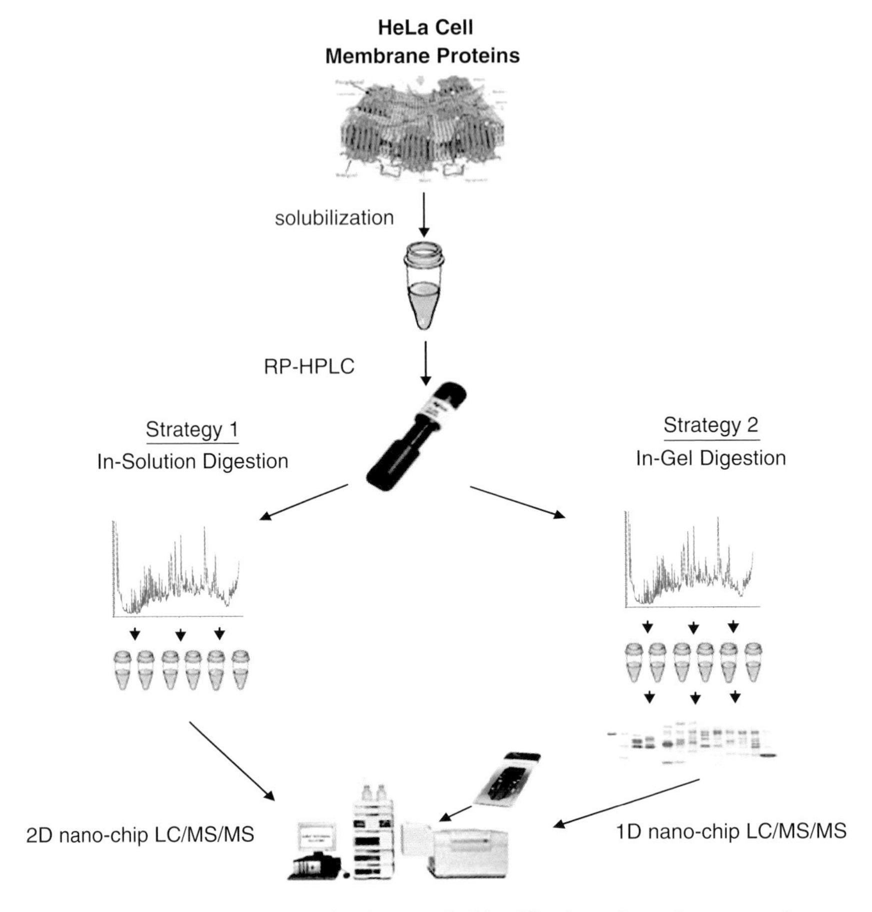

Fig. 8.17. Comparison of strategies for protein identification of membrane proteins.

ions can be a measure for relative quantitation. Even for more complex protein samples the profiling mode can be used to look at relative changes between samples as a result of an external change (e.g., drug treatment) or as a result of disease state [16].

Key to comparison of samples is reproducibility of the method and high resolution, which at the nanoflow scale can be achieved with the HPLC-chip system. Fig. 8.19 shows an example of the profiling of a mixture of glycosylated peptides. This mixture was obtained by affinity-based sample enrichment at the protein level using hydrazide beads [17]. In a first step the glycoprotein mixture is oxidized and subsequently covalently coupled to the hydrazide beads. The bound proteins are digested such that only the glycosylated peptide remains bound to the bead. By washing the beads the non-glycosylated peptides are removed. In the subsequent step the glycopeptides are enzymatically released and analyzed using HPLC-chip time-of-flight to obtain a profile of the mixture [18].

Fig. 8.19 shows the retention time versus *m/z* abundance plot of the analysis. On the left the map for run 1 and on the right the map of run 10 is compared. As can be seen the

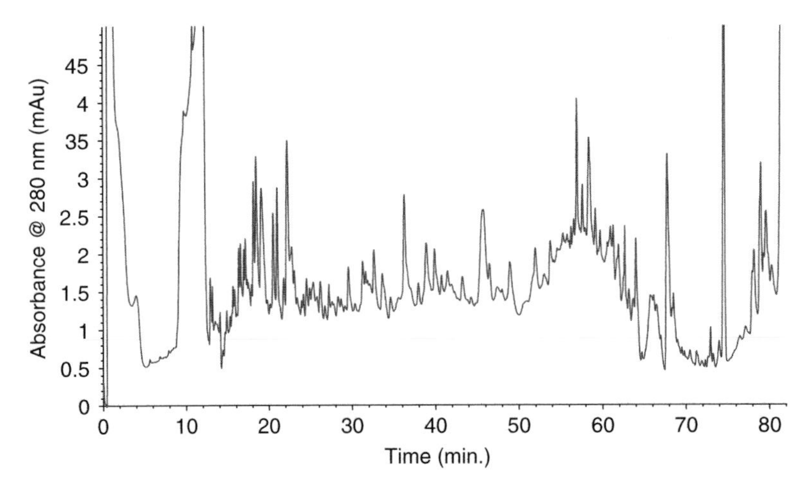

Fig. 8.18. High-resolution intact protein separation using macroporous reversed-phase chromatography.

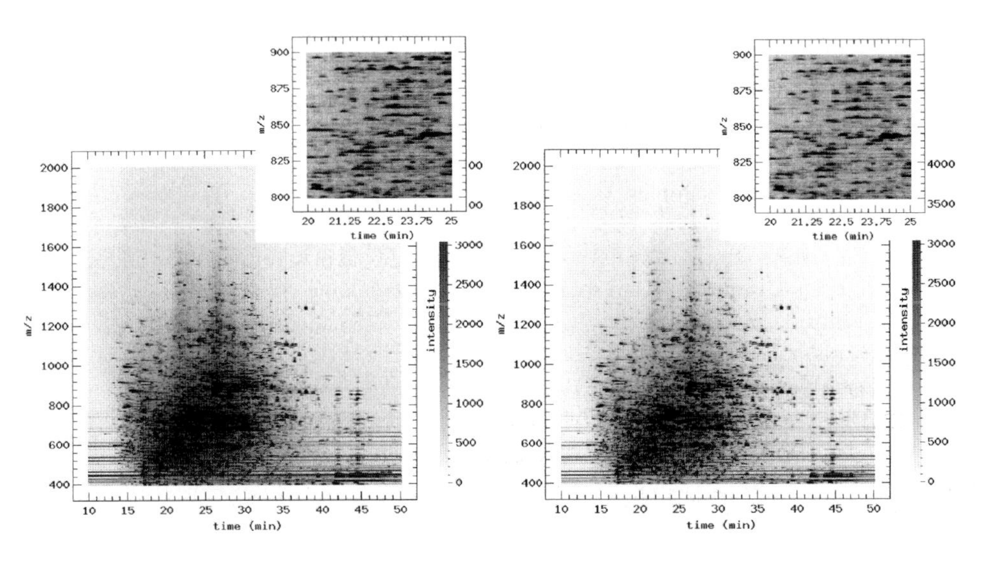

Fig. 8.19. Reproducibility of a profiling map of a glycopeptide mixture.

reproducibility is excellent, which shows even more clearly in the zoom area for each of the plots. This result compared very favorably with standard nano-LC/MS approaches both in terms of reproducibility as well as in the number of peptides that were resolved in the experiment. Over 5000 peptides could be seen in the map using a gradient of only 60 min as compared to about 2500 peptides using standard nano LC using even a gradient that was twice as long.

The ability to reproducibly profile such complex mixtures and the ability to resolve this many individual peptides makes the technique ideal for profiling applications and using the approach for QA/QC of batches of complex formulations.

8.4.3 Biomarker discovery

Biomarker discovery is of increased interest due to its clinical relevance. If specific metabolites or proteins can be identified that are characteristic for an early disease state, these compounds could be monitored as a clinical test for onset of disease or progression of disease [19]. Many researchers have used the Protein Identification Strategy based on data-dependent MS/MS to biomarker discovery using different strategies for relative quantitation during the discovery phase. Sample complexity and duty cycle limitations of this approach have however limited its success. A new approach uses MS only based profiling of the samples and controls and comparison of analytes based on accurate mass, retention time and abundance to biomarker discovery. This approach overcomes duty cycle issues and finds possible candidates but does not identify the potential markers in the MS experiment. Differential expression software is used to create target compound lists that are then identified in a subsequent run based on triggering MS/MS on the analytes recognized by the accurate mass/retention time tag. Retention time reproducibility, high mass accuracy and high sensitivity together with reliable nano-LC/MS are important requirements for this application. The HPLC-chip approach was used for this application by combining the profiling MS experiment using a time-of-flight mass spectrometer, followed by targeted MS/MS using an ion trap. The ease with which the HPLC-chip can be moved from one instrument to the other to obtain reproducible cross-platform results made this approach feasible.

To demonstrate the capability, 400 ng of an *E. coli* lysate (Biorad) was spiked with 100 fmol of bovine serum albumin (BSA) and 200 fmol of serotransferrin (Sigma) as a control sample. Two additional samples were prepared as the target samples. Sample A contained the same amount of *E. coli* proteins but had 200 fmol BSA and 100 fmol of serotransferrin spiked in, while sample B had 400 fmol BSA and 50 fmol serotransferrin added to the *E. coli* proteins. All samples were subsequently digested using trypsin and analyzed using an HPLC-chip–TOF system using a reversed-phase C18 chip with a 150 mm column. A chromatographic run of 100 min using an acetonitrile/water gradient from 3% to 45% with 0.1% formic acid was used for the analysis [20].

Fig. 8.20 shows the total ion chromatogram from the MS experiment, clearly showing the complexity of the sample. Even when extracting a single peptide at *m/z* 504.2507 using a narrow mass range of 1.9 ppm (Fig. 8.21) does not result in a single peak, again emphasizing the complexity of the sample. Fig. 8.22 shows the MS spectrum for the peak at

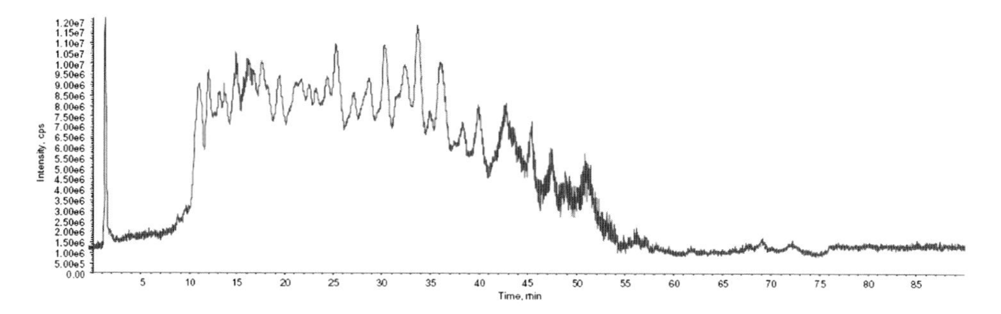

Fig. 8.20. Total ion chromatogram for sample B.

9.2 min. The target mass is not among the most abundant ions and would likely never be identified in a data-dependent experiment.

Using Mass Hunter software, features were extracted in both the control sample as well as sample A and B. Non-chromatographic chemical background was removed and related peaks such as isotopes, adducts, dimers, trimers and multiple charge states were clustered and combined and a neutral accurate mass, retention time and abundance for each feature exported to a feature list. Fig. 8.23 shows a contour plot for one of the samples showing the effectiveness of the feature extraction.

In a subsequent step differential expression profiling software in Mass Hunter was used to align and compare the control sample to sample A and sample B. Differentially expressed features were plotted as log 2 plots for both sample A versus the control as well as sample B versus the control. Fig. 8.24 shows these results clearly, finding the difference in the spike levels of the added proteins even with the highly complex background of the *E. coli* lysate.

MS data for the differentially expressed were submitted for database search using Spectrum Mill in peptide mass fingerprinting mode. Both BSA and serotransferrin were positively identified. Targeted MS/MS of the peptides using HPLC-chip ion trap and subsequent mass tag searching of the MS/MS spectra using Spectrum Mill versus the database gave the same results.

This example demonstrates the use of profiling on TOF combined with targeted MS/MS on trap using the HPLC-chip system as a powerful tool for biomarker discovery. Reproducible retention times and ion abundance for relative quantitation combined with high sensitivity and powerful software tools will increase the success rate in biomarker discovery. Discovery is of

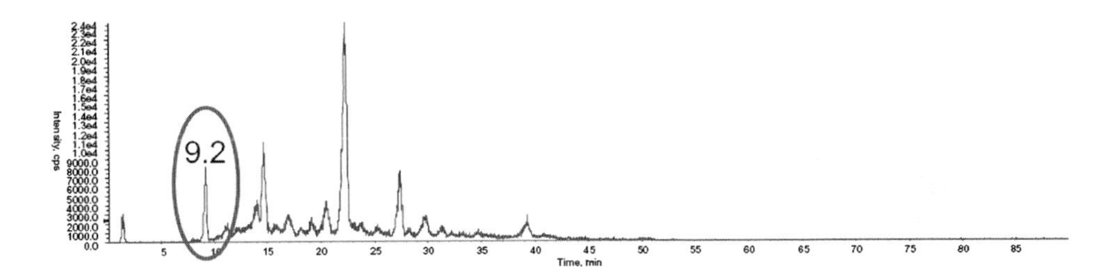

Fig. 8.21. Extracted ion chromatogram for $m/z = 504.2507$.

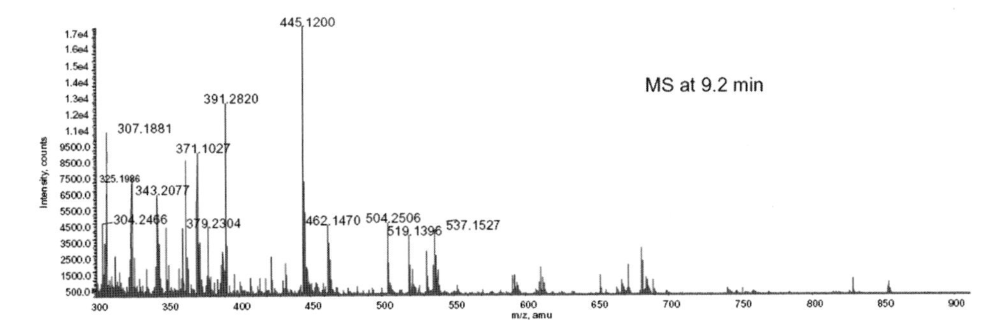

Fig. 8.22. Mass spectrum at 9.2 min for sample B.

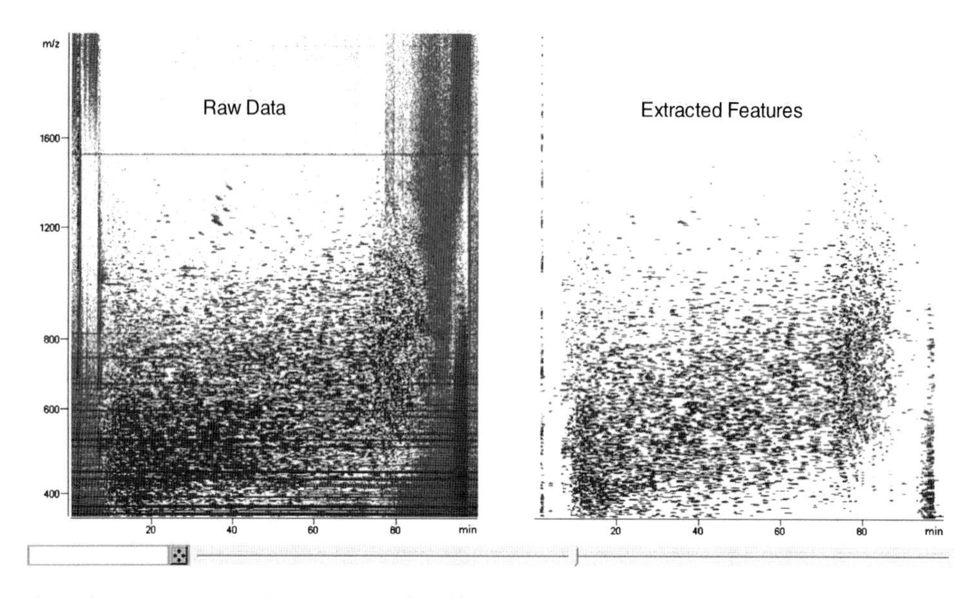

Fig. 8.23. Feature extraction using the Mass Hunter software.

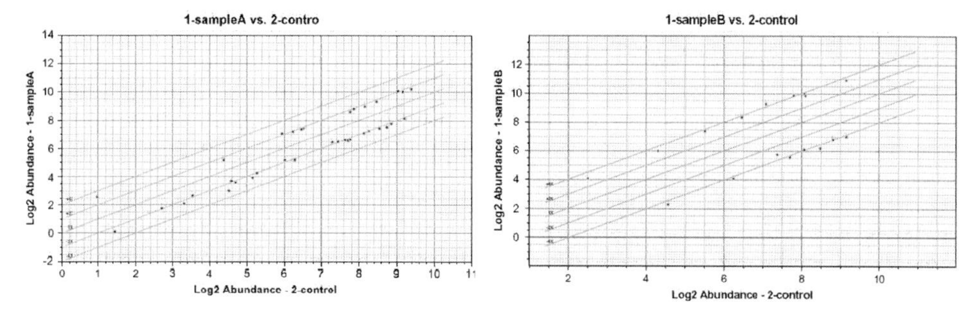

Fig. 8.24. Differential expression log/log plots for sample A and B versus the control. Sample A shows 2-fold up-regulation for BSA and 2-fold down-regulation for serotransferrin while sample B shows 4-fold up- and down-regulation.

course just a first step and potential markers would require further validation using absolute quantitation on large data sets, but this could become a targeted analysis using triple quadrupole MS/MS technology in multiple reaction monitoring (MRM) mode.

8.4.4 Oligosaccharide analysis

Oligosaccharides are the third largest component of human milk after lactose and lipids. The composition of the oligosaccharides varies widely from individual to individual and with stages of lactation. Among 900 possible structures only 86 have been elucidated. The function of oligosaccharides has been linked to prevention of bacterial adhesion to epithelial surfaces and has also been suggested as possible prebiotics. There is considerable interest

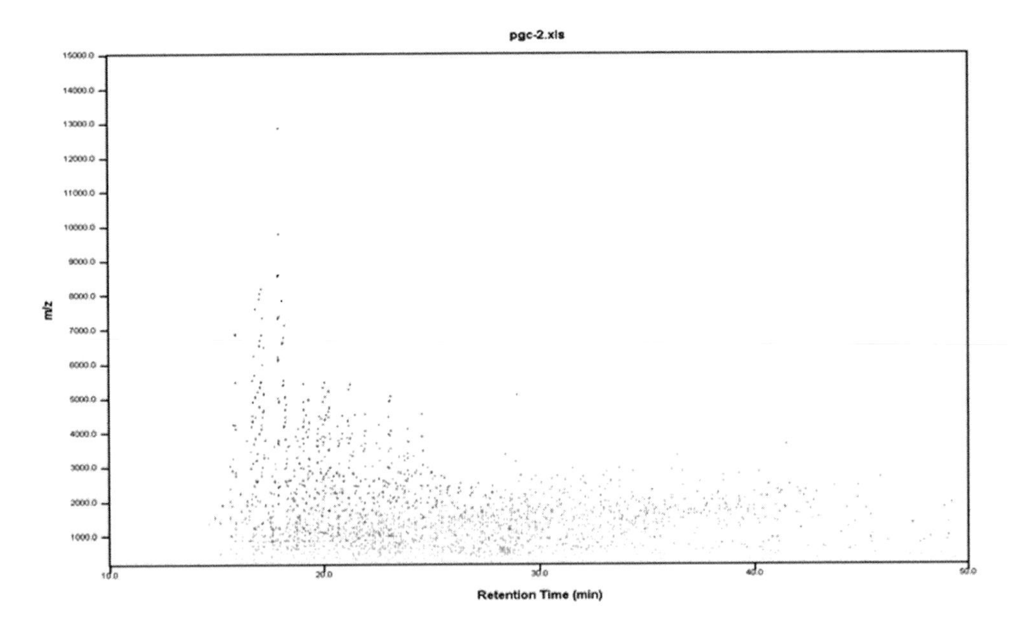

Fig. 8.25. Profiling map of the oligosaccharides present in human milk.

in further understanding the role of oligosaccharides and their use in formulated milk for additional health benefits [21].

Fig. 8.25 shows the oligosaccharide profile of human milk as analyzed using HPLC-chip–TOF using a graphitized carbon column. Around 2000 compounds could be resolved in this experiment but only 97 of them could be linked to previously identified compounds under *m/z* of 2200. Clearly many isomers of the same oligosaccharide can be found in the data [22,23].

As an example, the extracted ion chromatogram for $m/z = 1220.45$ is shown in Fig. 8.26. As can be seen the structure that consists of 4 hexose, 2, *N*-acetylhexosamines and 1 fucose is present at multiple locations in the chromatogram indicating that the isomers are separated using the graphitized carbon chip.

Some possible structures are shown in Fig. 8.27, but many more can be proposed. Identification of these compounds can be done using MS/MS fragmentation and subsequent database searching against a glycan database such as the GlycomIQ database offered by proteome systems.

This new approach to oligosaccharide profiling and identification is a powerful research tool that can help in further understanding the role these compounds play.

8.4.5 Small molecule application

Due to its high sensitivity and low sample consumption nano-LC/MS has gained interest in small molecule application areas like low-level metabolite identification. Several research teams have evaluated the potential of nanoscale-LC–MS on metabolite analysis using an in vitro human liver microsomal incubation assay. Buspirone was chosen as a test compound and LC/MS analysis was performed using a conventional nanocolumn system with a 75 μm ID C18 column. While the conventional nanocolumn format offered superior

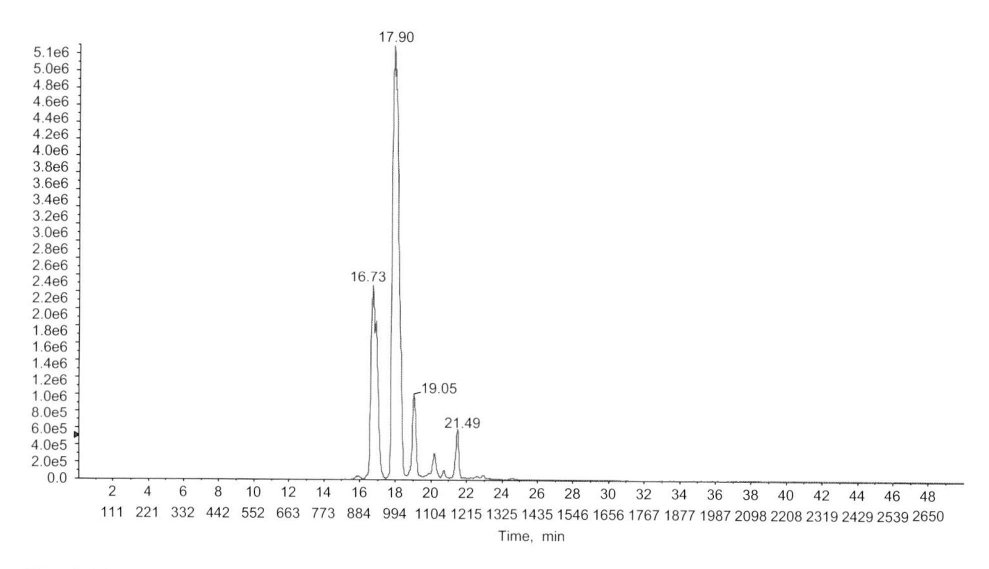

Fig. 8.26. Extracted ion chromatogram for $m/z = 1220.45$.

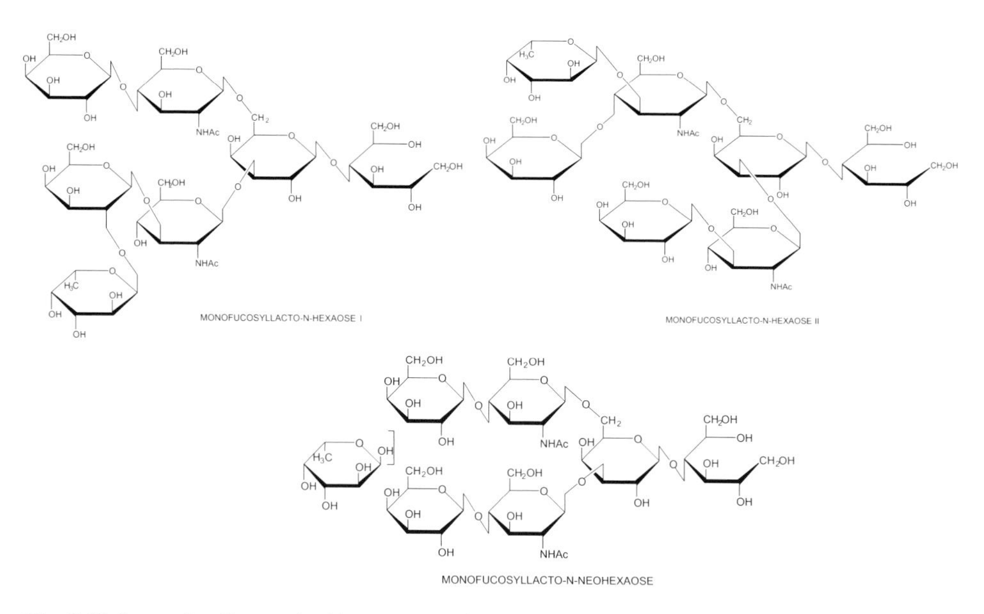

Fig. 8.27. Isomeric oligosaccharide structures for $m/z = 1220.45$.

sensitivity and separation performance compared to 2.1 or 4.6 mm ID columns, its use is hampered by the difficulty in setting up and using nanocolumns and its lack of robustness and reproducibility [24] (Fig. 8.28).

In this example, buspirone was incubated in rat liver S9 fractions for metabolite identification. The sample was first analyzed with a conventional nanocolumn set-up (top traces in blue) and then with an HPLC-chip/MS system (bottom trace in red) using the same ion

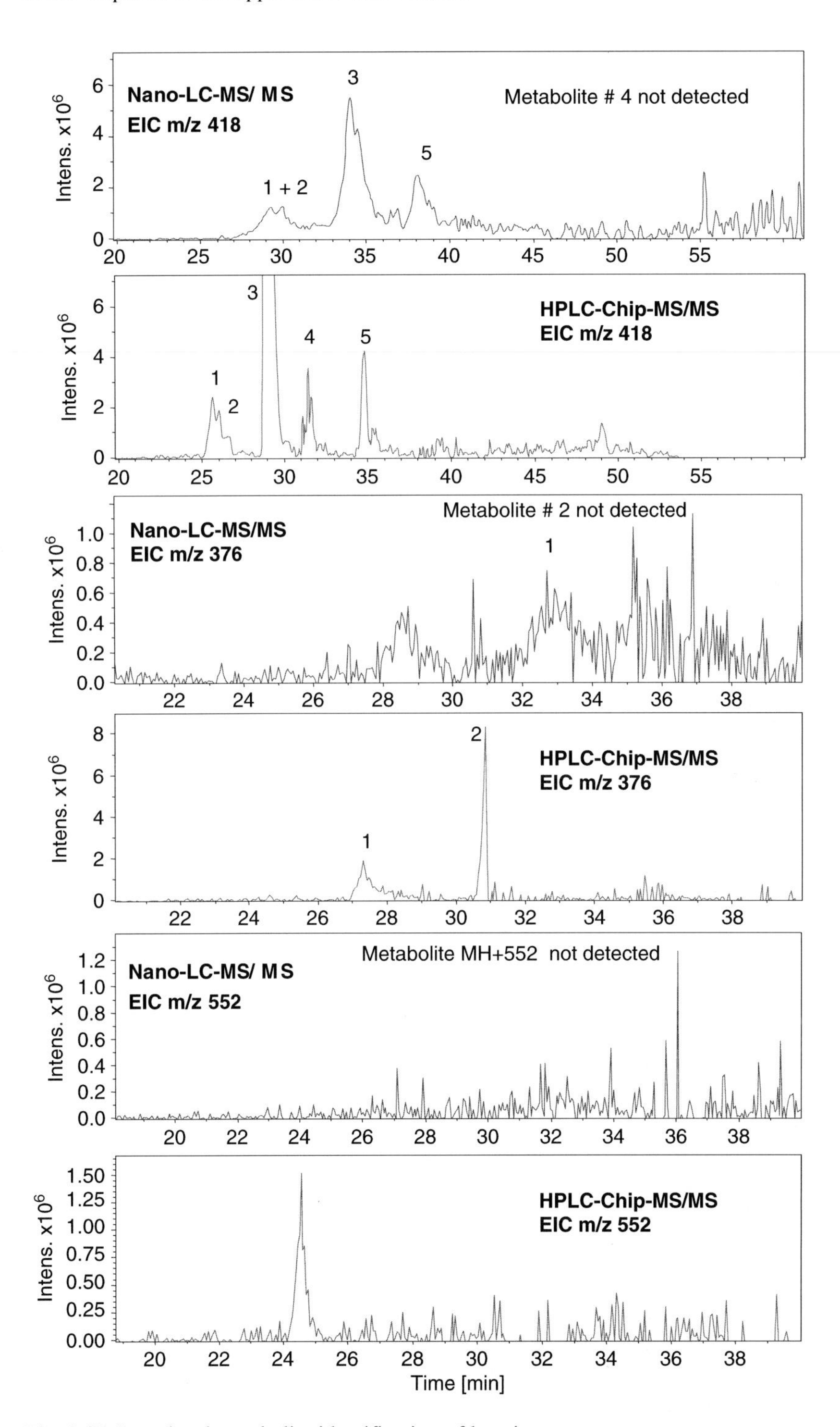

Fig. 8.28. Low-level metabolite identification of buspirone.

trap mass spectrometer. Overall, a 6× sensitivity increase was reported for the HPLC-chip system when compared to the conventional nanocolumn system. Compared to conventional nanocolumn LC/MS, the increased sensitivity and chromatographic performance of the HPLC-chip/MS system provided the detection of additional metabolites, namely a dihydroxylated metabolite (dihydroxybuspirone, MH + 418), *N,N*-desethyl hydroxybuspirone (MH + 376) and as a *N,N*-desethyl hydroxybuspirone glucuronide (MH + 552). In addition the HPLC-chip approach offers superior ease of use, which is an important attribute for implementation in a pharmaceutical environment.

8.5 FUTURE DIRECTIONS

8.5.1 Integration of functionality

The HPLC-chip platform is the ideal approach for further integration of functionality using microfluidics. Handling small volumes of sample without significant sample losses and retention of resolution requires such an integrated approach. In addition, as applications of these low-level high sensitivity analyses move from the research area into either the clinical lab, the pharmaceutical industry or into the broader biology community, standardization into an easy-to-use format is needed. Application-specific chips that integrate more steps of the workflow and that enable a broader user base can drive this capability.

As an example, Fig. 8.29 shows a prototype 2D LC/MS chip. This chip incorporates separation by both strong cation exchange and reversed phase, which is widely used for protein identification and is known as the multidimensional protein identification technology (MUDPIT) approach [25].

The chip is created by expanding the number of laminated layers and as such merging the strong cation exchange column with the reversed-phase column. Peptide samples are injected and trapped on the SCX column and subsequently fractionated and eluted using salt steps. Eluted peptides are trapped on the enrichment column and desalted after which they are released and separated on the reversed-phase column.

Although it is difficult to see the flow path in a top view, Fig. 8.30 shows a 3D image of the flow path.

The flexibility of the valve interface allows proper flow switching for the application and the use of the additional concentric valve in the interface can also enable applications such as fraction parking.

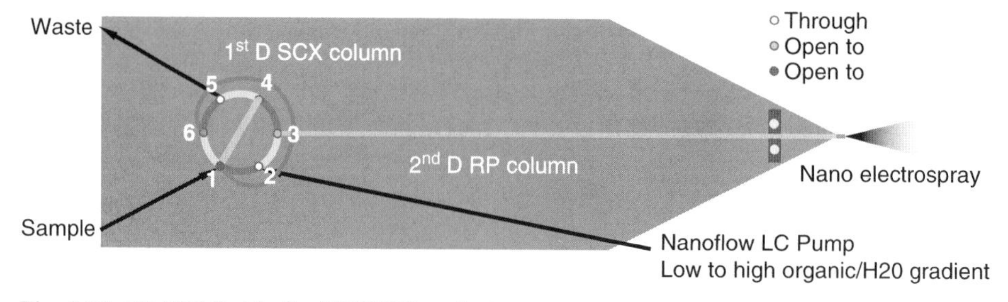

Fig. 8.29. 2D HPLC-chip for MUDPIT analysis.

8.5.2 Higher throughput

The rotor in rotor concept can also be used to further increase throughput by combining two separation columns that are switched between the enrichment column and the nanospray tip. This allows for reconditioning of one column while the other is being used for analysis. Since often reconditioning of the column can take almost as long as the analyses itself, these approaches can double the throughput of an experiment [26].

The layout of a dual column prototype chip is shown in Fig. 8.31. The valve configurations for the different positions are shown. Although the layout suggests 2 valves in different locations on the chip, this design can be made using the concentric rotors using the multilayer capability of the chip fabrication. A throughput improvement of about a factor of 2 can be achieved resulting in a more optimal utilization of MS acquisition time.

8.5.3 Application-specific chips

The integration of multiple functionalities on a single chip device is an attractive way of creating application-specific chips. Unlike electrophoretic microfluidic approaches, where channels are always in electrical contact with each other, the HPLC-chip approach with valving capability allows physical separation of such functions. This allows separate optimization and operation of the individual steps in the workflow and coupling them through valving. Application-specific chips will allow less experienced users to run applications in easy-to-use format.

An example of an application-specific chip is shown in Fig. 8.32. In this prototype design for intact protein analysis, an immobilized trypsin bed is integrated onto the chip for digestion prior to analysis and identification at the peptide level. The advantage of an immobilized trypsin approach over in-solution digestion is the speed with which digestion takes place. This increase in speed is achieved by the much higher concentration of trypsin used on the beads as compared to in-solution. Since the trypsin is immobilized, auto digestion is greatly reduced and thus peptides resulting directly from the trypsin itself are very low. In solution this would not be possible. Digestion times can be taken down from hours to minutes allowing more rapid analysis of intact proteins.

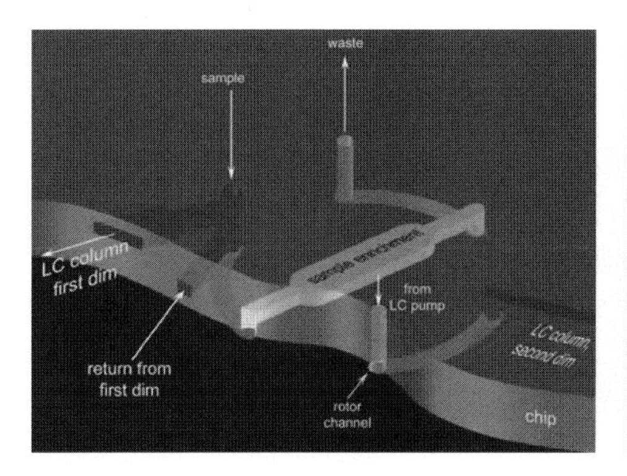

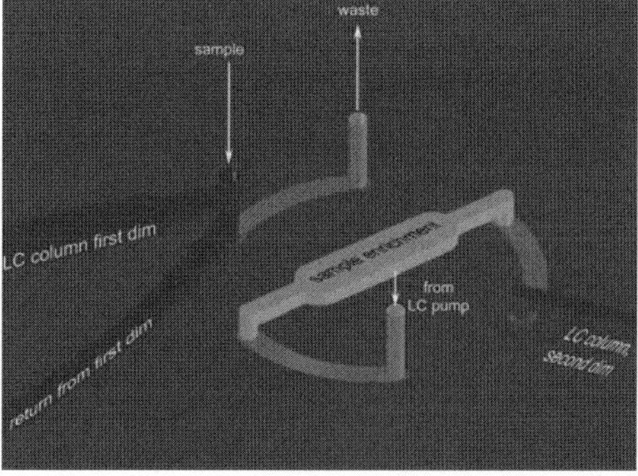

Fig. 8.30. Flow path of the 2D HPLC-chip prototype for MUDPIT analysis.

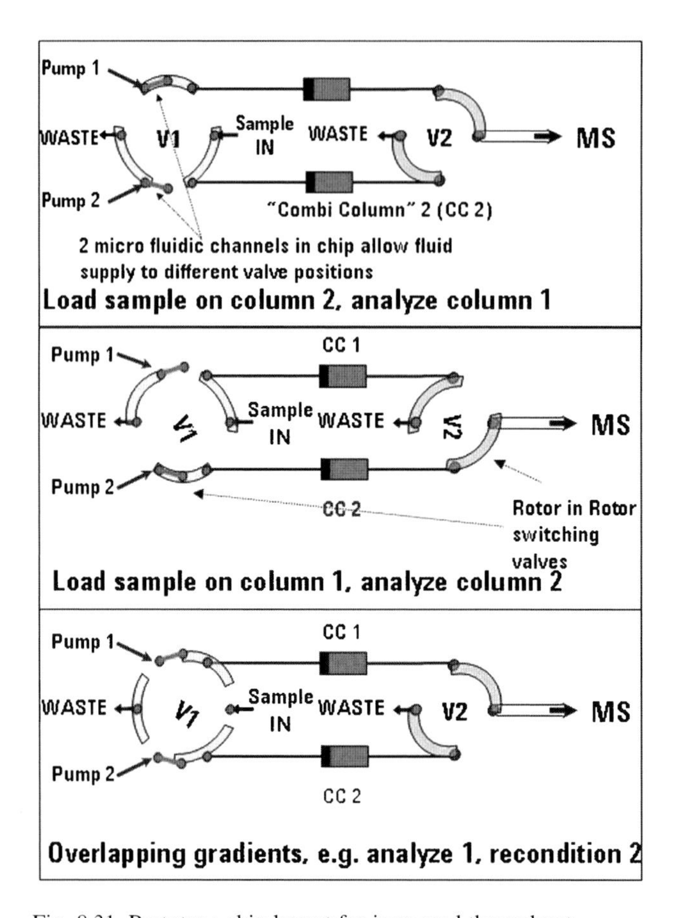

Fig. 8.31. Prototype chip layout for increased throughput.

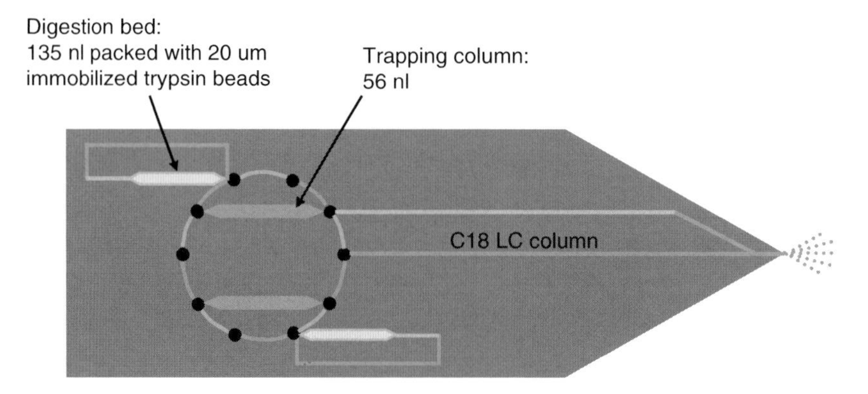

Fig. 8.32. Prototype chip for intact protein analysis.

8.5.4 Sample prep integration

In a similar fashion extra column chambers can be packed with bead-based affinity media such as IMAC beads for phosphopeptide enrichment, lectin beads for glycopeptide enrichment and antibody-based affinity media. The multilayer chip architecture with the on-chip valving again is the key to enabling these combinations. The sample prep steps can be executed independently of all the other functionality on the chip. Solvent compatibility and trapping of samples after enrichment are the key parameters that need optimization to move this type of chip approaches to a reality.

8.6 CONCLUSION

HPLC-chip technology is a new approach to nano-LC/MS. The integration of nano LC elements onto a single chip reduces the number of connections needed and thus reduces dead volume and the possibility of leaks. The reduced dead volume directly results in higher separation performance and sensitivity.

The implementation of the HPLC-chip technology into the LC/MS platform by using on-chip valving allows samples and solvents to be delivered to the chip device using standard nano LC components. Trapping of the sample on the chip allows large volume injection of samples, maximizing the sensitivity that can be achieved.

The valve configuration in the chip handler enables connecting fluids without removing any connections. The valve interface opens up and clamps the chip in a high-pressure leak tight clamp. This allows users to change chips in an easy-to-use way similar to inserting a credit card into an ATM machine.

The robotics in the chip handler positions the chip in an optimal location in front of the mass spectrometer optimizing performance of the nano electrospray. Manual adjustment is no longer needed taking a considerable amount of required expertise out of the process. These features make nano-LC/MS accessible to less experienced users.

Examples were shown that demonstrate the extremely high sensitivity that can be achieved with the HPLC-chip system, as well as the excellent reproducibility that results from this approach. For protein identification this offers identification of more and lower abundant proteins. For profiling applications, comparative studies are simplified due to the reproducibility of results. This leads in the Biomarker discovery area to a new approach of using profiling combined with targeted ID to a more efficient way of finding markers. The fact that chips can be moved easily from MS to MS offers the possibility of using multiple platforms to solve scientific questions in a reproducible way. Chips packed with different stationary phases expand the application area to many other applications.

The HPLC-chip platform with its multilayer structure and multi-valve switching capability allows multiple steps of a sample workflow to be integrated on a single device. Higher throughput and a more integrated, low sample loss solution are created by this functionality. Application-specific chips enable researchers to do complex analytical experiments in an easy-to-use way. This can lead to more standardization and easier comparison of results in complex biological problems and can bring nano-LC/MS as a routine technique to more laboratories.

REFERENCES

1 C. Dass, *Principles and Practice of Biological Mass Spectrometry*, Wiley, New York, 2001.
2 D. Liebler, *Introduction to Proteomics: Tools for the New Biology*, Humana Press, Inc., Totowa, NJ, 2002.
3 G. Marko-Varga and P. Oroszlan, *Emerging Technologies in Protein and Genomic Material Analysis*, Elsevier Science BV, Amsterdam, 2003.
4 H. Yin, K. Killeen, R. Brennen, D. Sobek, M. Werlich and T. van de Goor, *Anal. Chem.* 77 (2005) 527.
5 M. Hamdan and P. Righetti, *Proteomics Today: Protein Assessment and Biomarkers Using Mass Spectrometry, 2D Electrophoresis, and Microarray Technology*, Wiley, New Jersey, 2005.
6 G. Greg Gibson and S. Muse, *A Primer of Genome Science*, Sinauer Associates, Inc., Sunderland, USA, 2002.
7 M. Kinter and N. Sherman, *Protein Sequencing and Identification Using Tandem Mass Spectrometry*, Wiley, New York, 2000.
8 J. Sanchez, G. Corthals and D. Hochstrasser, *Biomedical Applications of Proteomics*, Wiley-VCH Verlag GmbH & Co. KGaA, Weinheim, 2004.
9 R. Ardrey, *Liquid Chromatography Mass Spectrometry: An Introduction*, Wiley, Chichester, 2003.
10 M. Vollmer, C. Miller and B. Glatz, Performance of the Agilent 1100 HPLC-chip/MS system, 2005, Agilent Application Note 5989-4148EN.
11 S. Minteer, *Microfluidic Techniques: Reviews and Protocols*, Humana Press, Inc., Totowa, NJ, 2006.
12 C. Sommer, *Non-Traditional Machining Handbook*, Advance Publishing Inc., Houston, TX, 2000.
13 U. Neue, *HPLC Columns: Theory, Technology, and Practice*, Wiley-VCH Inc., New York, 1997.
14 D. Rees, *Membrane Proteins* (Advances in Protein Chemistry, Volume 63), Academic Press, San Diego, CA, 2003.
15 J. Martosella, N. Zolotarjova and H. Liu, A proteomic strategy for increasing membrane protein identification with use of the Agilent high recovery mRP-C18 reversed phase HPLC column – A comprehensive survey of the HeLa membrane proteome, 2006, Agilent Application Note 5989-5020EN.
16 G. Siuzdak, *Mass Spectrometry for Biotechnology*, Academic Press, San Diego, CA, 1996.
17 H. Zhang, Z. Li, D. Martin and R. Aebersold, *Nat. Biotechnol.*, 21 (2003) 660.
18 H. Yin, K. Killeen, R. Grimm, X. Li, H. Zhang and R. Aebersold, ABRF Poster P150-T, 2005.
19 D. Liebler, *Proteomics in Cancer Research*, Wiley, Hoboken, NJ, 2004.
20 C. Miller and B. Miller, A highly accurate mass profiling approach to protein biomarker discovery using HPLC-chip/MA-enabled ESI–TOF MS, 2006, Agilent Application Note 5989-5083EN.
21 D.S. Newburg, *Bioactive Components of Human Milk* (Advances in Experimental Medicine and Biology), Kluwer Academic/Plenum Publishers, New York, 2001.
22 M. Ninonuevo, H. An, H. Yin, K. Killeen, R. Grimm, R. Ward, B. German and C. Lebrilla, *Electrophoresis*, 26 (2005) 3641.
23 H. Yin, K. Killeen, M. Ninonuevo, C. Lebrilla and R. Grimm, Screening of immunostimulatory oligosaccharides by using a new HPLC-chip/MS technology; FOCIS poster F2.25, 2005.
24 M. Vollmer, A. Fandino and G. Gauthier, Simultaneous assessment of drug metabolic stability and identification of metabolites using HPLC-chip/ion trap mass spectrometry, 2006, Agilent Application Note 5989-5129EN.
25 H. Yin, K. Killeen R. Brennen and T. van de Goor, Chip 2-D LC/MS: Seamless integration using a microfluidic multilayer structure, 2004, Agilent Application Note 5989-1296EN.
26 B. Glatz, P. Goodley, P. Hoerth, K. Kraiczek and M. Vollmer, Overlapped protein digest analysis using multiple nanospray columns on a microchip; ASMS poster ThPL 235, 2004.

Achille Cappiello (Editor)
Advances in LC–MS Instrumentation
Journal of Chromatography Library, Vol. 72

Chapter 9

Matrix effect, signal suppression and enhancement in LC–ESI–MS

LORIS TONIDANDEL and ROBERTA SERAGLIA

In the last decade various research fields (i.e. pharmaceuticals, toxicology, environmental and clinical) have changed the traditional one-dimension instrumentation liquid chromatography with ultraviolet detection (LC–UV) to hyphenated mass spectrometric technologies, based on atmospheric pressure ionization (API) methods coupled with liquid chromatography (LC) systems. Among the different API interfaces, the most widely used is the electrospray ionization (ESI) source. There are several reasons for this choice: it is relatively easy to use, its low solvent consumption and the large variety of analytes which can be analysed (e.g. high molecular weight compounds, polar and thermally labile species). However, its use is far from being problem-free.

Many papers that appeared in recent literature focused on the phenomenon of matrix effect in LC/MS, which leads to ion enhancement/suppression with a consequent error in the quantitation of an analyte of interest. This matrix effect originates from co-elution of endogenous and/or exogenous components present in biological, natural and synthetic matrix in which the analyte of interest is present, even at very low concentration. In particular, when detection limit drops to sub-ng/mL, matrix signal can produce a total ion suppression [1–6].

Moreover, signal suppression and enhancement in ESI also originate from several factors that can influence the ionization yield of the analyte(s) of interest, such as flow instability, background noise, sample contaminants and mobile phase composition.

An understanding of the nature of the mechanisms operating in the ESI is crucial in order to obtain an effective solution to this ion suppression/enhancement phenomenon.

Enke [7] studied, from the theoretical point of view, this problem and obtained a predictive model for matrix and analyte effects in ESI of singly charged ionic analytes.

The study of the basic aspect of ESI, described in Chapter 1, has led to two possible mechanisms for ion production from the sprayed solution droplets, i.e. the charge residue and the ion evaporation (CRM and IEM respectively). It is generally accepted that both mechanisms are operative: IEM is considered to be important for relatively small analyte species, while the CRM becomes important for macromolecules. How can these mechanisms be related to the analyte concentration in the droplet? Tang and Kebarle proposed that the ion evaporation rate from the droplets is proportional to the ion concentration in the droplet [8a]. This hypothesis was based assuming that IEM is the main mechanism for ion evaporation in gas phase, and that the fraction of the ion current of the analyte is dependent on the total ion formation rate I. Considering the co-presence of electrolytes other than the analyte (e.g. salts, impurities), I is given by

$$I = Pf \frac{k_A[A^+]}{k_E[E^+] + k_A[A^+]} \tag{9.1}$$

where P is the sampling efficiency of the systems; f the fraction of the charges brought by the droplet converted into gas-phase ions; ks the ratio constants; and [ ] the analytical concentration of the analyte A^+ and the electrolyte E^+.

The Enke model considers that the charge density is not homogeneous inside the droplet: the excess charge is necessarily localized on the droplet surface. The inside of the droplet is neutral and contains solvent and neutral analyte molecules and salts. Hence the surface excess charge state can be considered as a phase separate from that inside the droplet and ions are free to partition between interior and surface phases.

Assuming that the partitioning process is sufficiently rapid, an equilibrium state will be reached. The partition coefficient for the analyte ions A^+ and electrolyte ions E^+ will be:

$$(A^+X^-)_i \rightleftharpoons (A^+)_s + (X^-)_i \Rightarrow K_A = [A^+]_s[X^-]_i/[A^+X^-]_i \tag{9.2}$$

$$(E^+X^-)_i \rightleftharpoons (E^+)_s + (X^-)_i \Rightarrow K_E = [E^+]_s[X^-]_i/[E^+X^-]_i \tag{9.3}$$

where X^- is the associate anion.

The equilibrium can be considered as a displacement reaction, since A^+ and E^+ are both competing to supply a fixed number of surface charges. It can be expressed by the reaction:

$$(A^+X^-)_i + (E^+)_s \rightleftharpoons (A^+)_s + (E^+X^-)_i \text{ and } K_A/K_E = [A^+]_s[E^+X^-]_i/[A^+X^-]_i[E^+]_s \tag{9.4}$$

In an electrospray experiment the difference in the concentration of cations and anions can be considered the concentration of excess charge $[Q]$. $[Q]$ (in equivalents per litre) can be calculated as:

$$[Q] = I/F\Phi \tag{9.5}$$

where I is the current flowing from the ESI capillary to the counter electrode, F the Faraday constant and Φ the flow rate. $[Q]$ represents the sum of the concentration of all charged species present on the droplet surface:

$$[Q] = [A^+]_s + [E^+]_s \tag{9.6}$$

This equation provides evidence to an important point: when the total cation concentration is higher than $[Q]$, there will be a competition among the cations on the surface and those inside the droplet. If we assume that $[A^+]_s$ and $[E^+]_s$ are small (in other words only a small fraction of A^+ and E^+ is on the surface), Eqn. (9.2) and (9.3) become:

$$K_A = [A^+]_s[X^-]_i/C_A \text{ and } K_E = [E^+]_s[X^-]_i/C_E \tag{9.7}$$

where C_A and C_E are the concentration of analyte and electrolyte in the injected solution.
From these equations, by eliminating $[X^-]_i$ we obtain:

$$[A^+]_s = C_A K_A / C_E K_E [E^+]_s \tag{9.8}$$

Considering that Eqn. (9.6) can be written as $[E^+]_s = [Q] - [A^+]_s$ and substituting it in Eqn. (9.8), we obtain:

$$[A^+]_s = \frac{C_A K_A}{C_A K_A + C_E K_E}[Q] \tag{9.9}$$

and, if we assume that the ion current in the ESI spectrum is proportional to the concentration of the ion in the surface excess charge phase of the droplet, the instrumental response for A^+ (R_A) will be given by:

$$R_A = Pf\frac{C_A K_A}{C_A K_A + C_E K_E}[Q] \tag{9.10}$$

This equation is of the same form as that of Kebarle (Eqn. 9.1).
By introducing mass balance equations and in the case of C_A values lower than $[Q]$, $[A^+]_s$ can be calculated:

$$[A^+]_s = C_A\left(\frac{K_A/K_E}{K_A/K_E - 1 + C_E/[Q]}\right) \tag{9.11}$$

while in the more general case a quadratic equation in $[A^+]$ is obtained:

$$[A^+]_s^2\left(\frac{K_A}{K_E} - 1\right) - [A^+]_s\left([Q]\left(\frac{K_A}{K_E} - 1\right) + C_A\frac{K_A}{K_E} + C_E\right) + C_A[Q]\frac{K_A}{K_E} = 0 \tag{9.12}$$

Data from an experiment reported by Tang and Kebarle are plotted in Fig. 9.1. In this experiment, the response intensities of heroin hydrochloride ($HerH^+$) and Cs^+ were separately measured as function of analyte concentration. The electrolyte present was principally the NH_4CH_3O impurity present in the methanol solvent, estimated by the authors to be 10^{-5} M. The fit parameters were K_A/K_E and C_E. The values that fit were $C_E = 1.6\times10^{-5}$ M and K_A/K_E for Cs^+ and $HerH^+$, 0.32 and 4.5 respectively. These results demonstrate the ability of Eqn. (9.12) to fit experimental data in the low concentration ($<10^{-5}$ M) range where Eqn. (9.1) did not work. Not unexpectedly, the values of k obtained by Tang and Kebarle from fitting Eqn. (9.1) to the data do not exactly match the values of K_A/K_E obtained by fitting Eqn. (9.12). They do agree in the general trend of the k values but differ by a factor of 2.5 in the ratio of the K_s for Cs^+ and $HerH^+$.

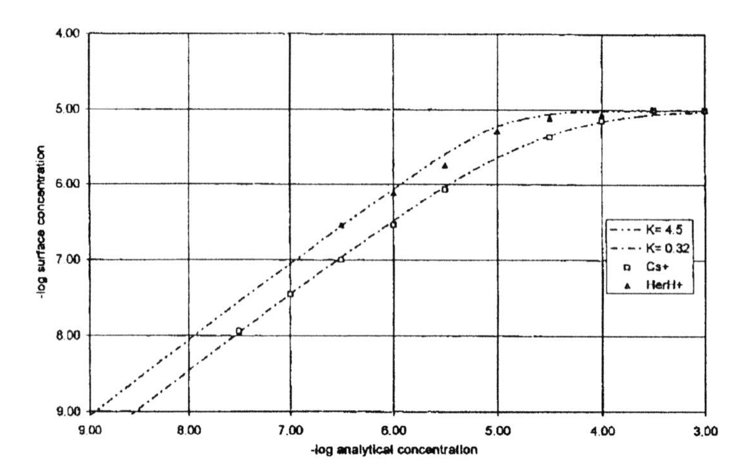

Fig. 9.1. Fit of Eqn. 9.12 to experimental data from Kebarle and Tang [8b]. (Reprinted with permission from Ref. [7], Copyright 1997 American Chemical Society).

In the case of ESI of multiple analytes the situation becomes more and more complicated. As an example we can consider the case of two analyte ions A^+ and B^+ and the electrolyte E^+. In this case the charge balance equation is $[A^+]_s+[B^+]_s+[E^+]_s = [Q]$ and, after a series of calculation, the result is a cubic equation:

$$[A^+]_S^3[K_B - K_A + K_E(1-K_B/K_A)]+[A^+]_S^2\{C_A(2K_A - K_B - K_E)+ C_B(K_A - K_E)+C_E K_E(1-K_B/K_A)+[Q][K_A - K_B - K_E(1-K_B/K_A)]\} - [A^+]_S C_A[[Q](2K_A - K_B - K_E)+C_B K_A + C_A K_A + C_E K_E]+C_A^2[Q]K_A = 0 \tag{9.13}$$

By considering different values of the various parameters, the results reported in Fig. 9.2, keeping C_A/C_B ratio equal to 1, and 3, keeping K_A/K_E constant, have been obtained. They give a rationalization of the matrix effect, i.e. a different response factor observed in ESI when different analytes are present in the solution injected in the source.

All that has been described until now represent processes occurring through an equilibrium competition among the ionic species present in the solution for the excess charge sites existing in the charged droplet. However another point can be relevant to explain different response factor for different analytes, i.e. the electrochemical processes mainly occurring at the spraying capillary tip as already described in Chapter 1.

As stated above, the matrix-induced signal suppression effects are believed to result from the competition of analyte ions and matrix component to access the electrosprayed droplet surface for gas-phase emission. In the presence of matrix components, the rate of analyte ion formation may differ significantly from that of a standard sample, leading to a variation in signal response. This implies that the quantitative data obtained cannot be fully validated unless a stable-isotope labelled analyte is used as an internal standard (ISTD). A full compensation of matrix effects is achieved only if the ionization yield of the analyte and ISTD are influenced by the matrix in the same manner, otherwise

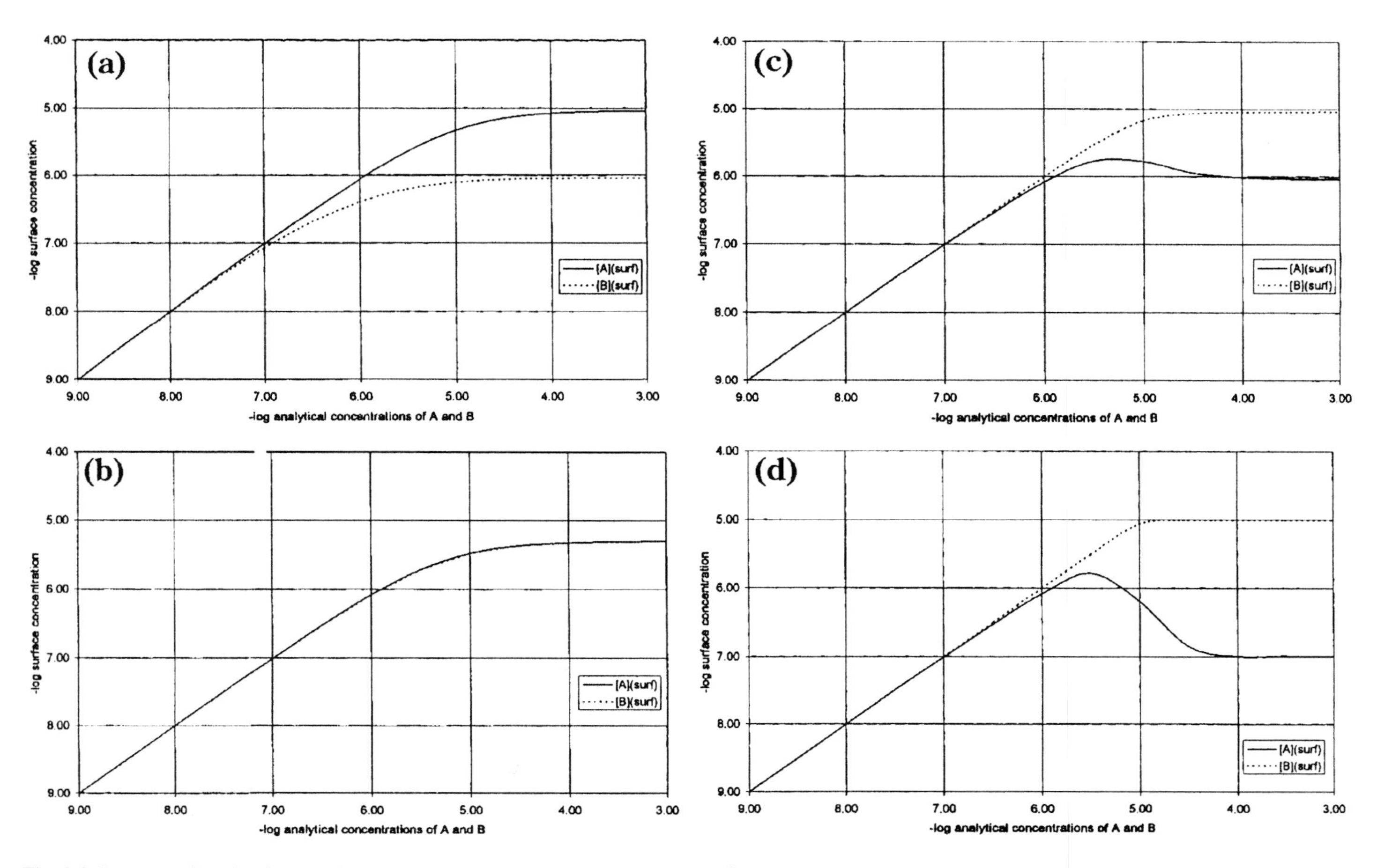

Fig. 9.2. Response function for two singly charged analytes with $C_E = [Q] = 10^{-5}$ M and $K_A/K_E = 1$: (a) $K_B/K_E = 0.10$; (b) $K_B/K_E = 1.0$; (c) $K_B/K_E = 10$ and (d) $K_B/K_E = 100$. (Reprinted with permission from Ref. [7], Copyright 1997 American Chemical Society).

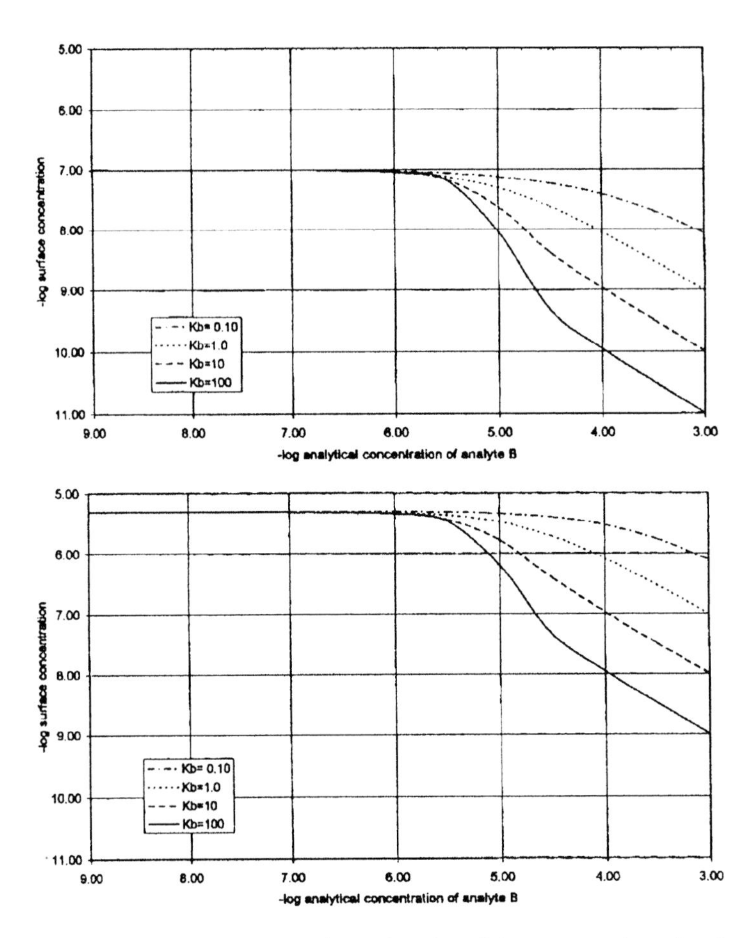

Fig. 9.3. Response to analyte A^+ as a function of the concentration of analyte B^+ with $[Q] = 10^{-5}$ M, $K_A/K_E = 1.0$ and $C_E = 10^{-5}$ M: (a) $C_A = 10^{-7}$ M and (b) $C_A = 10^{-5}$ M. (Reprinted with permission from Ref. [7], Copyright 1997 American Chemical Society).

different response of analyte and ISTD are formed leading to invalid and unreproducible quantitative data.

An approach to evaluate matrix effect is the method described by King *et al.* [4]. This approach gives a qualitative view of suppression effects during the course of a chromatographic separation of a blank sample extract (zero analyte) using an added (with constant flow) infusion of the target analyte solution (fixed concentration) via post-column tee. In the absence of any interferences eluting from the column, the post-column infusion of analyte will give a constant signal in a multiple reaction monitoring (MRM) channel. The resulting chromatogram will show a reduction of this constant baseline signal due to suppression of the analyte signal by elution of endogenous material from the sample (i.e. extract of urine, plasma, etc.). By this method valuable information is obtained on any potentially high suppression regions in the chromatogram. In this way the chromatographic conditions can be modified to reduce or increase the retention time of the analyte of interest such that its retention time will be outside of these suppression zones. Obviously, the change of the chromatographic conditions can shift also the elution time of the suppression region and so the test must be repeated.

The post-column infusion method was employed for establishing the matrix effects observed in rat plasma samples due to endogenous materials (i.e. salts, amines, fatty acids, etc.) from the vehicles (or mixtures of vehicles) used in dosing solution for administration of drugs at discovery stage. These vehicles are present at high concentration (>1 mg/mL) in the early time-point samples, in particular for intravenous (IV) dose. Shou and Naidong [9] studied by post-column infusion the effect of different vehicles [polyethylene glycol 400 (PEG 400), polysorbate 80 (Tween 80), hydroxypropyl-β cyclodestrine (HPBCD) and *N,N*-dimethylacetamide (NN)] on the LC/ESI response of eight model analytes, using a fast gradient LC/MS/MS method. The eight model analytes were metoprolol (MET), α-hydroxymetoprolol (OHMET), midazolam (MID), 1′-hydroxymidazolam (OHMID), bufuralol (BUF), 1′-hydroxybufuralol (OHBUF), propranolol (PRO) and fexofenadine (FEX). The authors observed a decrease of analyte response when rat plasma samples that originated from drug administration using vehicles were analysed. To elucidate if the matrix effect originates from endogenous component of plasma or from the presence of high amount of vehicles in plasma at early time-point sample, the same concentration of analytes (250 mg/mL) was added to mobile phase solution, vehicle-free rat plasma, and rat plasma containing different vehicles. Plasma sample was extracted using the protein precipitation method. The samples so obtained were injected into the LC/MS/MS system in triplicate. The average peak areas were plotted against the respective matrices as shown in Fig. 9.4. It is noted that for all the compounds the peak areas in vehicle-free plasma sample are very close to that formed in mobile phase solution. This is a good indication that the decrease in the ESI response of these analytes in rat plasma is not attributable, in this case, to endogenous component of plasma. Moreover, the observed suppression is compound-different as well as vehicle-dependent: PRO, MET, and OHBUF seemed to be the most influenced by ion suppression, whereas BUF, MID, OHMID and FED were less suppressed. The polymeric vehicles PEG 400 and Tween 80 were the most prone to induce signal suppression on the analyte of interest. Taking into account that polymers have a good surface activity in the ESI ion evaporation processes, they cause ionization suppression when co-eluting with the analyte of interest. To demonstrate this hypothesis a solution of the analyte of interest is continuously infused post-column, while blank plasma extract and plasma extract fortified with different vehicles were injected into the LC column. The "valley" in the resulting infusion chromatograms indicated the retention windows of signal suppression. Since a gradient method has been employed the analyte responses changed with the mobile phase composition; the highest response was observed at the end of the run where the organic content was the highest. In Fig. 9.5 (top) the overlaid infusion chromatograms for OHBUF from injection of vehicles-free plasma extract and a plasma extract fortified with PEG 400 are reported, while in Fig. 9.5 (bottom) similar chromatograms are reported for BUF. As it can be seen, a valley, corresponding to induced ion suppression by PEG 400, is present in the retention time region of OHBUF, while BUF eluted after this valley. Taking into account that OHBUF is one of the analyte more influenced by ion suppression (Fig. 9.4) and that this effect is negligible for BUF, these results confirmed the

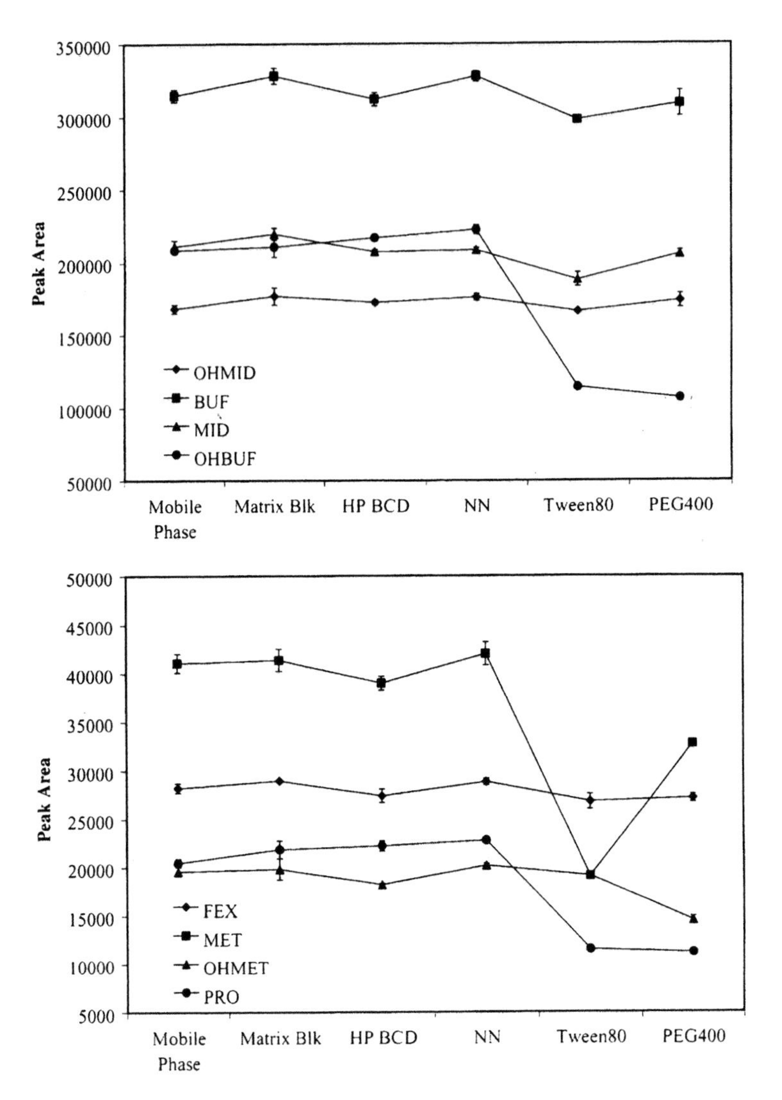

Fig. 9.4. Analyte peak areas from protein precipitation extracts of rat plasma containing various dosing vehicles. See text for abbreviation of analytes and vehicles. ("Post-column infusion study of the 'dosing vehicles effect' in the liquid chromatography/tandem mass spectrometric analysis of discovery pharmacokinetic samples" by W.X. Shou and W. Naidong, Copyright 2003 John Wiley. Reproduced with permission).

assumption that the suppression effect was due to the co-elution of analytes with polymeric vehicles. On real sample the ion suppression effect involve both analyte and its internal standard, perhaps to different extent. The improvement of both chromatographic condition and sample clean-up lead to better reproducibility of the results and increase response of the analyte as shown in Fig. 9.6.

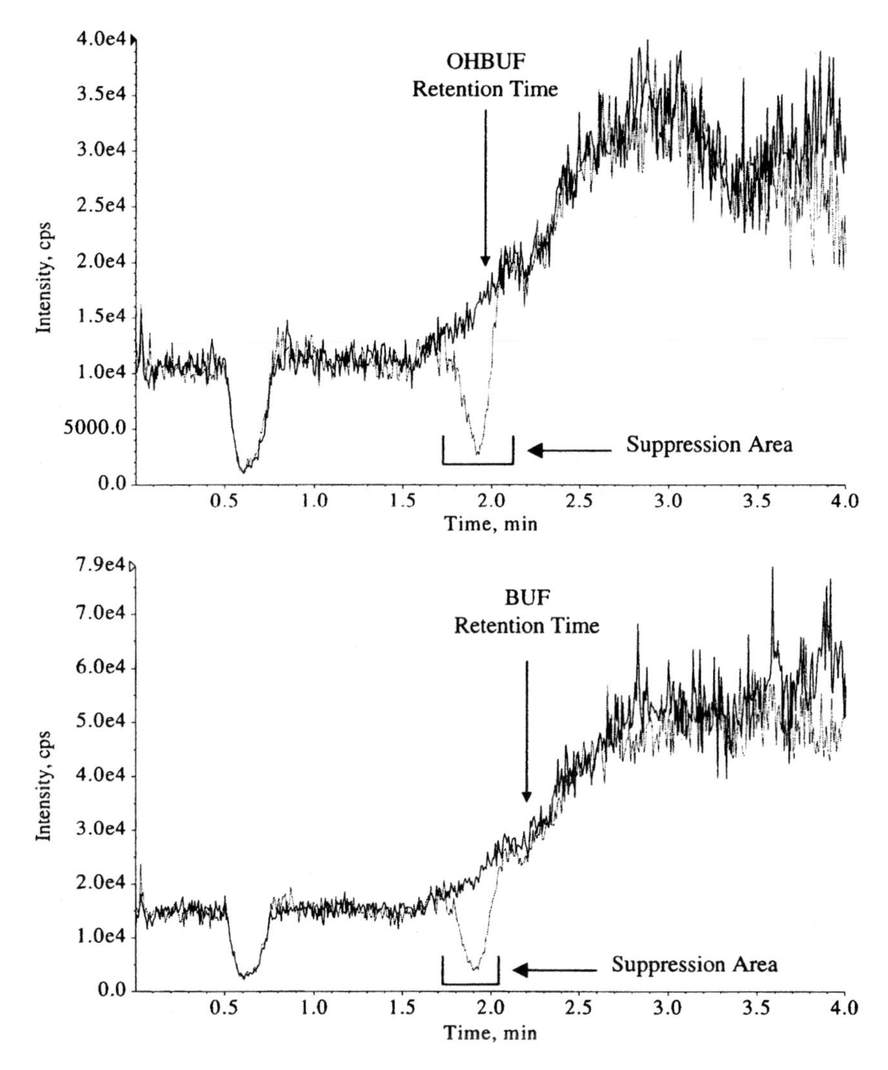

Fig. 9.5. Overlaid post-column infusion chromatograms for OHBUF (top) and BUF (bottom). The darker traces were from injections of a regular plasma blank protein precipitation extract and the lighter traces (showing the suppression areas) from injections of an extracted plasma blank fortified with 5 mg/mL of PEG 400. ("Post-column infusion study of the 'dosing vehicles effect' in the liquid chromatography/tandem mass spectrometric analysis of discovery pharmacokinetic samples" by W.X. Shou and W. Naidong, Copyright 2003 John Wiley. Reproduced with permission).

Another factor which can influence the potential matrix effects of dosing vehicles is the ESI source design as pointed out by Xu *et al.* [10]. After optimization of the purification procedure of rat plasma (by protein precipitation) and of chromatographic separation, the authors observed that different extent of ion suppression/enhancement depends on both dosing vehicles and ion source design. As an example the pseudo-ephedrine (PSE) time-dependent

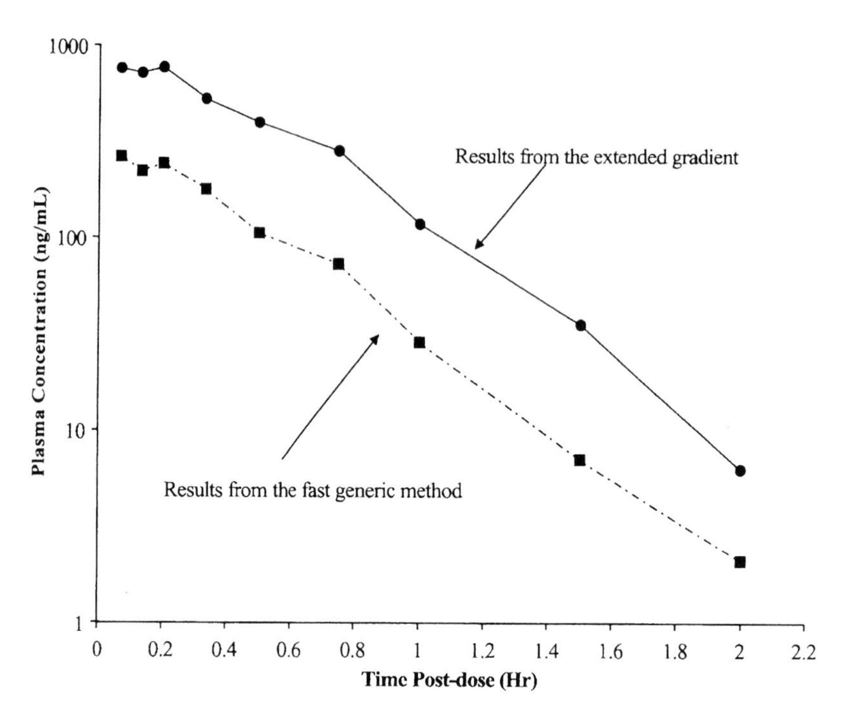

Fig. 9.6. Comparison of analysis results obtained from the same set of dog IV plasma samples containing PEG 400 when two different gradient LC/MS/MS methods were used. ("Post-column infusion study of the 'dosing vehicles effect' in the liquid chromatography/tandem mass spectrometric analysis of discovery pharmacokinetic samples" by W.X. Shou and W. Naidong, Copyright 2003 John Wiley. Reproduced with permission).

MS response using Tween 80 0.1% as vehicles in IV and oral (PO) administration is reported in Fig. 9.7. Taking into account that mass response of PSE in control plasma (no vehicles in the administered solution, only water) was set as 100%, a clear different response has been obtained with the three ion source employed (Thermo Finnigan Quantum, Waters-Micromass model Quattro Ultima and AB Sciex model API 3000), with a 50% ion suppression observed in the case of Thermo Finnigan Quantum instrument. Using PEG 400 as dosing vehicles (Fig. 9.8) the strongest ion suppression takes place in the Waters-Micromass Quattro Ultima instrument, especially in PO-dosed PSE. The matrix effect of these dosing vehicles is dependent also on the ionization mode (ESI or APCI) employed. In this case the APCI mode resulted less sensitive to matrix effect than the ESI one.

In another work by Mei *et al.* [11], the Waters-Micromass Quattro Ultima instrument equipped with APCI ion source was shown to be the most sensitive apparatus for revealing matrix effect caused by exogenous materials, such as polymer plasticizers, contained in different brands of plastic tubes, or Li-heparin, a commonly used anti-coagulant. The use of the instrument arrangement more sensitive against matrix effect can help to eliminate the source of sample contamination during the processing raw sample procedure.

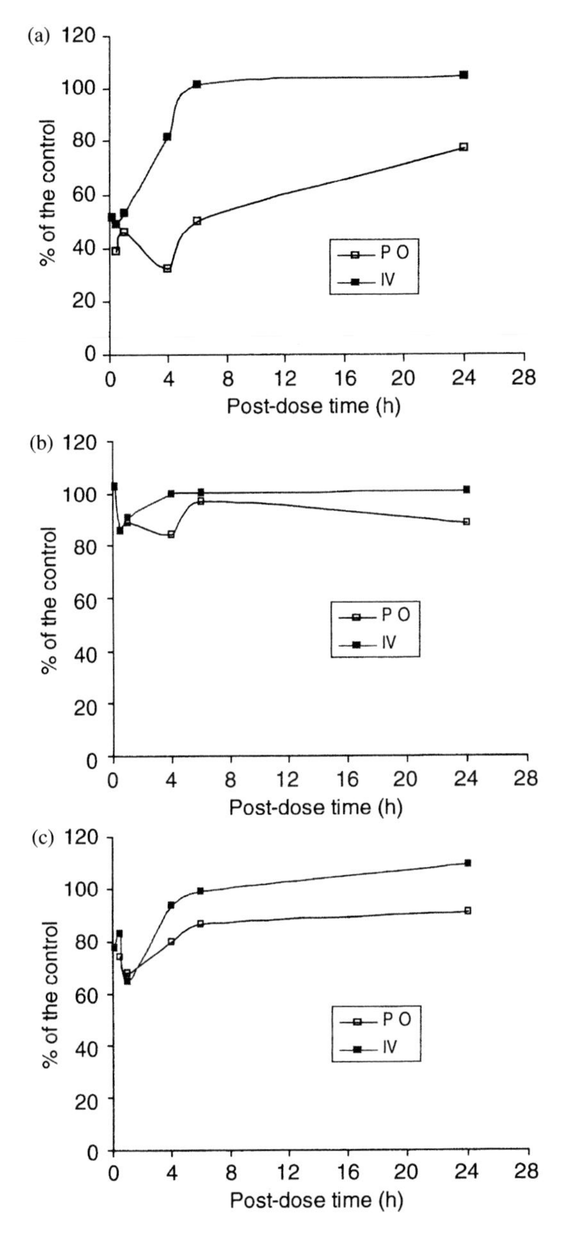

Fig. 9.7. PSE time-dependent MS response after 0.1% Tween 80 used as vehicle in IV and PO routes: (a) Finnigan ESI mode, (b) Sciex ESI mode and (c) Micromass ESI mode. ("A study of common discovery dosing formulation components and their potential for causing time-dependent matrix effects in high-performance liquid chromatography tandem mass spectrometry" by X. Xu *et al.* Copyright 2005 John Wiley. Reproduced with permission).

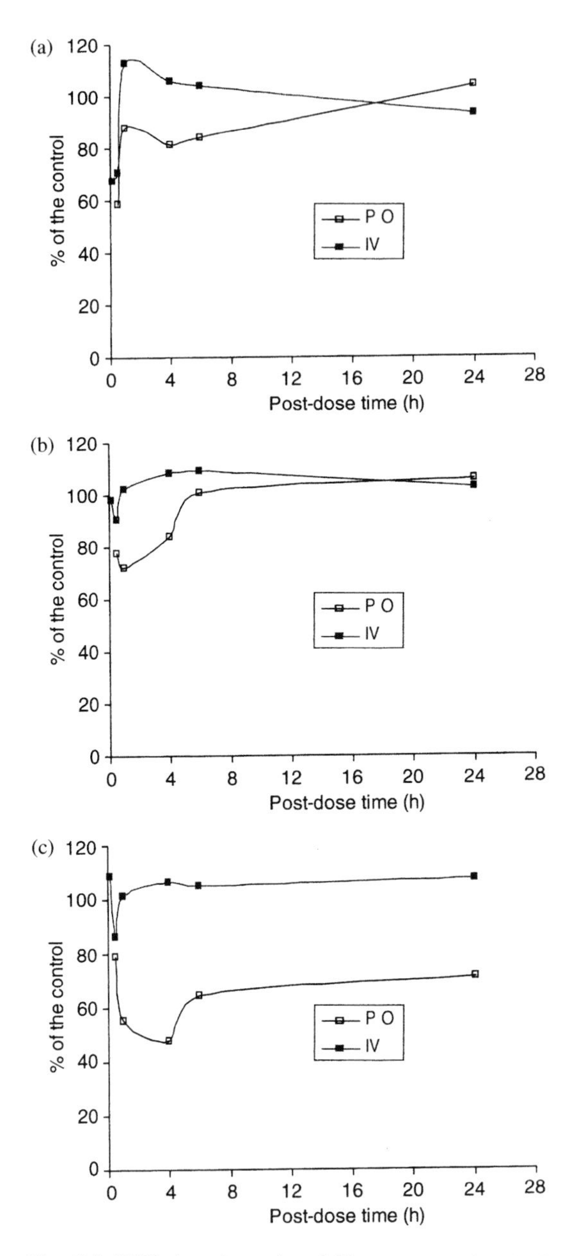

Fig. 9.8. PSE time-dependent MS response after PEG 400 used as vehicle in IV and PO routes: (a) Finnigan ESI mode, (b) Sciex ESI mode and (c) Micromass ESI mode. ("A study of common discovery dosing formulation components and their potential for causing time-dependent matrix effects in high-performance liquid chromatography tandem mass spectrometry" by X. Xu *et al.* Copyright 2005 John Wiley. Reproduced with permission).

The ion suppression/enhancement in LC/ESI/MS strongly depend also on the additives of the mobile phase employed. Additives and buffers in LC mobile phase are primarily employed to achieve reproducible retention time due to their buffering capacity and, secondly, to produce sharper peak shape if used as an ion pairing agent capable of creating neutral species with the target analyte. In LC/ESI/MS the additives and buffer can serve to protonate basic molecules when operating in positive ion mode and vice versa for acidic molecules. Generally in LC/ESI//MS analysis, volatile buffers and additives (i.e. ammonium acetate, formate or hydroxide, formic or acetic acid or trifluoroacetic acid) are preferred to the non-volatile ones (i.e. phosphates, sulphates, borates and citrates). In fact the non-volatile eluent components can be deposited on the ion surface, leading to contamination of the electrodes inside the ion source. This contamination reduces the effective charge density at the electrode surface, emulating a lower voltage for the electrode: a reduction in the signal intensity will be observed over time.

However, volatile buffer and additive must be used at a concentration as low as possible, consistently with chromatographic resolution. In fact buffer and additive species compete for surface site on the electrosprayed droplets and an increase in their concentration leads to decrease in the analyte signal: the competition for surface sites of the droplets favours the higher concentration species and, in this case, analyte species become depleted. This effect is compound-dependent: some analytes show only a small decrease of signal while for others the response can be decreased much more.

An example of buffer effect on ion suppression is reported in Fig. 9.9. While the UV chromatogram (Fig. 9.9b) shows well-resolved peaks, in the TIC chromatogram (Fig. 9.9a) a low sensibility is present due to ion source contamination induced by the high concentration of buffer present in the mobile phase (0.1 M ammonium acetate). Taking into account that the analytes of interest had retention time larger than 10 min, it was possible to obtain TIC chromatogram with no ion suppression effect connecting the LC column to the ESI source after 10 min from the beginning of the run (Fig. 9.10).

An interesting study on the ion suppression/enhancement due to additives and the buffer of mobile phase was carried out by Mallet *et al.* [12]. They used a variation of the post-column infusion method [4] described above for analysing the effect of additives and buffers of the mobile phase on the ESI analysis of acidic and basic drugs (Table 9.1), choiced for their differences in molecular mass, polarity and structure. In this study the LC column was omitted to eliminate any possible contribution on ion suppression from column itself (column bleeding). The solution of mobile phase additives is mixed through an infusion pump line into a constant flow of a target solution (containing either basic or acidic analytes in 50/50 methanol/water solvent), obtained by connecting the LC system directly to ESI source. The reference point to evaluate the ion suppression/enhancement was measured mixing target solution with mobile phase (50/50 methanol/water) with no adjective added. In Fig. 9.11 the ion suppression and ion enhancement effects on terfenadine is shown. The partial mass spectrum reported in Fig. 9.11B clearly detected the $[M+H]^+$ of terfenadine at m/z 472.6, acquired with 50/50 methanol/water. When the analyte flow is mixed with an equivalent flow of 0.5% ammonium hydroxide in methanol/water 50/50 (Fig. 9.11A) an enhancement effect of 41% was observed. On the contrary, a suppression effect of 75% was found in the case of the same solvent with 0.5% of trifluoroacetic acid (TFA) (Fig. 9.11C). In Table 9.2 the results obtained for the test solutions of both basic and

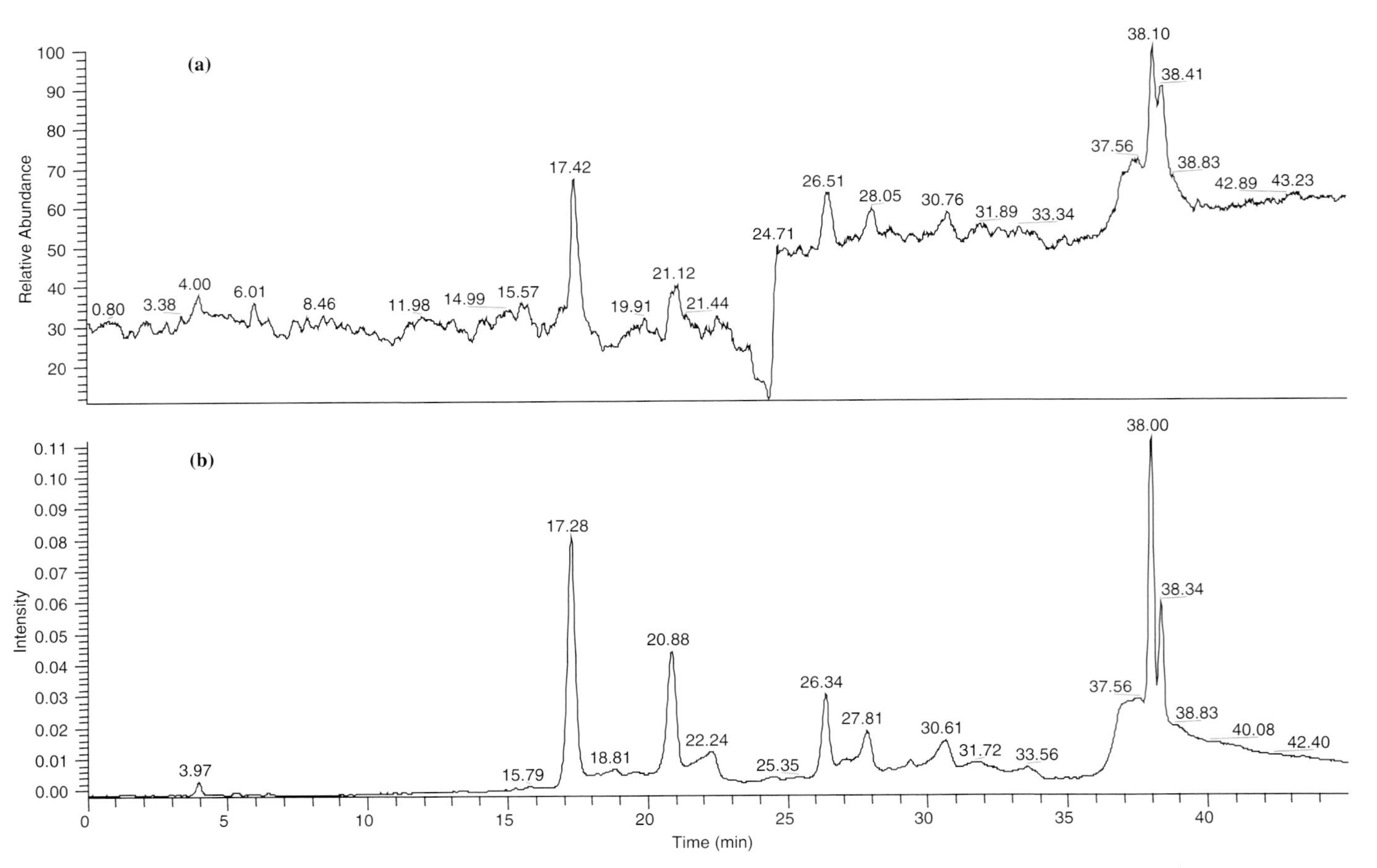

Fig. 9.9. LC–ESI/MS chromatograms obtained with mobile phase containing 0.1 M ammonium acetate: (a) TIC chromatogram and (b) UV trace.

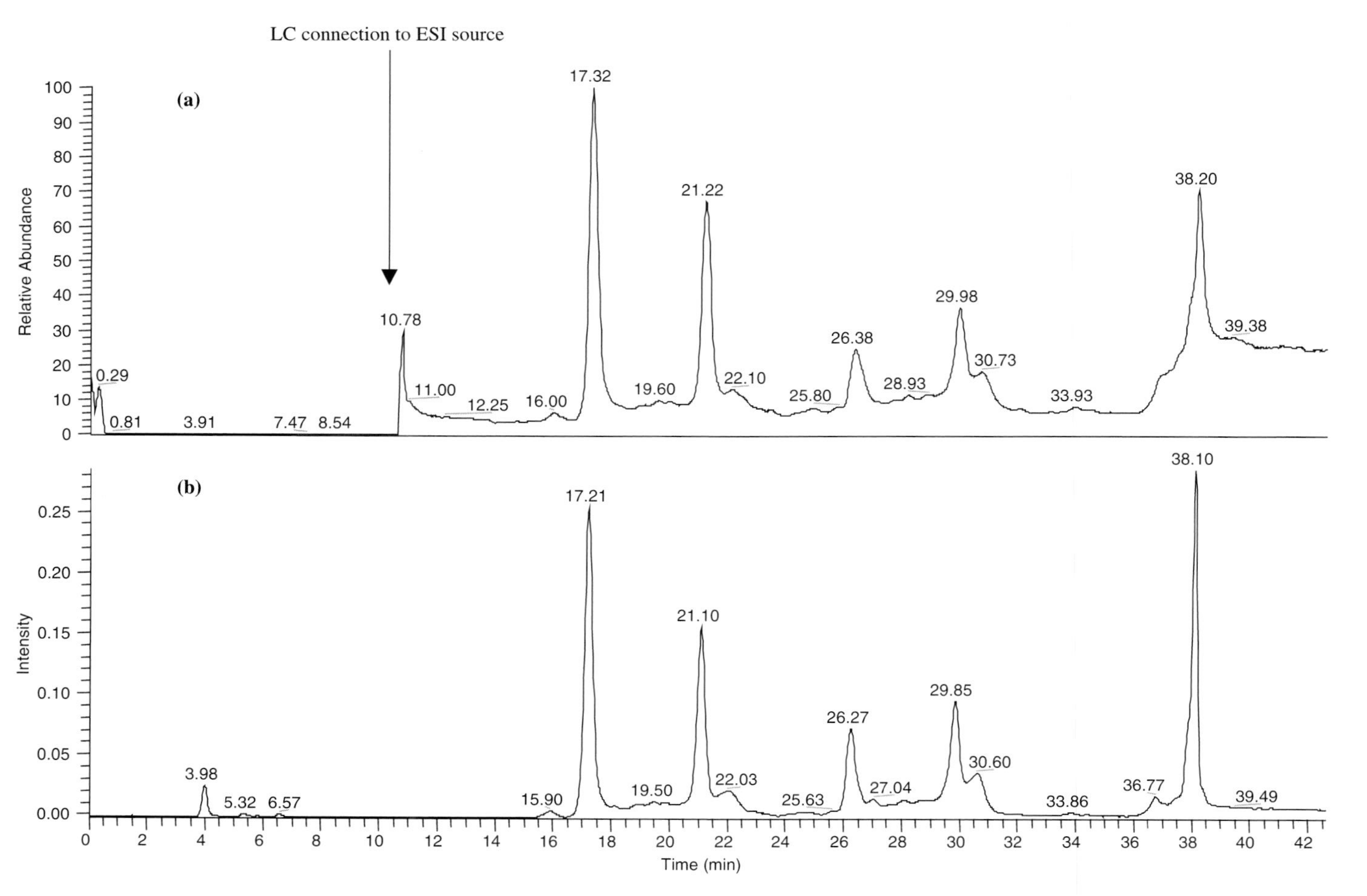

Fig. 9.10. LC–ESI/MS chromatograms obtained with mobile phase containing 0.1 M ammonium acetate with delayed LC connection to ESI source: (a) TIC chromatogram and (b) UV trace.

TABLE 9.1

CONCENTRATIONS OF ACIDIC AND BASIC DRUGS USED AS TEST ANALYTES FOR SUPPRESSION/ENHANCEMENT EVALUATION

Compound	$[M-H]^-$	[ ] (ng/μL)	Compound	$[M+H]^+$	[ ] (ng/μL)
Fumaric acid	115.0	1.0	Propranolol	260.2	1.0
Malic acid	133.1	1.0	Trimethoprim	291.3	1.0
Etidronic acid	205.2	8.0	Pipenzolate	354.4	1.0
Clodronic acid	243.2	1.0	Resperidone	411.4	0.5
Niflumic acid	281.4	6.0	Terfenadine	472.6	1.0
Canrenoic acid	357.6	6.0	Methoxyverapamil	485.6	1.0
Cholic acid	407.7	10.0	Benextracmin	591.6	7.0
Raffinose	503.4	10.0	Reserpine	609.6	5.0

Source: From Ref [12]. A study of ion suppression effects in electrospray ionization from mobile phase additives and solid-phase extracts. Copyright 2003 John Wiley. Reproduced with permission.

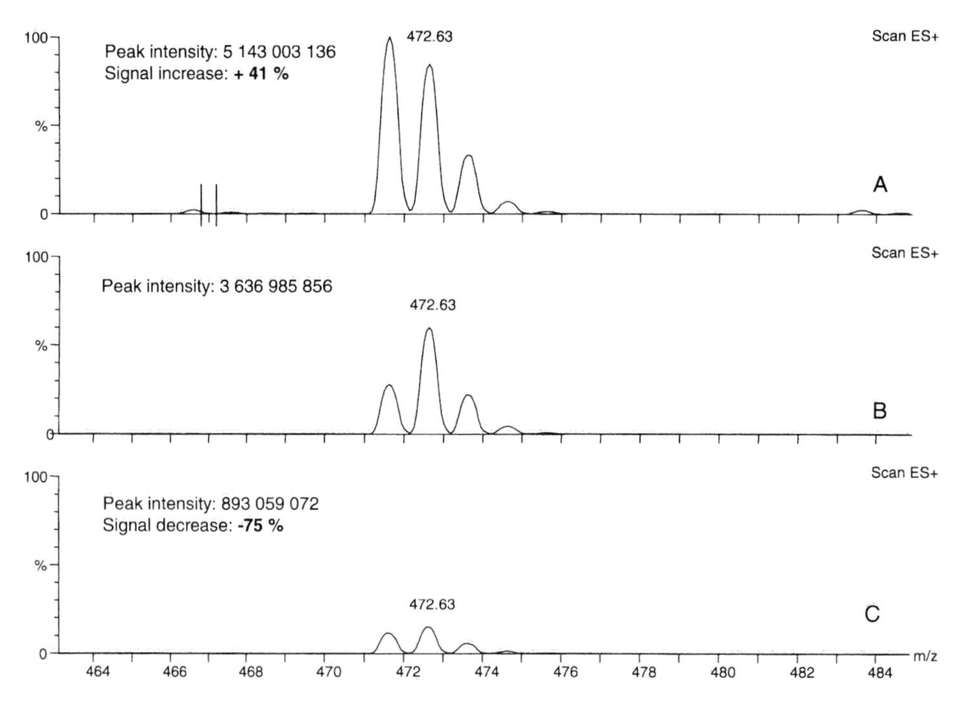

Fig. 9.11. Examples of ion suppression and enhancement on terfenadine. (A) Terfenadine mixed with 50/50 $MeOH/H_2O$ + 0.5% NH_4OH; (B) terfenadine mixed with 50/50 $MeOH/H_2O$ + no additive and (C) terfenadine mixed with 50/50 $MeOH/H_2O$ + 0.5% TFA. ("A study of ion suppression effects in ESI from mobile phase additives and solid-phase extracts" by C.R. Mallet *et al.* Copyright 2003 John Wiley. Reproduced with permission).

TABLE 9.2

SUPPRESSION AND ENHANCEMENT EFFECTS FOR THE pH ADDITIVES

	0.05%	0.10%	0.50%	1.00%	0.05%	0.10%	0.50%	1.00%
	Formic acid				Ammonium hydroxide			
Positive test solution								
Propranolol	36.5	28.8	4.5	−8.3	−2.2	2.02	10.2	11.4
Trimethoprim	41.7	30.1	−5.3	−17.5	−5.4	−5.4	4.2	8.9
Pipenzolate	−0.1	−0.2	−5.5	−9.5	0.02	0.02	0.02	0.01
Resperidone	−27.5	−37.1	−54.2	−59.4	6.1	9.6	16.1	16.8
Terfenadine	17.3	11.6	−7.9	−16.5	10.8	21.3	57.9	66.6
Methoxyverapamil	22.8	17.1	−1.8	−10.7	38.8	41.1	46.6	49.1
Benextramine	−39.77	−44.1	−52.7	−52.8	22.1	30.7	37.9	38.3
Reserpine	21.4	21.4	17.2	8.9	−12.1	−11.9	−6.2	−3.2
Negative test solution								
Fumaric acid	−11.9	−29.5	−64.7	−68.1	−38.4	−41.1	−45.8	−57.8
Malic acid	−11.2	−27.9	−62.2	−63.9	−35.5	−38.8	−42.4	−53.4
Etidronic acid	29.8	17.8	−17.2	−30.9	−61.9	−63.5	−75.9	−70.3
Clodronic acid	5.7	−15.7	−58.3	−66.6	0.3	1.3	−5.3	−27.7
Niflumic acid	−0.28	−21.4	−60.9	−64.5	14.1	11.1	5.3	−11.6
Canrenoic acid	13.8	−11.1	−51.6	−57.6	196.1	202.5	201.9	127.3
Cholic acid	31.9	3.7	−40.8	−44.7	420.5	454.9	403.1	352.8
Raffinose	−4.6	−26.3	−39.4	−43.7	60.9	61.9	66.6	32.1
	Trifluoroacetic acid				Acetic acid			
Positive test solution								
Propranolol	−54.8	−62.8	−74.7	−77.1	25.5	25.8	17.3	−0.2
Trimethoprim	−40.1	−58.1	−73.9	−76.6	18.3	10.4	−0.4	−7.1
Pipenzolate	−27.5	−37.4	−43.9	−43.7	0.01	−0.01	−0.4	−1.7
Resperidone	−53.7	−62.3	−68.2	−69.3	−2.1	−16.8	−37.7	−44.2
Terfenadine	−24.4	−44.6	−61.5	−64.8	15.9	11.9	7.5	−2.8
Methoxyverapalmil	−59.9	−57.3	−70.2	−72.6	19.5	16.6	8.9	−4.8
Benextramine	−29.4	−41.8	−42.7	−38.7	−21.9	−28.9	−29.9	−27.8
Reserpine	−32.5	−52.8	−71.7	−75.7	19.3	15.6	12.4	11.1
Negative test solution								
Fumaric acid	−87.4	−89.7	−91.1	−91.2	−15.1	−29.1	−51.3	−59.5
Malic acid	−84.1	−86.9	−88.4	−88.1	−14.5	−27.6	−48.3	−58.04
Etidronic acid	−71.9	−73.1	−71.6	−65.9	29.2	26.8	10.1	−17.8
Clodronic acid	−95.6	−97.4	−98.8	−98.8	4.9	−4.9	−36.5	−49.6
Niflumic acid	−91.7	−94.8	−98.2	−98.1	38.5	20.6	−22.1	−34.7
Canrenoic acid	−93.8	−96.1	−96.8	−96.1	−16.7	−33.8	−67.4	−59.4
Cholic acid	−95.2	−97.5	−99.5	−99.6	−18.9	−33.1	−48.5	−63.9
Raffinose	−84.1	−91.2	−96.5	−97.6	−5.7	−19.4	−26.3	−63.6

Source: From Ref. [12]. A study of ion suppression effects in electrospray ionization from mobile phase additives and solid-phase extracts. Copyright 2003 John Wiley. Reproduced with permission.

acidic drugs in the presence of different amounts of formic acid, ammonium hydroxide, TFA and acetic acid are reported. As it can be seen at higher acid concentration the ESI response is drastically reduced and this suppression effect depends on the analyte. The well known strong suppression effect of TFA is well demonstrated in both positive and negative modes for all the compounds. This behaviour is explained by the neutralization of charge in positive ion mode. In fact in the presence of stronger acids such as TFA, perfluoroacids and hydrochloric acid the formation of ion pairs within the electrosprayed droplet is a favourite reaction:

$$[M+H]^+ + [A]^- \rightarrow [M+H+A]$$

This leads to charge neutralization and consequently to reduced ion signal. However, TFA is of crucial importance in liquid chromatography: its use in the mobile phase improves peak shape and chromatographic resolution. The post-column addition of a weaker, less volatile acid (i.e. propionic acid) [13] can displace the stronger, more volatile acid based on their different volatility. During ion evaporation, most of the stronger acid is removed from the droplet and the equilibrium shown below favours dissociation of the ion pair with the consequent release of the required $[M+H]^+$ ions.

REFERENCES

1 B.K. Matuszewcky, M.L. Constanzer and C.M. Chavez-Eng, *Anal. Chem.*, 70 (1998) 882–889.
2 M. Jemal and Y.-Q. Xia, *Rapid Commun. Mass Spectrom.*, 13 (1999) 97–106.
3 R. Bonfiglio, R.C. King, T.V. Olah and K. Merkle, *Rapid Commun. Mass Spectrom.*, 13 (1999) 1175–1185.
4 R. King, R. Bonfiglio, C. Fernandez-Metzler, C. Miler-Stain and T. Olah, *J. Am. Soc. Mass Spectrom.*, 11 (2000) 942–950.
5 D.T. Rossi and M.W. Sinz (Eds.), *Mass Spectrometry in Drug Discovery*, Marcel Dekker, New York, 2002.
6 C. Chi, L. Liang, P. Padovani and S. Unger, *J. Chromatography B.*, 783 (2003) 163–172.
7 G.C. Enke, *Anal. Chem.*, 69 (1997) 4885–4893.
8 (a) L. Tang and P. Kebarle, *Anal. Chem.*, 63 (1991) 2709–2715.
(b) P. Kebarle and L. Tang, *Anal Chem.*, 65 (1993) 972A–986A.
9 W.Z. Shou and W. Naidong, *Rapid Commun. Mass Spectrom.*, 17 (2003) 589–597.
10 X. Xu, E.H. Mei, S. Wang, Q. Zhou, G. Wang, L. Broske, A. Pena and W.A. Korfmaker, *Rapid Commun. Mass Spectrom.*, 19 (2005) 2643–2650.
11 H. Mei, Y. Hsieh, C. Nardo, X. Xu, S. Wang, K. Ng and W.A. Korfmaker, *Rapid Commun. Mass Spectrom.*, 17 (2003) 97–103.
12 C.R. Mallet, Z. Lu and J.R. Mazzeo, *Rapid Commun. Mass Spectrom.*, 18 (2004) 49–58.
13 R.D. Voyksner, In R.B. Cole (Ed.), *Electrospray Ionisation Mass Spectrometry*, John Wiley, 1997, p. 337.

Achille Cappiello (Editor)
Advances in LC–MS Instrumentation
Journal of Chromatography Library, Vol. 72

Chapter 10

Differential mobility spectrometry (FAIMS): a powerful tool for rapid gas phase ion separation and detection

RAANAN A. MILLER, ERKINJON G. NAZAROV and DAREN LEVIN

10.1 INTRODUCTION AND HISTORY

Differential mobility spectrometry (DMS), often referred to as high-field asymmetric waveform ion mobility spectrometry (FAIMS) is a rapidly advancing technology for gas phase ion separation. It is a technique that holds promise to dramatically improve or rival a number of mainstream ion analysis technologies. While the DMS can be used as a stand-alone analyzer, when combined with other analytical instruments such as mass spectrometry (MS) or gas chromatography (GC) it offers significant enhancement to the quality of the overall analysis. This is because DMS separation is rapid and substantially orthogonal to the separations provided by these other analyzers. In DMS ion separation is based on changes in ion mobility, relating to how the molecule structure and clustering levels change, in response to applied electric fields.

Over the years, DMS has been called by many names. These include ion drift non-linearity spectrometry [1,2], field ion spectrometry (FIS) [3], and radio frequency ion mobility spectrometry [4]. In this chapter we refer to all of these as differential mobility spectrometry, since measurements of differential mobility are core to this method.

When coupled to the front end of a mass spectrometer, the DMS provides a means of enhancing mass spectrometer (MS) selectivity as well as improving sensitivity. The DMS can continuously separate ions based on ion conformation rather than simply on mass-to-charge ratio. In addition the ability to eliminate molecular and isobaric mass interference problems in quantitative applications is invaluable since metabolites can produce the same ion in the mass spectrometer source. Furthermore, DMS has the potential to eliminate the need for doing liquid chromatography and thus dramatically increasing analytical throughput [5].

The enabling aspect of the differential mobility technique is that ions have field-dependent mobilities when exposed to electric field strengths above approximately 10,000 volts/cm at atmospheric pressure (field to gas density $E/N = 40\,Td$). Mason and McDaniel [6] were among the first to describe this phenomenon in their book *The Mobility and Diffusion of Ions in Gases*, published in 1973. This field dependence is significant, as it provides information about a molecule that is fundamentally different from, and richer than, low field techniques such as time-of-flight ion mobility spectrometry [7]. In DMS, changes in the molecule cross section, alignment, and structure in response to these strong electric fields induce changes in the mobility of the ions. These changes are used as the basis for filtering and identifying the ion species.

The method to harness this differential mobility was first proposed in the Soviet Union in the 1980s by Groshkov [9]. Buryakov and his colleagues were the first to develop a practical implementation of this technique and they presented their findings via journal publications in the early 1990s [1,2]. Since its introduction, DMS technology has been adopted by various researchers resulting in multiple instrumentation designs. Commercialization of the method began in the mid-1990s by Mine Safety Appliances Company (MSA), who investigated the technology for portable field detection of explosives and illicit drugs, and as a front end to MS [3,8]. Sionex Corporation was founded in May 2001 to commercialize work done at The Charles Stark Draper Laboratory, and at New Mexico State University, to develop micro-fabricated (MEMS) DMS sensors and filters using mass production, semiconductor and related fabrication technologies (see Figs. 10.1 and 10.2) [4,7]. At approximately the same time Ionalytics Inc. was also founded to commercialize a cylindrical DMS configuration that was being developed at the National Research Council of Canada (NRC) [10,11]. In August of 2005, Ionalytics was sold to Thermo Electron Corporation further validating the commercial importance of this technology.

10.2 PRINCIPLE OF OPERATION OVERVIEW

Operationally, the differential mobility spectrometer functions as a tunable, or programmable, ion filter as illustrated in Fig. 10.3. Ions permitted to pass through the filter without being neutralized are selected based on voltages, or electric fields, applied across the filter electrodes. These voltages, called compensation voltages (Vc), correspond to particular differential mobility values for the ions. From a mass spectrometric perspective, its operation can be likened to that of a quadruple mass filter, with the critically important caveat that in

Fig. 10.1. Sionex micro-fabricated DMS spectrometer chip, Sionex microDMx™, including fluidic channel next to a US dime for scale. Smaller embodiments of the device have also been realized.

the DMS, ion separation is based on mobility and not just mass. Ion mobility is a measure of how "easily" ion species move through an atmosphere in response to an applied electric field. A mobility coefficient, *K*, is usually described for each species. This is a proportionality coefficient relating the net velocity of the ions with the applied driving electric field [12]. The mobility coefficient is different for different ion species and depends on ion shape, charge, and mass. Additionally, it is also dependent on the environment through which the ions travel, the composition and density of the drift gas, as well as its pressure and temperature. In other words, ion mobility is not entirely an intrinsic property of the molecular ion (such as mass, size and shape), but is also dependent on the analytical environment of the ion. This environmental dependence of the ion has some advantages and disadvantages. This subject will be discussed in detail later in this chapter.

To analyze a sample in the DMS it is first ionized and then transported through the differential mobility ion filter as shown in Fig. 10.3. A combination of an asymmetric RF

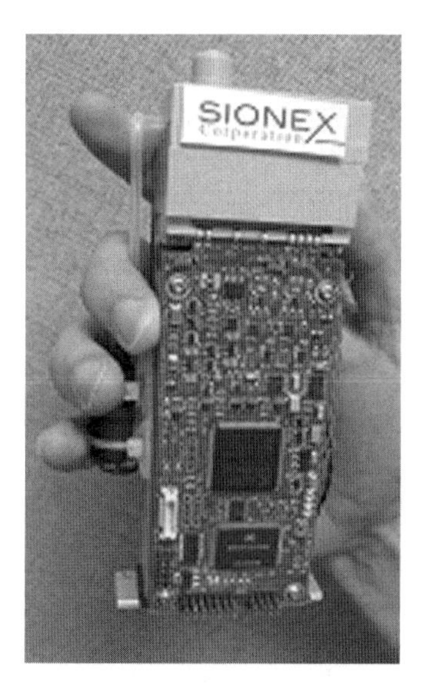

Fig. 10.2. Photograph of Sionex microDMx system including all drive electronics, transport gas conditioning and recirculation system, for a handheld system.

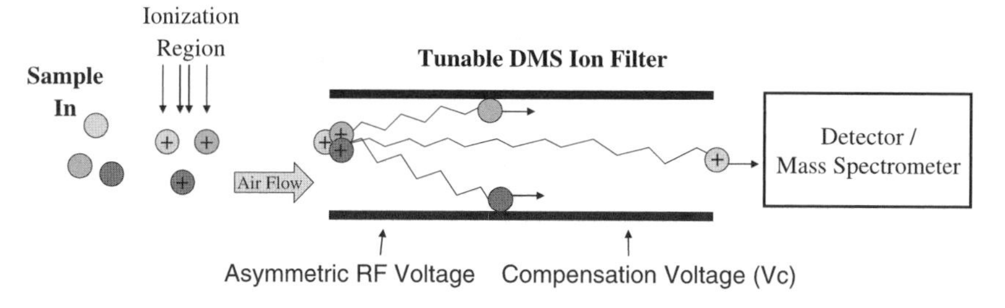

Fig. 10.3. Conceptual schematic of the DMS ion filter system.

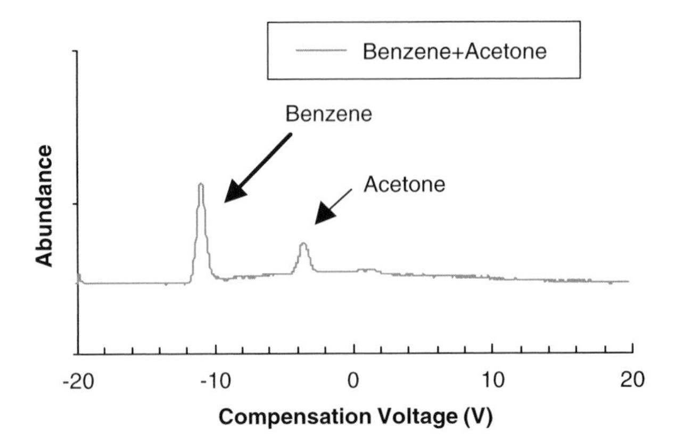

Fig. 10.4. DMS response for a mixture of acetone and benzene showing compensation voltage versus ion intensity or abundance measured at a fixed RF voltage.

and a DC compensation electric field are applied across the filter to select the ion species allowed to pass through the filter and reach a detector. The ions can be formed by a multitude of means ranging from electro or nano-spray ionization to matrix assisted laser desorption and ionization (MALDI), photoionization, to the use of radioactive ionization sources such as ^{63}Ni or ^{241}Am. A discussion of some of these ionization methods and their use in practical applications will also be covered in this chapter.

In some applications the detector is a series of faraday plate electrodes that measure the ion current while in others the detector can be a mass spectrometer. Figs. 10.4 and 10.5 illustrate typical DMS spectra obtained from scanning the applied voltages to the filter, and detecting the ions on a faraday plate detector [4,7]. In Fig. 10.4 the asymmetric RF voltage (V_{RF}) is kept constant and only the compensation voltage (Vc) applied to the filter is tuned. The result is a spectrum showing ion current intensities as a function of the compensation voltage required to pass particular ion species through the filter. In Fig. 10.5, both the RF and compensation voltages are in a predetermined manner adjusted providing a three-dimensional plot of DMS response. This three-dimensional representation, or a similar two-dimensional plot, is called a dispersion plot, and we will show how it is a powerful tool for analyzing DMS response.

10.3 HOW IT WORKS: FUNDAMENTALS OF THE ION SEPARATION METHOD

10.3.1 Field-dependent ion mobility

The DMS operation principle utilizes high-strength electric fields to separate ions in a regime where ion mobility is not constant but rather is dependent on the electric field strength at ambient gas density. The DMS principle is based on pioneering work by McDaniel and Mason [6], who found that the mobility of an ion is affected by the applied electric field strength at a constant gas density. Above an electric field to gas density ratio (*E/N*) of 40 Td ($E > 10{,}700$ V/cm at atmospheric pressure) the mobility coefficient *K(E)* has a non-linear

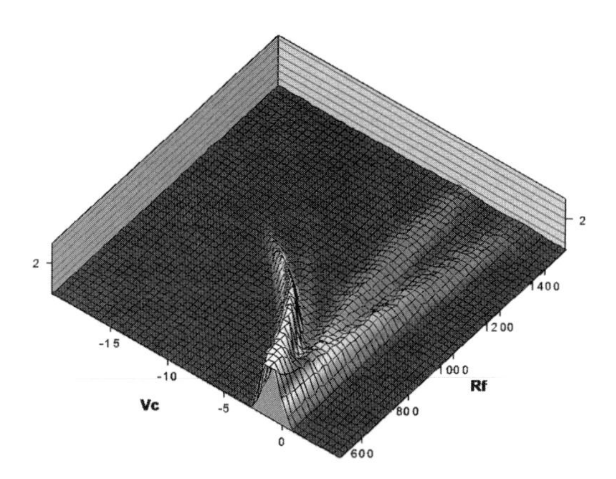

Fig. 10.5. 3-Dimensional DMS dispersion plot showing ion intensity measured at detector electrodes when both the RF (V_{RF}) and compensation (Vc) voltages are systematically scanned.

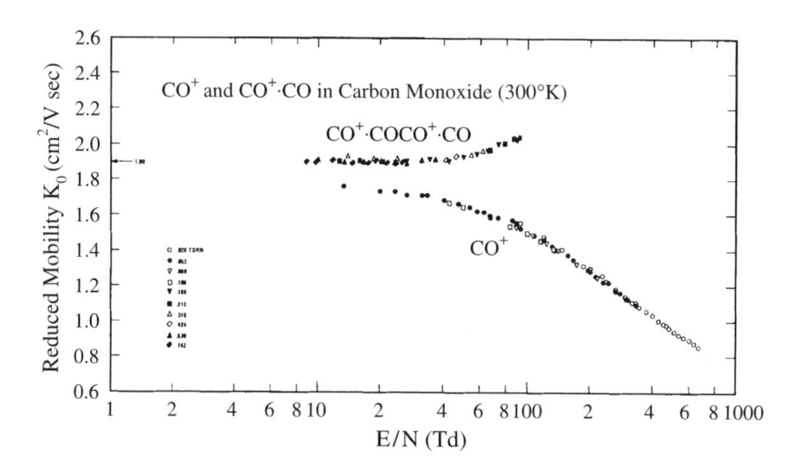

Fig. 10.6. Ion mobility as a function of electric field and gas density for carbon monoxide ions. Reprinted from Figure 7-1-K-1, E.W. McDaniel and Edward A. Mason, *The mobility and diffusion of ions in gases*, copyright 1973, with permission from John Wiley & Sons, Inc., 1973.

dependence on the field. This dependence is believed to be specific for each ion species. Below are some examples from McDaniel and Mason [6]. The mobility for the cluster ion CO^+CO increases with increasing field strength while that of CO^+ actually decreases with increasing field strength (Fig. 10.6). For some molecular and atomic ions the coefficient of mobility can change in a more complex way. For example, for atomic ions K^+, the mobility coefficient in carbon monoxide gas increases with increasing field by as much as 20%, but above $E/N \sim 200$ Td the coefficient starts to decrease (Fig. 10.7). For some other ions, for example N^+, N_3^+, and N_4^+, the mobility changes very little [6,13].

Fig. 10.8 illustrates schematically three possible ion mobility dependencies on electric field and gas density for three different ion species (A, B, and C). For simplicity we

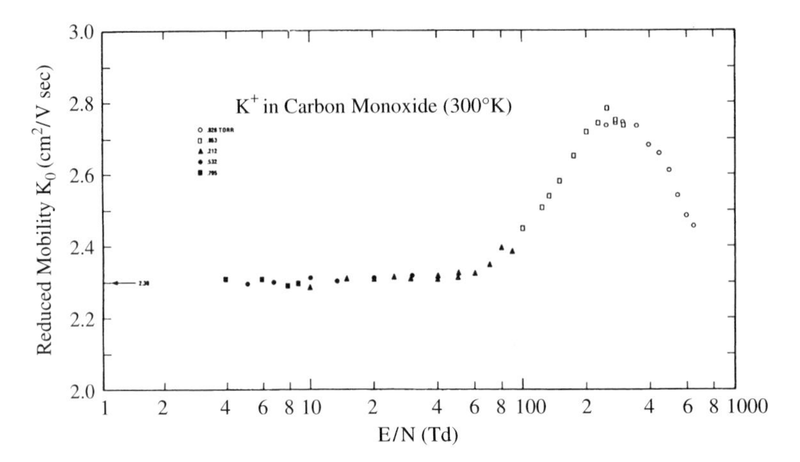

Fig. 10.7. Field-dependent mobility for potassium ion as a function of electric field and gas density. Reprinted from Figure 7-1-K-3, E.W. McDaniel and Edward A. Mason, *The mobility and diffusion of ions in gases*, copyright 1973, with permission from John Wiley & Sons, Inc., 1973.

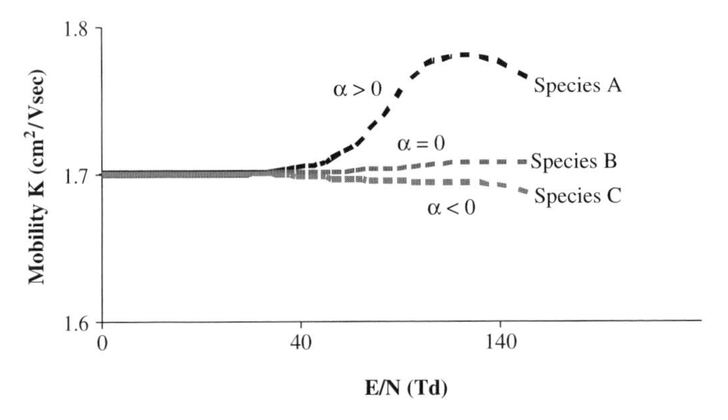

Fig. 10.8. Mobility dependence on electric field for three hypothetical ion species for the same gas density.

assume that at low *E/N* values (*E/N* < 40 Td) the mobility is the same for all three ion species. However, at *E/N* values greater than 40 Td the mobility coefficients for the different ion species become distinct from one another and the plots of mobility as a function of *E/N* have become unique for the different ion species. These unique plots can be used to fingerprint the ion species and enhance ion species identification.

The field dependence of the mobility coefficient *K(E/N)*, shown in Fig. 10.8, can be represented by a series expansion of even powers of *E/N* [14]

$$K(E/N) = K(0)\,[1 + \alpha_1 (E/N)^2 + \alpha_2 (E/N)^4 + \ldots\ \ldots] \qquad (10.1)$$

where $K(0)$ is the coefficient of mobility of the ion in a weak electric field (E < 1000 V/cm at atmospheric pressure, *E/N* < 40 Td), and α_1, α_2 are coefficients of the expansion and relate the dependence of K to the electric field. This equation can be simplified by using an effective $\alpha(E)$ as shown in Eqn. 10.2 [13].

$$K(E/N) \approx K(0)\,[1+\alpha\,(E/N)] \tag{10.2}$$

$\alpha(E)$ is used in place of $\alpha(E/N)$ throughout this text for convenience, when the gas density is held constant at a particular operating pressure, but the applied electric field strengths are varied. According to this expression when $\alpha(E) > 0$ the mobility coefficient $K(E)$ increases with field strength, when $\alpha(E) \sim 0$ the mobility $K(E)$ does not change, and when $\alpha(E) < 0$ then $K(E)$ decreases with increasing field strength. The physical meaning of the alpha parameter becomes clear after rearranging Eqn. 10.1 into the form of Eqn. 10.3.

$$\alpha(E) = \frac{K(E) - K(0)}{K(0)} = \frac{\Delta K(E)}{K(0)} \tag{10.3}$$

where the alpha parameter shows the relative change of the coefficient of mobility with applied electric field strength. The sign of the alpha parameter depends on whether mobility increases with increasing electric field (positive alpha) or decreases with increasing electric field (negative alpha). In general, the alpha parameter also inversely depends on the ambient gas density if this is allowed to change. Further details on this effect can be found elsewhere [16].

10.3.2 Mechanisms for differential mobility

Until now we simply stated that the mobility of an ion changes with electric field at high field strengths, and we have defined a function called the alpha parameter which describes how the mobility changes with field strength. In the following sections, we will discuss some of the mechanisms that produce the changing mobility with field strength.

To better understand these mechanisms it is easier to start with a more physical expression for the field-dependent mobility coefficient provided by the Mason–Schamp equation, Eqn. 10.4. This equation can be derived from momentum and energy balance considerations when mean ion-neutral collision energy is written as $\varepsilon = (3/2)\kappa T_{\mathrm{eff}}$ [13–15].

$$K(E/N) = \frac{V}{E} = \frac{q}{N}\left(\frac{1}{3\mu\kappa T_{\mathrm{eff}}}\right)^{1/2} \frac{1}{\Omega(T_{\mathrm{eff}})} \tag{10.4}$$

Eqn. 10.4 can then be used to explain how the alpha parameter can decrease with increasing electric field $\alpha(E) < 0$. In this case one must assume that the value of the ion neutral cross section $\Omega(T_{\mathrm{eff}})$ does not change significantly for rigid-sphere interactions [13,14] and the reduced mass μ is constant. Under these conditions one finds that the mobility $K(E)$ will decrease if the effective temperature, or energy, of the ion increases. Physically this effect has a simple explanation. When the electric field strength is increased the ions are driven harder through the neutral gas. This increases the ion neutral collision frequency, which leads to an increased T_{eff} and a reduced ion mobility coefficient (Eqn. 10.4).

The rigid-sphere model, however, does not explain the experimental results which show that with many ions the mobility increases with increasing electric field $\alpha(E) > 0$. One of the possible explanations for the increased mobility at elevated values of *E/N* is offered when one considers that ions can de-cluster at high-strength electric fields. Ions in ambient conditions in a weak electric field generally do not exist in a free state. They are usually associated with small neutral molecules such as water to form loosely bound clusters for example, $MH^+(H_2O)_n$ with n polar molecules such as water attached. As the electric field strength is increased the kinetic energy and consequently the effective temperature (T_{eff}) of the ion increases due to the energy imparted by frequent collisions. This can lead to a reduction in the level of ion clustering (reduction in n) resulting in a smaller ion cross section, $\Omega(T_{eff})$, and a smaller reduced mass, μ, for the ion. Then, according to Eqn. 10.4, if due to de-clustering the cross section and reduced mass decrease in a sufficient manner to overcome the increase in T_{eff}, the case where $\alpha(E) > 0$ can be explained.

The third case when $\alpha(E) \sim 0$ can be explained by a decrease in ion cross section due to de-clustering which is offset by an increase in the effective temperature of the ion. The result is no net change to the mobility coefficient of the ion, and is predicted for ion-induced dipole interactions.

While the model presented above provides a first-order understanding of the mobility dependences on electric field strength, the reality is more complex. Listed below are a number of additional mechanisms which can have an effect on mobility.

1. *Elastic scattering due to polarization interaction.* This mechanism plays a significant role at low electric field conditions in non-polar gases. When ions move slowly in an initially non-polarized gas they can induce polarization of nearby drift gas molecules and consequently attract them. Even longer range interactions occur in gases containing trace amounts of polar molecules, for example drift gasses with intentionally added dopants, or ubiquitous water molecules. With increasing ion velocity the range of the polarization forces decrease, resulting in a decreasing effective ion cross section. For the polarization interactions, the effective cross section $\Omega_{pol} \sim 1/\varepsilon_i$ of the ion is inversely proportional to the velocity or energy of the ions. So, scattering due to the polarization effect can explain an increase of the mobility coefficient with increasing electric field *E*, through changes in the effective cross section from a large polarization cross section to a smaller rigid-sphere value $\Omega_{pol} > \Omega$.
2. *Resonant charge transfer.* When ions move in a gas which contains neutral molecules with a similar structure to the ions themselves, the neutral molecule and ion may exchange charges. The previously ionized molecule slows while the neutral molecule that just became charged experiences a force and reaches a characteristic velocity. This electron jumping process has a resonant nature and results in an effective ion charge transfer cross section that is often larger than the elastic scattering cross section. This effect results in a decrease in ion mobility with increasing electric field strength.
3. *Changing cross section in a strong RF electric field.* This effect plays a significant role for big molecules. The strength of the asymmetric RF electric field varies from 1000 V/cm up to 30,000 V/cm, inducing changes in the dipole moments of ion complexes or re-alignment of these dipoles with the field. As a consequence the effective cross sections and the resultant mobility depends on the applied electric field.

10.3.3 Differential mobility method for ion separation

In order to better understand the ion separation principle of the DMS we use a simplified model as described in [7]. Consider three kinds of ions under the same constant gas density operating conditions but different mobility coefficient dependencies on electric field i.e. $\alpha(E) > 0$, $\alpha(E) < 0$, $\alpha(E) \sim 0$. All three ion species have the same zero-field mobility, $K(0)$, which are formed due to local ionization of neutral molecules at the same location in a narrow gap between two electrodes as shown in Fig. 10.3. A stream of carrier gas transports these ions longitudinally down the drift tube between the gap. If an asymmetric RF electric field is then applied to the electrodes the ions will oscillate in a perpendicular direction to the carrier gas flow, in response to the RF electric field, while moving down the drift tube with the carrier gas. A simplified asymmetric RF electric field waveform (Fig. 10.9a) with maximum field strength $|E_{max}| > 10{,}000$ V/cm and minimum field strength $|E_{min}| << |E_{max}|$ is used here to illustrate the operation principle of the DMS. The asymmetric RF waveform is designed such that the time average electric field is zero and

$$|E_{max}|t_1 = |E_{min}|t_2 \tag{10.5}$$

where t_1 is the portion of the period where the high field is applied and t_2 is the time the low field is applied, and β is a constant corresponding to the area under the curve in the high field and low field portions of the period. The ion velocities in the y direction are given by

$$V_y = K(E)E(t) \tag{10.6}$$

Here K is the coefficient of ion mobility for the ion species and E is the electric field intensity, in this case entirely in the y direction. If the amplitude of the positive polarity RF voltage pulse (during t_1) produces an electric field of strength greater than 10,000 V/cm then the velocity toward the top electrode

$$V_{up} = K_{up}|E_{max}| \tag{10.7}$$

will differ for each of the ion species (Fig. 10.9b) since, as shown in Fig. 10.8, the coefficient of mobility K_{up} for each ion at the high field condition is different. The ions with $\alpha(E) > 0$ will move faster and ions with $\alpha(E) < 0$ will have the smallest velocity, therefore, the slope of each ion trajectory will also differ. In the next portion of the period (t_2), once the polarity of the RF field has switched to the low field part, all three ion types will begin moving with the same velocity

$$V_{down} = K(E_{min})|E_{min}| \tag{10.8}$$

down toward the bottom plate. In this low field strength condition (see Fig. 10. 8) all three ion types will have the same mobility coefficient K_{down}. Therefore, all three ion trajectories will have the same slope in this portion of the period (Fig. 10.9).

The ion displacement from its initial position in the y direction is the ion velocity in the y direction, V_y, multiplied by the length of time, Δt, the field is applied

$$\Delta y = V_y \Delta t \tag{10.9}$$

In one period of the applied RF field the ion moves in both the positive and negative y directions. By substituting Eqns. 10.6, 10.7, and 10.8, into Eqn. 10.9 the average displacement of the ion over one period of the RF field can be written as

$$\Delta y_{\mathrm{RF}} = K_{\mathrm{up}}\left|E_{\max}\right|t_1 - K_{\mathrm{down}}\left|E_{\min}\right|t_2 \tag{10.10}$$

Using Eqn. 10.5 this expression can be re-written as

$$\Delta y_{\mathrm{RF}} = \beta(K_{\mathrm{up}} - K_{\mathrm{down}}) = \beta\Delta K \tag{10.11}$$

Since β is a constant determined by the applied RF field, the y-displacement of the ion per period of the RF field $T = t_1 + t_2$ depends on the change in mobility of the ion between its high and low field conditions. Assuming the carrier gas only transports the ion in the z direction, the total ion displacement Y (in the y direction) from its initial position (due to the electric field) during the ions residence time t_{res} between the ion filter plates can be expressed as

$$Y = \frac{\Delta y_{\mathrm{RF}}}{(t_1 + t_2)} t_{\mathrm{res}} = \frac{\beta\Delta K}{T} t_{\mathrm{res}} \tag{10.12}$$

The average ion residence time inside the ion filter region is given in Eqn. 10.13, where A is the cross-sectional area of the filter region, L the length of the ion filter electrodes, V the volume of the ion filter region ($V = AL$), and Q the volume flow rate of the carrier gas.

$$t_{\mathrm{res}} = \frac{AL}{Q} = \frac{V}{Q} \tag{10.13}$$

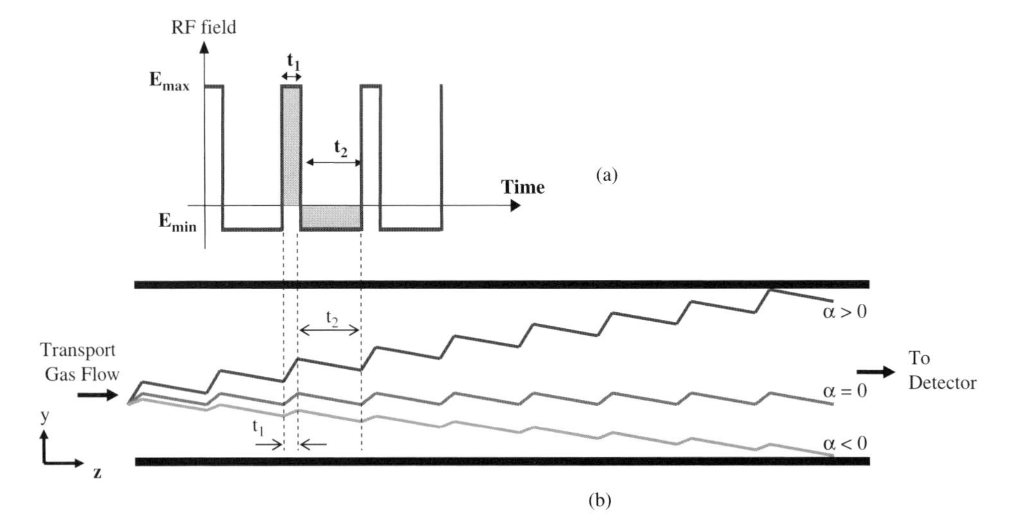

Fig. 10.9. (a) Simplified asymmetric RF electric field waveform used for ion filtering. (b) Trajectories of ions in the gap between the upper and lower parallel plate electrodes of the ion filter, under the combined influence of the carrier gas flow and an asymmetric RF electric field waveform. Reprinted from Miller *et al.* [7], with permission from Elsevier.

Substituting Eqn. 10.13 into Eqn. 10.12, noting from Eqn. 10.1 that $\beta = |E_{max}|\ t_1$ and defining the duty cycle of the RF pulses as $D = t_1/T$, the equation for displacement of the ion species, Eqn. 10.12, can be re-written as

$$Y = \frac{\Delta K E_{max} V D}{Q} \tag{10.14}$$

where Y is now the total displacement of the ion in the y direction based on the average ion residence time in the ion filter region. From Eqn. 10.14 it is evident that the vertical displacement of the ions in the gap are proportional to the difference in coefficient of mobility between the low and high field strength conditions. Different species of ions with different ΔK values will displace to different values of Y for a given t_{res}. All the other parameters including the value of the maximum electric field, the volume of the ion filter region, the duty cycle and the flow rate, are the same for all ion species.

When a low strength DC field, E_c, where ($|E_c| < |E_{min}| \ll |E_{max}|$) is applied in addition to the RF field, in a direction opposite to the average RF-induced (y-directed) motion of the ion, the trajectory of a particular ion species can be "straightened" (see Fig. 10.10). This allows the ions of a particular species to pass unhindered between the ion filter electrodes while ions of all other species are deflected into the filter electrodes and subsequently neutralized. The DC voltage that "tunes" the filter and produces a field which compensates for the RF-induced motion is characteristic of the ion species and is called the compensation voltage (Vc). The compensation voltage is usually said to increase with value when it is shifted away from a zero volts value. A complete spectrum for the ions in the gas sample can be obtained by ramping or sweeping the DC compensation voltage applied to the filter. The ion current versus the value of the sweeping voltage forms the DMS spectra. If instead of sweeping the voltage applied to one of the ion filter electrodes, a fixed DC voltage (compensation voltage) is applied, the spectrometer will work as a continuous ion filter allowing through only ions of a particular differential mobility.

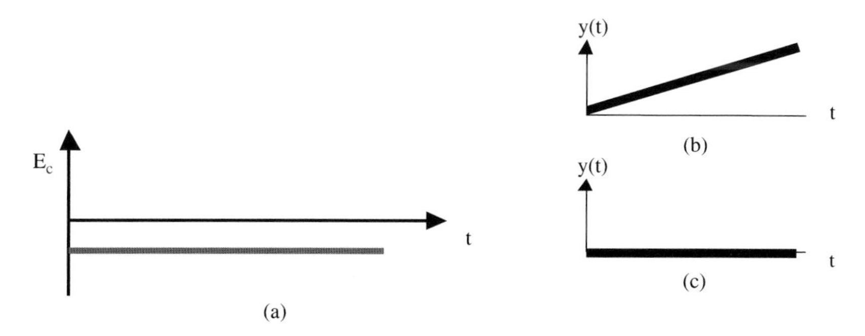

Fig. 10.10. (a) Ion trajectory "correction" through the application of a dc compensation voltage to the ion filter electrodes. The compensation voltage applied to cancel out displacement produced by the asymmetric RF field. (b) Average trajectory of ion from starting position at entrance to filter with only the RF field applied. (c) Trajectory of ion with both RF and compensation field applied. Reprinted from Miller *et al.* [7], with permission from Elsevier.

10.3.4 Ways to improve or control ion separation

10.3.4.1 Transport gas modifiers (doping)

The addition of gas modifiers to the ionization region of time-of-flight ion mobility spectrometers (TOF-IMS), generally trace levels of gases or vapors, often referred to as dopants, has been common practice for several decades. The dopants are used to adjust the ion formation chemistry in the TOF-IMS in order to enhance ionization efficiency and to suppress interferences [16]. Generally, ambient gas conditions in the ion separation of the TOF-IMS are controlled much more tightly than in the ionization region, since the mobility coefficient of an ion is a function not only of its intrinsic properties, but is very sensitive to the bath gas through which it travels [16]. This control is required to provide spectral stability and the desired ion chemistry. However, the dependence of the ion on its environment can also be used advantageously. When controlled amounts of dopant are added to the drift region (transport gas stream) ion separation and spectrometer resolution can be regulated. For time-of-flight ion mobility spectrometry, the benefit of adding dopants to affect resolution is relatively small. At higher dopant concentrations, resolution actually degrades due to ion cluster formation, since at high dopant levels the ion clusters have very similar effective cross sections. In contrast, the addition of a dopant in DMS has a significant affect on instrument resolution. The dopant concentration and transport gas composition can be used almost like a "knob" which can be adjusted to control DMS ion separation. Recently, gas modifiers, at percent down to trace levels, have been added to the DMS transport gas to enhance analytical separation. These modifiers change the differential ion mobility by the mechanisms discussed in Section 10.3.2, and act to shift the peak positions, i.e. compensation voltages, of given compounds in a way that enhances peak separation between analytes. Figs. 10.11 and 10.12 illustrate the dramatic effect the addition of a controlled amount of methylene chloride dopant (1000 parts

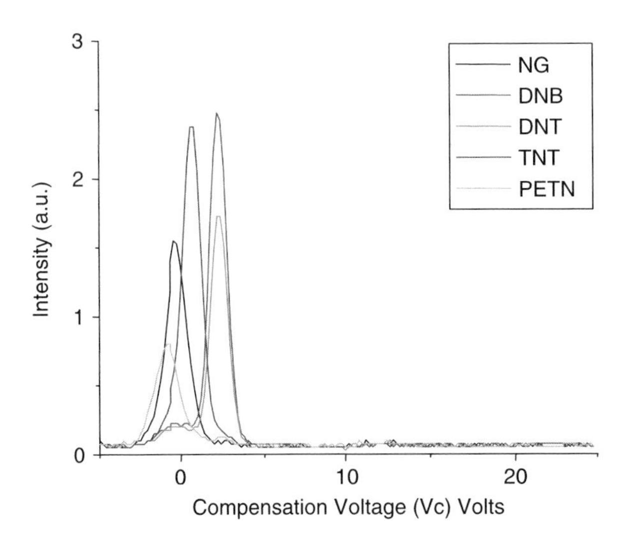

Fig. 10.11. DMS response for five explosives compounds with no dopant added.

per million) can have on DMS spectra. As shown in Fig. 10.12, the addition of the dopant spreads out the spectral peaks of the explosives compounds (TNT, DNT, PETN, etc.) resulting in enhanced analytical separation and improved compound detection.

The ability to affect peak separation, or influence the differential mobility of ions, was recognized in the early works of Buryakov *et al.*, he observed that the differential mobility could be modified based on the concentration of moisture and other organic compounds added to the drift gas [2,48]. Krylova *et al.* further studied the effect of moisture on the field dependence of mobility and proposed models related to enhanced low field clustering and de-clustering as an explanation for the increased differential mobility with increased water concentration for organophosphate compounds [17]. Eiceman *et al.* further explored the effects of various dopants including acetone, water, and methylene chloride on DMS peak positions (compensation voltages) for explosives compounds (Fig. 10.13). They proposed a model and showed mass spectrometric validation of the clustering/de-clustering mechanism as being dominant for enhanced ion separation of the explosives [18].

Guevremont *et al.* investigated changing mixtures of carrier gases (N_2, O_2, CO_2, NO_2, and SF_6) in the DMS to affect compound separation. They showed the separation of three positional (ortho-, meta-, para) isomers of phthalic acids, using a mixture of 95% nitrogen and 5% CO_2 [19,20]. Recently Shvartsburg *et al.* described a model for the DMS separation of mixtures of gasses that uses physical chemical fundamentals that determine the high-field mobilities in heteromolecular gases [21]. In the applications section we will illustrate further how the addition of drift gas modifiers can be used advantageously to enhance DMS separation for mass spectrometric ion pre-separation.

10.3.4.2 Operating pressure effect on ion separation

Another means of affecting ion separation is through the control of DMS operating pressure. Ion mobility has an inverse relationship with gas density according to the

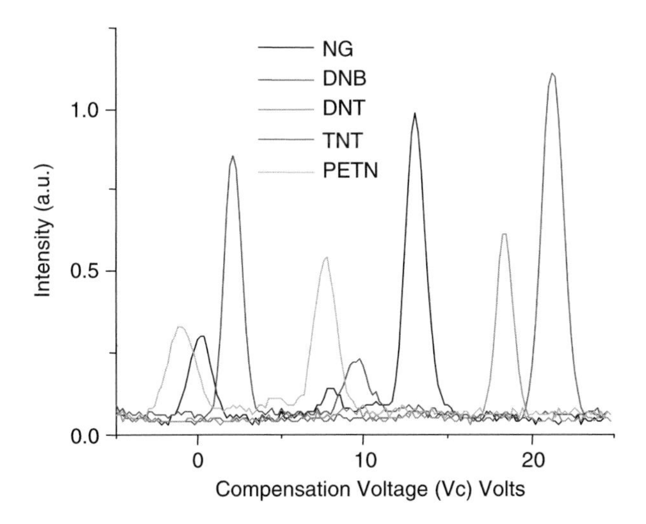

Fig. 10.12. DMS response for the same explosives with the addition of 1000 ppm methylene chloride.

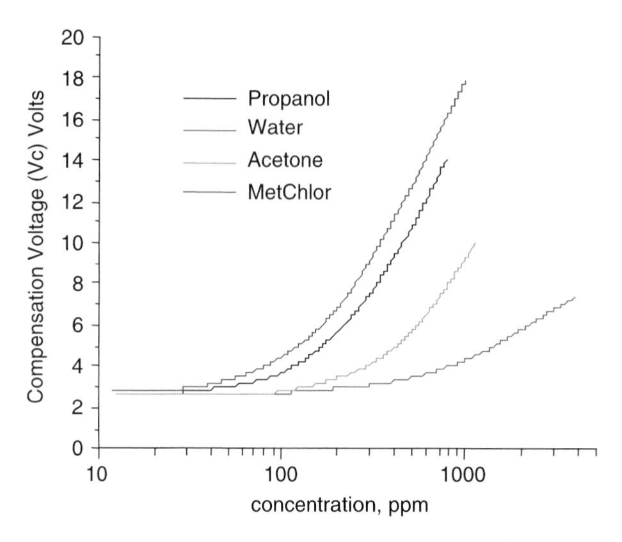

Fig. 10.13. DMS spectral peak position (Vc) as a function of solvent concentration for 50 ppb 2,6-DNT in purified air at 150°C and at a RF field strength of 20 kV/cm. Reprinted with permission from Eiceman *et al.* [18]. Copyright 2004 American Chemical Society.

Mason–Schamp equation (Eqn. 10.4, and Figs. 10.6 and 10.7). The mobility curves can be treated as either being a function of electric field for a constant transport gas density, or a function of gas density for a constant maximum RF electric field strength or voltage (V_{RF}) applied. This means that pressure, like RF electric field strength, can also be used as a "knob" to affect differential mobility and allow the optimization of DMS performance. Operation of a DMS under reduced pressure conditions can be especially useful for applications, when the DMS serves as an ion pre-filter for mass spectrometry, or as an interface device to enhance the efficiency of ion introduction from ambient pressure to vacuum (e.g., in atmospheric pressure MALDI). It is also useful to reduce power requirements for portable DMS systems, as the maximum RF voltage intensities can be reduced.

Fig. 10.14 shows three dispersion plots for dimethyl methylphosphonate (DMMP), at a concentration of about 4.6 parts per billion, at three different DMS sensor operating pressures. The dispersion plot (shown here in 2 dimensions) shows the ion intensity (density of traces) as a function of compensation voltage versus the maximum asymmetric RF voltage. The three traces evident in each of the Fig. 10.14a–c correspond to the reactant ion peak (RIP) the left-most trace, the monomer (MH^+) ion middle trace, and the dimer (M_2H^+) ion the right-most trace. The variation between Fig. 10.14a–c illustrates that pressure does have a significant effect on DMS spectra. Fig. 10.14b shows that at 0.76 atm the dimer and cluster ion level decreases, enhancing the monomer peak intensity relative to Fig. 10.14a. At reduced pressures the compensation voltage (Vc) reaches a maximum value (see crosses in Fig. 10.14) at significantly lower RF voltages (V_{RF}) compared with measurements at atmospheric pressure (at 1.55 atm, $V_{RF} = 1500$ V; at 0.76 atm, $V_{RF} = 1200$ V; at 0.42 atm, $V_{RF} = 1000$ Vfor the monomer ion).

Fig. 10.15 summarizes the effect of pressure on DMS spectral response. It shows the effect of pressure on DMS peak width (FWHM), area, resolving power and compensation voltage for a particular V_{RF} voltage ($V_{RF} = 1000$ V). This figure clearly shows that there is

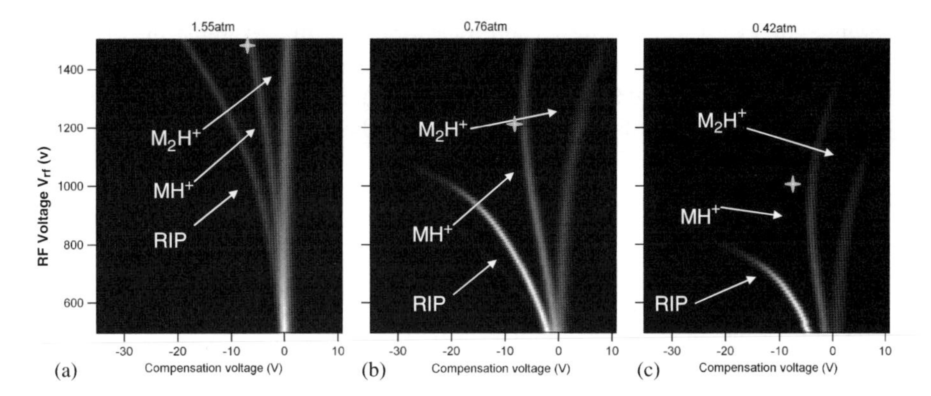

Fig. 10.14. Comparison of the traces for reactant ion peaks (RIP), monomer (MH^+), and dimer (M_2H^+) ions shows that pressure has a significant effect on DMS spectra. Traces for all ion species are different at different pressures.

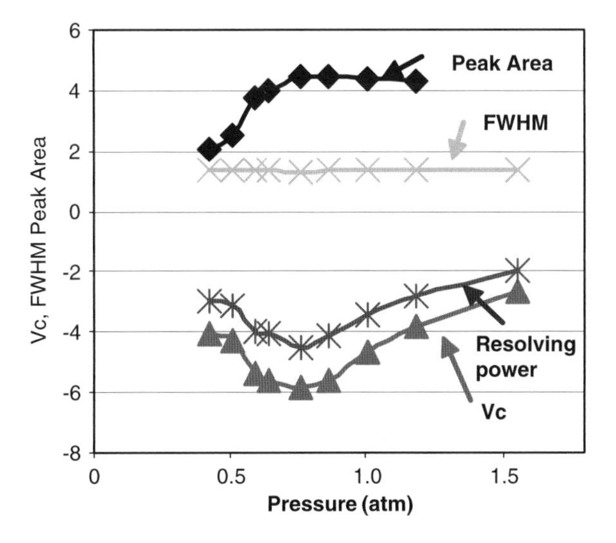

Fig. 10.15. Effect of pressure on the DMMP monomer peak at a V_{RF} of 1000 V.

an optimal DMS operating pressure to obtain the highest response and largest compensation voltage value. For $V_{RF} = 1000$ V an optimal pressure is between 0.6 and 0.8 atm.

From Fig. 10.15, one can see that the peak width does not change with changing pressure under conditions of constant mass flow. This is an expected result and can be explained as follows. The net effective trajectories of ions traveling through the analytical gap of the DMS ion filter are a function of the superposition of two independent forces. The ions are carried along the length of the ion filter by the transport gas flow which produces a longitudinal ion movement in the z direction, see Fig. 10.9. A perpendicular force due to the applied RF and compensation voltages moves the ions transverse to the direction of motion induced by the transport gas flow (in the y direction, as defined in Fig. 10.9).

To determine the effect of pressure on ion trajectories we first examine the effect of pressure on ion residence time in the analytical gap. Based on the ideal gas law equation $PV = N\kappa T$; we can estimate the amount of drift gas molecules inside the DMS sensor with volume V at two pressures under isothermal conditions (where temperature T is constant):

$$P_1 V = n_1 \kappa T \tag{10.15a}$$

$$P_2 V = n_2 \kappa T \tag{10.15b}$$

There are several ways the pressure can be varied in the DMS. One way is by controlling the mass flow of the transport gas. Another way is to control the volume of the transport gas under different pressure conditions. Controlling mass flow is experimentally more convenient $F = n/t$ while allowing the pressure to change. Standard mass flow controllers can be used to provide flow rate information. In this case, the relationship between pressure and residence time τ can be written as follows:

$$\frac{n_1}{\tau_1} = \frac{n_2}{\tau_2} \tag{10.16}$$

And by using Eqns. 10.15 and 10.16 the relationship between ion residence time and pressure can be written as:

$$\tau_2 = \frac{P_2}{P_1}\tau_1 \tag{10.17}$$

So, under these experimental conditions the residence time changes proportionately to the pressure in the DMS filter. And consequently, the velocities of the ions along the longitudinal axis of the filter under the first and second pressure conditions are

$$\vartheta_{\|1} = \frac{L}{\tau_1} \tag{10.18a}$$

$$\vartheta_{\|2} = \frac{L}{\tau_2} = \vartheta_{\|1} = \frac{N_1}{N_2} \tag{10.18b}$$

where L is the length of the DMS ion filter region. The velocity of the molecules moving through the analytical gap increases proportionately to the change in pressure (i.e. density) N_1/N_2.

Next the effect of pressure on ion movement along the y axis induced by the asymmetric RF field will be considered.

Based on Eqn. 10.12, the total displacement in the y direction is $Y = (\beta\Delta K/T)t_{\text{res}}$. We can determine the drift velocity of the ions in this perpendicular direction at the first pressure P_1 as

$$\vartheta_{\perp 1} = \frac{h}{\tau_1} = \frac{\beta\Delta K}{T} \tag{10.19}$$

where h is the height of the analytical gap. At the second pressure, P_2, the ion velocity will be as follows:

$$\vartheta_{\perp 2} = \frac{h}{\tau_2} = \frac{hP_2}{\tau_1 P_1} = \vartheta_{\perp 1}\frac{N_2}{N_1} \tag{10.20}$$

From Eqn. 10.20 one can see that pressure affects ion velocity. However, the relationship between ion velocities in the y axis direction does not contain a pressure-related term:

$$\frac{\vartheta_{\perp 2}}{\vartheta_{\parallel 2}} = \frac{\vartheta_{\perp 1}}{\vartheta_{\parallel 2}} \tag{10.21}$$

From Eqn. 10.21 one can see that the ions will have similar effective trajectories under different pressure conditions, if the value of $\Delta K \sim \alpha$ does not change with changing pressure. Based on Fig. 10.15, we see that the peak width at FWHM does not change with pressure between 0.4 and 1.5 atm and therefore the alpha parameter is the same. Therefore, the result applies independent of the way the value of E/N is changed whether through varying pressure, varying the RF voltage V_{RF}, or by adjusting both experimental parameters simultaneously. This conclusion is supported by Fig. 10.16 which plots the DMMP monomer spectra at different pressures. To unify the axes in this figure we used the units recommended by E.W. McDaniel and E.A Mason and converted the axis units to Townsend where $1\ \text{Td} = 10^{-17}\ \text{V cm}^2$.

Plotting the data on a Townsend scale enables graphing of all the experimental data on a single chart for a particular RF waveform shape, Fig. 10.16. Interestingly, all the graphs obtained at different pressures fall on one another, and indicate that a particular combination of RF field strength and gas density, i.e. E/N value, provides an optimal DMS operating point. For the DMMP monomer ions plotted in Fig. 10.16 DMS provides the best resolution for an E_{RF}/N of around 140 Td. The optimal separation of DMMP ions in air occurs at around $E/N \sim 140$ Td.

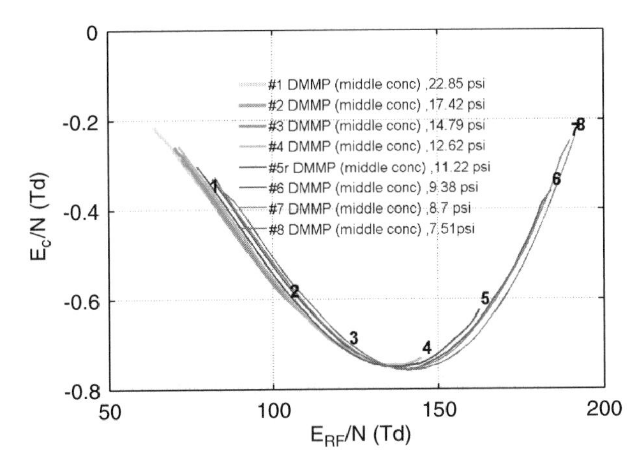

Fig. 10.16. Compensation field versus RF field for DMMP at different pressures, normalized to operating gas density.

Any combination of E_{RF} and gas density that produces a ratio of around 140 Td will result in optimal DMS resolution.

The non-monotonic behavior of the plots can be explained as follows: Below 140 Td the coefficient of ion mobility increases with increasing ion energy. Above 140 Td the coefficient of ion mobility decreases with increasing ion energy. The asymmetric behavior of these curves suggests the possibility that an ion neutral interaction transition occurs from elastic scattering, due to polarization interaction, rigid-sphere ion neutral interaction or resonance charge transfer.

10.4 PHYSICAL IMPLEMENTATION: PLANAR VERSUS CYLINDRICAL ION FILTER GEOMETRIES

The first DMS embodiments used planar ion filter electrodes [1,2]. In this configuration, the electric field between the electrodes is essentially of constant strength along the *y* axis, see Fig. 10.17. Buryakov, Krylov and Soldatov were the first to demonstrate a DMS with a coaxial-cylindrical electrode design. In this configuration the electric field strength varies with radial distance from the central electrode [49,50,51]. This design was explored due to its ability to provide for ion focusing in the DMS ion filter region which was thought would provide enhancement for ion transmission through the filter. The focusing of the ions into radial bands is due to the interaction of the radially varying electric field strength and the field-dependent mobility of the ions [49].

MSA was the first company to build a commercial prototype based on this cylindrical electrode configuration [3,52]. The cylindrical design was adopted and modified by Guevremont *et al.* [10,11]. In this embodiment the inner electrode tended to be a solid rod with the end closest to the mass spectrometer shaped to have a hemispherical tip as shown in Fig. 10.18. A more detailed schematic can be found in [6]. A more detailed investigation of ion focusing was carried out by Guevremont *et al.* [11,19,20,22]. They showed that resultant spectral peak shape in the cylindrical DMS is a balance between the ion focusing between the concentric cylindrical electrodes and diffusion and ion–ion repulsion effects spreading the ions out radially [22–24].

Much has been discussed about the advantages of ion focusing in a cylindrical DMS, and for certain ion species this is clearly advantageous. Guevremont observed that when the ion focusing conditions are met, the ion transmission efficiency of a coaxial DMS can be greater than for the planar DMS [22,24]. However, there are a number of significant

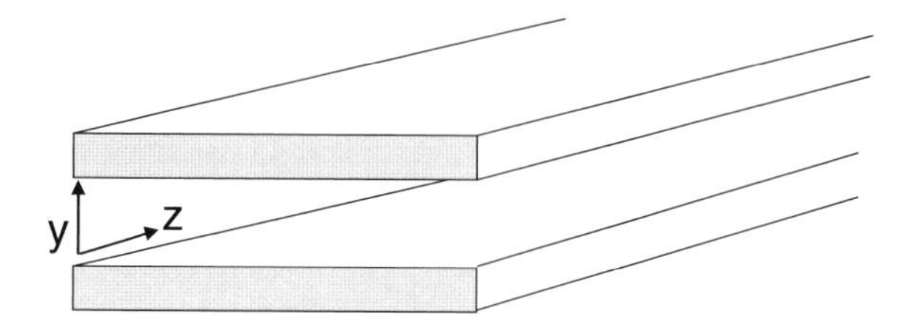

Fig. 10.17. Planar DMS filter electrodes with uniform electric field as a function of *y* in the gap.

disadvantages to ion focusing and the cylindrical construction that should be considered as well. Krylov [23] compared the planar and coaxial DMS performance. His conclusions were that for some ions the transmission efficiency of the coaxial DMS can indeed be higher than for the planar DMS but with a much longer response time. Krylov found that the response time of a coaxial DMS with ion focusing is about two orders of magnitude slower than for a planar DMS. This is due to the fact that the ion focusing process requires the ions to spend a certain time in the filter, and that this focusing time is significantly longer than the time required for actual ion filtering. Usually the ion focusing time is greater than 100 milliseconds. DMS response time is an important consideration for high-speed analysis, such as in high-speed gas chromatography, or even high-throughput screening with MS. In comparison to the coaxial DMS, the planar spectrometer typically has an ion residence time of less than 2 milliseconds in the ion filter. This short residence time helps to minimize diffusion losses. So while the coaxial configuration does compensate for diffusion losses, which is necessary due to the long residence times required for ion focusing, the very small, fast, and significantly less complex planar DMS device does not require the ion focusing to get excellent performance and ion transmission. The DMS can also be made low cost and even disposable for applications which are intolerant of any potential sample-to-sample carry over or cross contamination.

Another aspect of the cylindrical device is that only ions with substantial field dependencies on mobility (large positive or negative α values) can be effectively focused. The focusing conditions in the coaxial DMS are not the same for all ion types. This is significant for quantitative analysis of different ion species in mixtures. Unfortunately, the coaxial cylindrical DMS has different ion transmission efficiencies for the different ion species (alpha values), and thus will not provide an accurate representation of the relative concentrations of the various analytes in the sample. In contrast, the coefficient of ion transmission in the planar DMS does not depend on the alpha parameter, or the polarity of

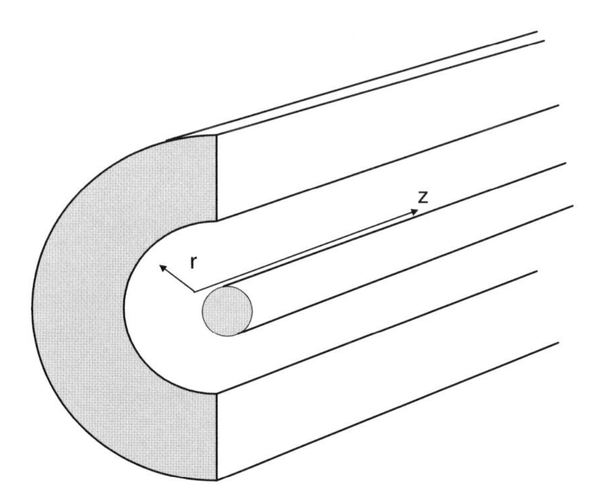

Fig. 10.18. Coaxial cylindrical DMS filter electrodes, cross section shown for clarity, with radially varying electric field strength.

the ions, and thus provides a significantly more uniform ion transmission characteristic for different ion species. This uniform transmission allows much better quantitation of the relative concentrations of analytes in a sample. As a result, the planar configuration allows a single calibrant to be used to calibrate the system for a wide range of ion species, the cylindrical configuration requires that the coefficient of transmission be calibrated for each ion species individually, in order to determine relative concentrations in a sample.

Figs. 10.19 and 10.20 respectively show DMS responses for planar and cylindrical DMS analyzers for the same water molecule cluster ($(H_2O)_nH^+$) RIP peak. Fig. 10.19 shows the planar Sionex microDMx response for an RIP peak formed by a ^{63}Ni ionization. The response in Fig. 10.19 for the planar microDMx shows that the RIP peak intensity is relatively uniform over a V_{RF} range from 500 V to 1020 V. In contrast Fig. 10.20 shows that the RIP peak intensity continuously increases as a function of V_{RF} up to 2600 V for the cylindrical, Ionalytics Selectra, DMS. In the cylindrical device the RIP peak ions are formed by corona discharge. We show in Fig. 10.21 that virtually the same ion species are formed with both ionization sources, since the normalized plots of Figs. 10.19 and 10.20 fall on top of each other. The two DMS analyzers also have different ion filter gaps. The planar DMS has a gap of 0.5 mm while the cylindrical DMS has a gap of 2 mm. The different gaps are also normalized out in the chart of Fig. 10.21. Fig. 10.21 plots the alpha parameters as a function of *E/N* using the data from Figs. 10.19 and 10.20. As expected very similar results were obtained for analyzers (since we are detecting the same ion species). However, even though the applied RF voltages used in the cylindrical DMS ranged up to 2600 V, the cylindrical DMS only covered a relatively small *E/N* range. This is due to a number of limitations of this embodiment. First, it is technically difficult to generate high V_{RF} voltages at the required frequencies. Second, as the RF field applied is increased in the cylindrical DMS the spectral peak width increases and spectrometer resolution is decreased. Under the same conditions, the range of *E/N* that the planar DMS can cover is significantly wider even though the voltages applied are only up to 1500 V. The intensity of the peaks in the planar DMS are of almost constant up to about 1100 V (90 Td) and the peak width does not vary significantly between the low and high field strength.

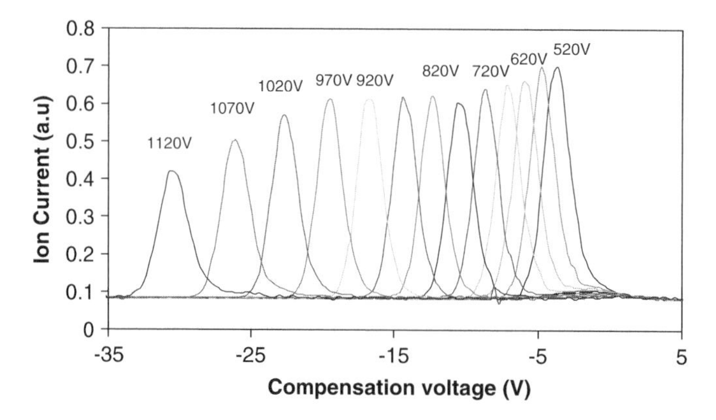

Fig. 10.19. Planar DMS (microDMx) CV spectra for protonated water ions $(H_2O)_nH^+$. ^{63}Ni ion source.

What this means is that the planar DMS appears to provide higher accuracy, selectivity, and sensitivity, for quantitative analysis applications compared with the cylindrical DMS device. Even though, for some select ion species with high alpha dependencies the ion transmission efficiency in the cylindrical device may be higher. A. Shvartzburg and K. Tang of the Pacific Northwest National Laboratory (PNNL) have also advocated the benefits

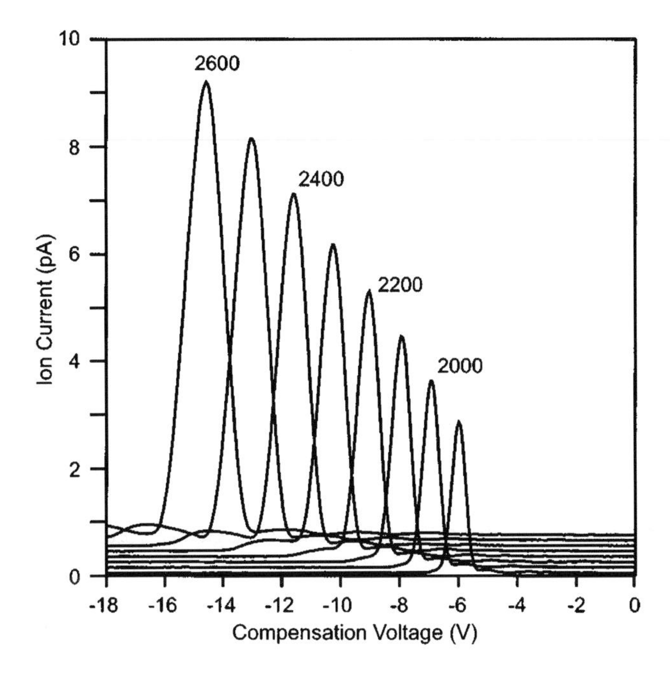

Fig. 10.20. Cylindrical DMS (FAIMS-E) CV spectra showing the $(H_2O)_nH^+$ ion. Corona Ion source. Reprinted with permission from Guevremont and Purves, [22], copyright 1999.

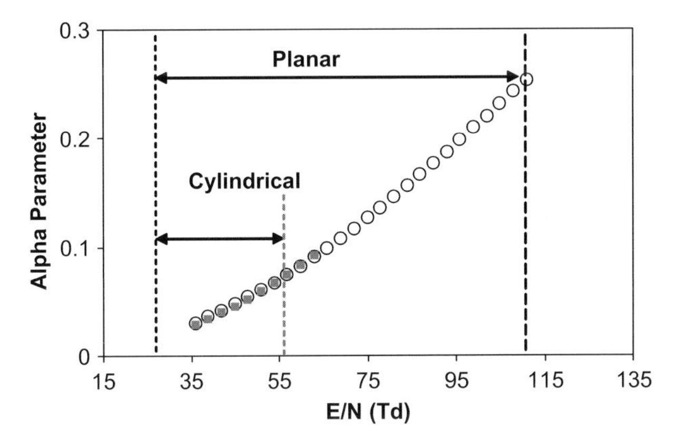

Fig. 10.21. Comparison of alpha parameters as a function of *E/N* for RIP peaks from cylindrical and planar DMS systems.

of planar DMS approach for quantitative analysis over the cylindrical configuration, and have demonstrated this for heavy ion species such as proteins [53,54].

One additional advantage of the planar DMS design is that for the same RF voltage waveform one can simultaneously separate and detect both positive and negative ions. This is very useful for enhanced compound identification (e.g., fast gas chromatography), as relative intensities of both positive and negative ions for exactly the same sample can be analyzed simultaneously. In the cylindrical embodiment, on the other hand, only one polarity of ions at a time can be filtered in the DMS. In order to switch between separation of positive or negative ions the polarity of the RF voltage must be switched.

10.5 THE IMPORTANCE OF IONIZATION: FUNDAMENTALS OF ATMOSPHERIC PRESSURE ION FORMATION AND ION CHEMISTRY

Ion formation is a prerequisite step to ion filtering and compound analysis in the DMS. It operates at or near atmospheric pressure and therefore utilizes various ionization sources that form ions under atmospheric pressure conditions. In this section we review a number of the principal mechanisms and methods for ion formation and contrast them with impact ionization performed under vacuum conditions.

Ionization can be induced through the collision of highly energetic accelerated particles (e.g., electrons, photons, ions) with neutral molecules. Under vacuum conditions, see Fig. 10.22, product ions are formed by direct impact ionization when these highly energetic particles collide with analyte molecules. During the collision process the energy is transferred to the analyte molecules causing extensive fragmentation of the molecules resulting in, molecular, quasy-molecular (M^+, M^-, $(M-R_i)^+$, $(M-R_i)^-$) and fragment (R_i^+, R_i^-) ions. The greater the impacting particle energy the greater the amount of fragment ions formed.

In contrast, at atmospheric pressure, the ionization of analyte molecules occurs in the gas phase and is generally a "soft" multi-step process. The ionization at atmospheric pressure is often termed atmospheric pressure chemical ionization (APCI) when ionization sources such as radioactive (^{63}Ni, a beta emitter, or ^{241}Am an alpha emitter), corona discharge, or plasma ionization are employed. The ionization process is not a direct one from energized particle to analyte ion, but generally involves formation of intermediate ions,

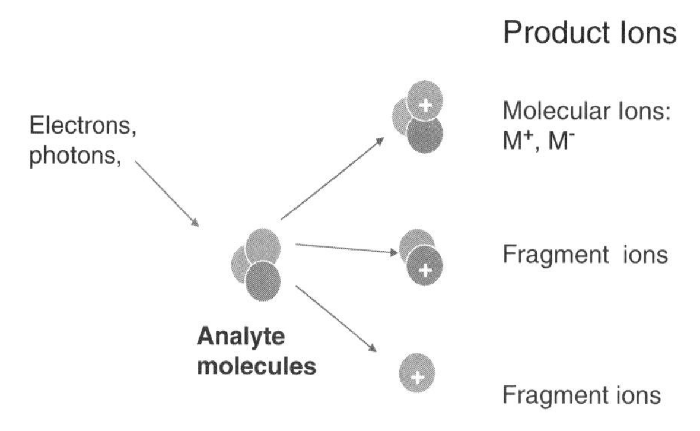

Fig. 10.22. Formation of analyte ions in vacuum conditions by impact ionization.

often termed reactant ions, which ultimately transfer an electron, a proton, or other charged species to the neutral analyte to form the analyte ion. When photons are used to generate the reactant ions (e.g., by UV PID or laser photoionization), the ionization process is often termed atmospheric pressure photo ionization (APPI). APCI is also a widely used technique for liquid chromatography–mass spectrometry applications using sample nebulization and vaporization. The APCI ionization process does not directly apply to electrospray ionization (ESI) or MALDI as ions are generally formed in solution or by interacting with a matrix prior to volatilization by extraction into the gas phase by an electric field. The ESI process will be described further in the next section. Listed below are some of the principal steps involved in ion formation at atmospheric pressure by APCI processes.

1. *Ionization of ambient gas molecules.* At atmospheric pressure the analyte molecules generally tend to be at trace concentrations (ppm, parts per million or ppb, parts per billion levels) relative to the ambient gas which is often air or nitrogen. As a result, the probability of direct ionization of analyte molecules is negligible, and the ionizing particles will ionize ambient, non-analyte, gas molecules with approximately a 100% probability. This will result in the formation of some positive ions and free electrons. In the case where the ambient gas is very dry air, mostly nitrogen and oxygen ions will form.
2. *Ion "cooling" and ion-neutral collisions.* Under ambient conditions ($P = 1\,\text{atm}$, $T = 273\,\text{K}$) the density of molecules is $n = N_a/0.0224 \sim 3\times10^{25}\,\text{m}^{-3}$. Consequently the mean free path for air molecules (O_2, N_2) with a typical diameter $d \sim 2 \times 10^{-10}\,\text{m}$ are estimated at $\lambda = 1/\Omega n \sim 3 \times 10^{-7}\,\text{m}$. The average molecule velocity in ambient conditions is $\langle \vartheta \rangle = \sqrt{3kT/M} \sim 500\,\text{m/s}$. The time between collisions, therefore is a very short $\Delta t = \lambda/\bar{\vartheta} = 6\times10^{-10}\,\text{s}$. As a result newly formed excited ions will immediately collide with other gas molecules and will lose their energy i.e. be "cooled" very rapidly.
3. *Reactant ion (charge reservoir) formation.* In this step gas phase chemical reactions occur between the initially formed ions and other ambient gas molecules. Thermodynamic and traditional theoretical kinetic approaches can be used to predict the reactant ions formed. Ions which have the longest lifetime in the ambient gas will generally define the reactant ion species formed. The positive reactant ions are generally formed from molecules with the highest proton affinity and the negative reactant ions are formed from molecules which have the highest electron affinity. For example, in air with moisture levels around 10 ppm, the reactant ions are formed from the protonated water molecules $H^+(H_2O)_n(N_2)_m$ and the negative reactant ions will be composed of oxygen ions and water molecules $(H_2O)_mO_2^-$. These ions generally form the reactant ion peaks or RIP.
4. *Formation of analyte ions.* When the density of the reactant ions becomes sufficiently high ($>10^7$ ions/cc) the interaction between analyte molecules and reactant ions is facilitated, see Fig. 10.23. The result of the interaction depends on the energetic properties of the reactant ions and analyte molecules and follows expected thermodynamic considerations. When the energetic properties of the analyte molecules relative to the reactant ions are favorable, for example, when analyte ions have higher proton or electron affinity, analyte ion formation can take place. The

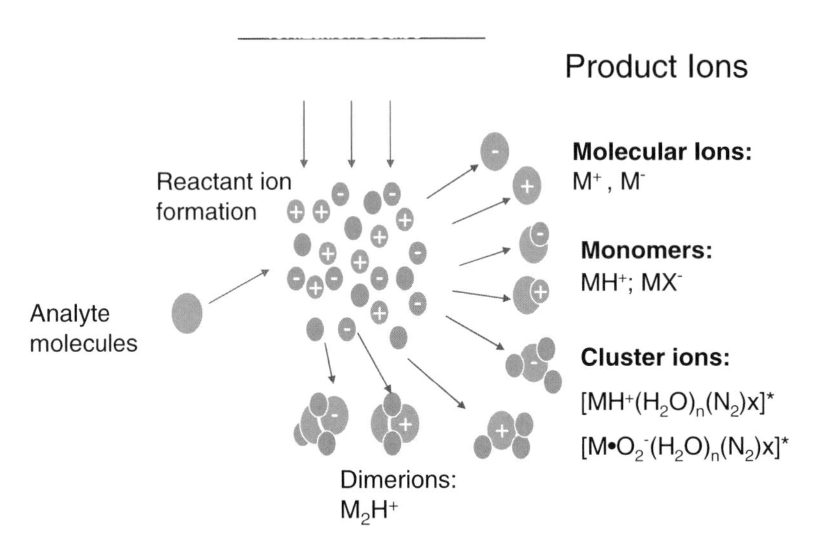

Fig. 10.23. Formation of analyte ions at atmospheric pressure.

efficiency of analyte ion formation can be as high as 100% if the collision frequency between analyte molecules and reactant ions is high. This is the reason why detectors which used APCI exhibit extremely high sensitivity toward chemical warfare agents (organo-phosphonates) and explosives.

5. *Formation of analyte cluster ions.* As the concentration of analyte is increased proton bound dimers may form due to an increase in the probability of analyte ion and analyte molecule interactions. For positive ions, the process of cluster formation is shown in equation

$$M + MH^{+}(H_2O)_n \rightarrow M_2H^{+}(H_2O)_{n-1} + H_2O \qquad (10.22)$$

To summarize then, APCI has the following characteristics: Ions are formed through mild energetic interactions; there is little molecular fragmentation; analyte ion formation involves charge transfer and competitive distribution of available charge among analyte molecules and there is an inherent tendency to form ion clusters.

APCI provides for highly efficient ion formation, and the ability to advantageously control the gas phase chemistry by using special dopants. It also allows the possibility of predicting the ions formed based on the well-developed kinetic and thermodynamic principles of gas phase chemistry.

Figs. 10.24 and 10.25 highlight the principal difference between electron impact ionization in vacuum and APCI for acetone analyte molecules. The electron impact spectrum shows a molecular acetone peak $m/z = 58$ with a relative abundance of about 60% and a number of peaks with smaller masses: $m/z = 43$ (relative abundance 100%), 42 (relative abundance 5.5%), 15 (relative abundance 12.2%) which are related to fragmentations of the acetone molecules.

In the APCI mass spectrum, Fig. 10.25, the dominated peaks are protonated molecular ion peak ($m/z = 59$, (MH^{+}) with a relative abundance of 100%), and proton bound dimers

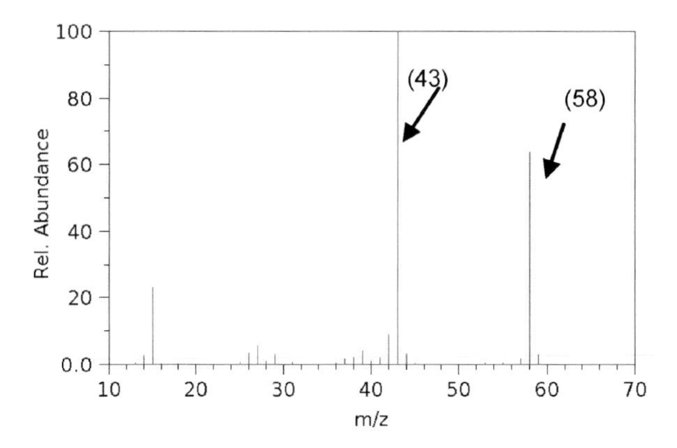

Fig. 10.24. Mass spectra of acetone ionized by electron impact ionization [55].

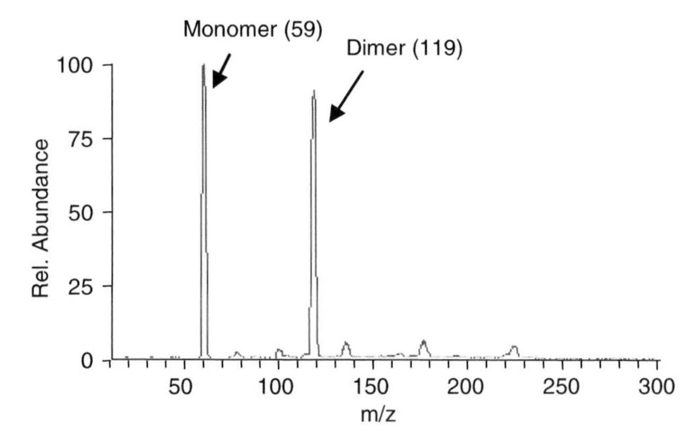

Fig. 10.25. Mass spectra for acetone ionized by APCI [56].

($m/z = 117$, (M_2H^+) relative abundance 90%). The spectrum also contains a number of low intensity ions with $m/z = 135$ [$(H_2O)M_2H^+$], 176[M_3H^+], peaks which are products of clustering of protonated acetone ions with water molecules.

10.5.1 Electrospray ionization

Electrospray ionization (ESI) was first developed by Dole *et al.*, in the 1960s [25], as a means to transfer ions from solution into the gas phase. In 1984, John Fenn incorporated the use of ESI with MS [26]. Since then, ESI has become one of the most widely used ionization techniques for MS. This is primarily due to its soft ionization characteristics, and compatibility with upstream analyte separation techniques such as liquid chromatography and capillary electrophoresis. In addition, ESI is capable of ionizing a wide variety of compounds from solution, including biological compounds such as peptides, proteins, oligosaccharides, and oligonucleotides. An ESI mass spectrum usually consists of a series of multiply-charged ions, e.g., $[M+nH]^{n+}$, for a single compound. The sample is dissolved in a suitable electrospray solvent such as 50:50 methanol:water with 2% acetic acid. Of particular interest has been the capability of ESI to successfully transfer non-covalent interactions between compounds in solution into the gas phase for analysis by MS [27–29].

In general, ESI is performed by pumping a solution of ions at atmospheric pressure through a metal capillary, to which a high-voltage is applied (typically 3–5 kV). In positive mode ESI, the high voltage applied to the capillary will be positive and the negative terminal is connected to a metal plate a few millimeters away from the tip of the capillary. Fig. 10.26 illustrates schematically the typical positive mode ESI ionization process. As can be seen, the applied voltage creates ion separation in solution, whereby cations move away from the capillary into the liquid drop extending away from the capillary tip and anions are attracted to the positively charged metal capillary. This process is termed electrophoretic charging [30,31]. The process of electrophoretic charging, for positive mode ESI, results in a high concentration of cations at the liquid surface from the capillary. The electro hydrodynamics of the charged liquid surface result in the formation of a liquid cone, referred to as the *Taylor cone*. Extending from the Taylor cone is a thin filament jet of solution which extends out until it breaks into droplets. The droplets are all positively charged due to the high concentration of cations at the liquid surface extending from the Taylor cone [31–34]. The formation of negative ions via ESI would simply be the reversal of applied voltage polarities.

One key difference between electrospray and APCI of liquid samples by nebulization is that many of the ions formed are believed to be formed in solution before the electrospray process or created by simple changes in concentrations as the aerosolized droplets shrink [35].

The charged droplets which are emitted from the filament jet are the result of the charge repulsion of the ions in solution exceeding the surface tension of the liquid. Once the charge at the surface of the liquid droplet exceeds the Rayleigh limit [36], the droplet breaks apart as part of the fission process into smaller droplets. This fission process is often referred to as coulombic expansion or explosion due to the process of exceeding the coulombic forces

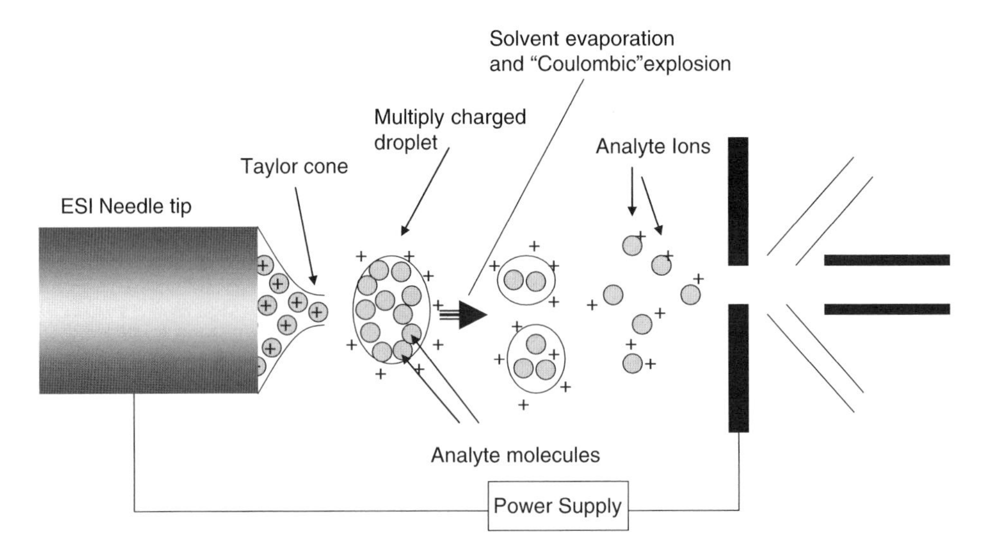

Fig. 10.26. Schematic of the electrospray ionization process.

maintaining the intact droplet [25]. In ESI, as the droplets collide with the ambient gas a drying effect occurs, evaporating off solvent from each droplet, and in turn increasing the charge concentration at the liquid surface. This process results in the repeated fission process until ions suitable for analysis are produced [25,37]. Examples where an electrospray ionization source is coupled with DMS and MS will be discussed in the applications section.

10.6 APPLICATIONS OF DMS

DMS has previously been demonstrated as a stand alone detector, in a number of applications [3,38]. In this section we focus on applications where DMS is used in conjunction with other analytical instruments to enhance the capabilities of these instruments and the quality of their overall analysis. First, applications using the DMS as a back-end detector for a gas chromatograph will be presented. Then, applications where the DMS is used as a front-end ion pre-separator for MS will be described.

The hyphenated GC–DMS system has a number of advantages. When used as a detector for GC, the DMS provides a second degree of compound separation, based on differential mobility. This is complimentary to the GC retention time based separation. The hyphenated system can provide separation for co-eluting compounds that cannot be resolved by the GC alone. Additionally, the burden of analytical separation can be shared between the GC and the DMS, so that poorer chromatographic separation can be tolerated but the quality of the original analysis preserved. This can enable faster chromatographic runs while still maintaining the quality of the overall analysis.

The DMS can be used to pre-separate complex mixtures into simpler fractions when placed between the ionization source and inlet of an atmospheric pressure ionization mass spectrometer. This differential mobility based pre-separation is sufficiently orthogonal to the mass spectrometric separation to provide a significant enhancement to the overall analysis. The DMS can be used to reduce the complexity of the sample presented to the mass spectrometer. This allows for enhanced composition analysis, can be used to significantly improve signal to noise, and elucidate structural information, and even to resolve between isomers [11].

10.6.1 Applications to gas chromatography

Gas chromatographic analysis is often hampered by the quality of peak or analyte separation in the chromatographic column. If two compounds have the same retention time, called co-elution, the resultant chromatographic peaks overlap. This separation can usually incorporate a unique tunable ion filter, capable of selecting specified components out of any matrix. This innovative approach to GC analysis adds separation power to the detector portion for greatly enhanced selectivity.

Fig. 10.27 illustrates the enhancement provided by the DMS to a chromatographic analysis for a homologous series of ketones from acetone to decanone (C3–C10). A UV photoionization lamp ($\lambda = 116\,\text{nm}$) was used to generate the ions. Fig. 10.27a shows a one-dimensional chromatogram of the form one might obtain if a GC were outfitted with a flame ionization detector (FID) or thermal conductivity detector (TCD). Fig. 10.27b is called a topographic plot and shows information from both the GC and DMS separations. The topographic plot

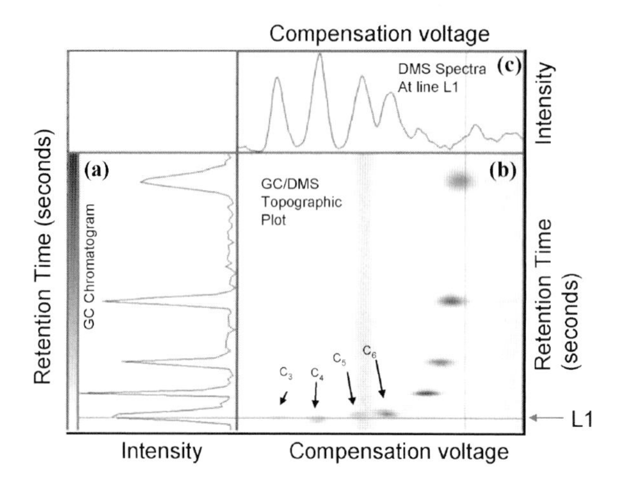

Fig. 10.27. GC–DMS spectra where the chromatographic runtime has been decreased leading to co-eluting species, showing that the DMS is able to resolve the co-eluted species.

provides three levels of information: retention time, compensation voltage, and ion intensity (indicated by the density of the spots) all of which can be used to better fingerprint or identify the analytes. Fig. 10.27c is a one-dimensional plot of the DMS intensity as a function of compensation voltage taken along line L1. The chromatogram in Fig. 10.27a was in fact also generated by the DMS and produced by summing the ion intensities over the entire compensation voltage scan range for each retention time, i.e. effectively a "total ion current" measurement. If chromatographic conditions (e.g., fast temperature ramps to reduce analysis time) result in degraded or co-eluted peaks, the DMS can be used to resolve and identify the co-eluted compounds. From Fig. 10.27b one can determine that the first chromatographic peak of the homologous series of ketones, Fig. 10.27a, actually contains four unique co-eluted compounds which can be identified in the DMS by their particular compensation voltages. Furthermore, the wealth of information provided by the GC–DMS can eliminate the need of external calibration through standards. Fig. 10.20 shows the DMS spectra only for positive ion species, however, DMS spectra can be obtained simultaneously for both positive and negative ions.

The performance of the DMS was benchmarked by comparing it with that of a FID, one of the most widely used GC detectors. Figs. 10.28 and 10.29 show a comparison of the concentration dependence and detection limits for a homologous series of ketones measured with the FID and the DMS equipped with a UV ionization source (116 nm), further details are described in Ref. [39]. The experimental results for these compounds showed that the average DMS detection limits were approximately an order of magnitude better than those of the FID. The reproducibility of the DMS spectrometer was also investigated and was found to compare very well to that of the FID [39].

10.6.1.1 Industrial applications of DMS (natural gas analysis)

Controlling and monitoring the levels of sulfur-containing compounds (e.g., odorants) in natural gas is an important commercial problem as these compounds contribute to corrosion

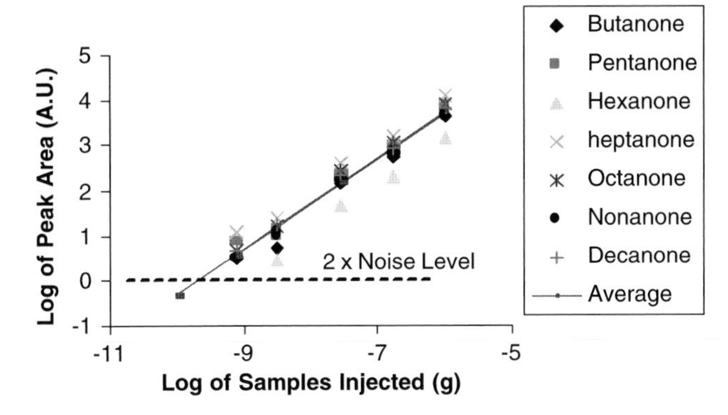

Fig. 10.28. FID and DMS response as a function of compound concentration for a homologous series of ketones. Note average FID detection limit is 2×10^{-10} g.

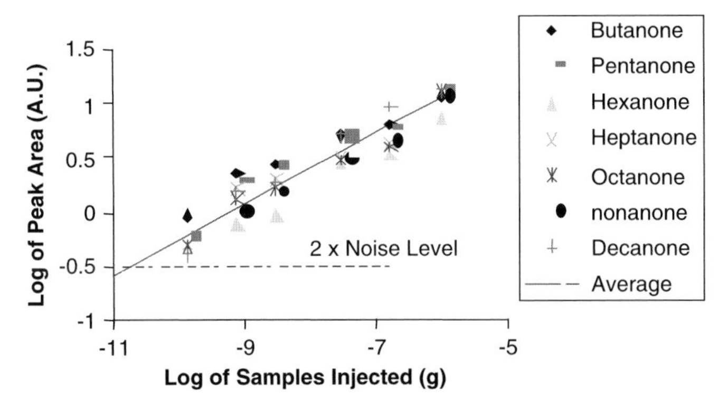

Fig. 10.29. Average DMS detection limit is 2×10^{-11} g for a homologous series of ketones.

of pipes and equipment, degrade catalysts, and have environmental, safety, and health implications. For many applications, the ability to detect sulfur-containing compounds at 1 ppm levels or less is needed. Current detectors either suffer from quenching effects in the presence of the hydrocarbon backgrounds (e.g., flame photometric detector), or the hydrocarbon peaks co-elute with the sulfur odorant peaks [57]. DMS has been successfully used as a highly sensitive and selective GC detector in the detection of trace levels of sulfur-containing compounds in hydrocarbon gas mixtures. Fig. 10.30a shows a typical GC–TCD chromatograph for a sample of natural gas containing the odorant methyl ethyl sulfide (MES). MES elutes on the tail of a hydrocarbon peak and as a result, limits the detection limit for MES to around 10 parts per million. In comparison, Fig. 10.30b shows a GC–DMS chromatogram with atmospheric pressure UV ionization with the DMS tuned and fixed to the compensation voltage for MES. The UV ionization provides inherent selectivity over hydrocarbons. So simultaneously we perform selective ionization of odorants and selective detection of particular ions related to sulfur odorants. In this case, the MES peak is baseline resolved with

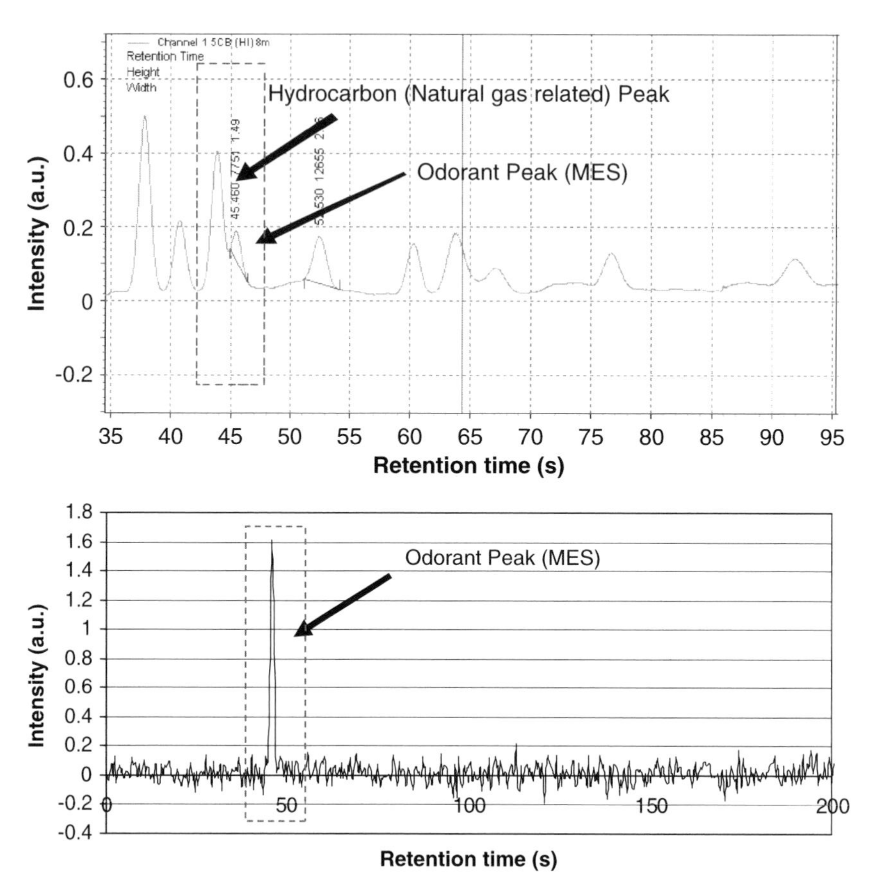

Fig. 10.30 (a) GC response with TCD detector. (b) GC response with DMS detector tuned to the compensation voltage for MES (Vc∼ −9.5 V). Reprinted with permission from Jos Curvers, Varian Inc., Copyright 2002.

detection limits down around 100 ppb. Fig. 10.31a illustrates how the DMS spectral separation helps separate the odorant from the hydrocarbon background using a non-selective radioactive ^{63}Ni source. The x axis in Fig. 10.31a corresponds to the DMS compensation voltage (Vc) and the y axis is the chromatographic retention time in the GC–DMS topographic plot. The z axis is the intensity or sample concentration in a direction out of the page. In addition to the MES there are actually two other sulfur odorants (TBM, THT) present in the natural gas sample. From this topographic plot, one can see that the odorants are spatially separated from the hydrocarbons, demonstrating the power of the DMS. Specialized software integrated with DMS control allows the collection of GC–DMS responses at select spatial locations in the topographic plot. This enables the presentation of the data a traditional chromatographic format. Thanks to the DMS separation, the three odorants at 100 ppb concentrations can be completely baseline resolved from the hydrocarbon background.

A productized GC detector based on the Sionex microDMx detector has been incorporated at into the Varian Instruments, Inc. CP-4900 μGC and is called the μDMD, see Fig. 10.35.

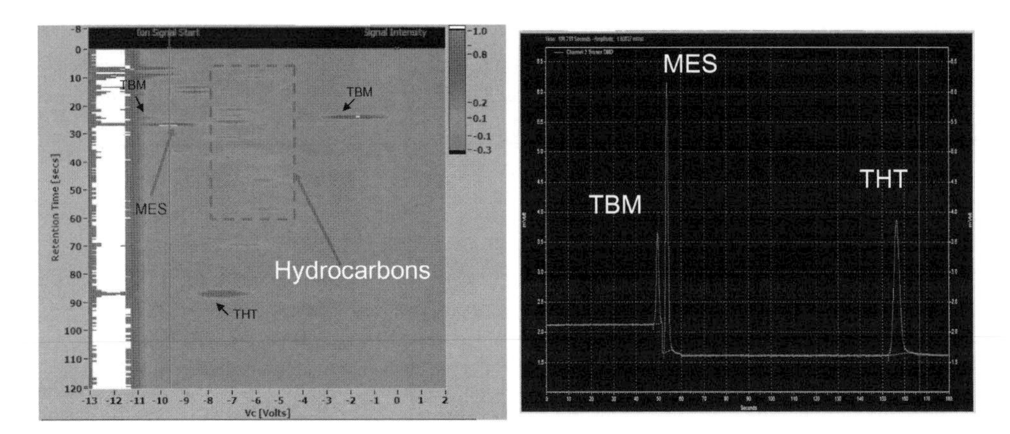

Fig. 10.31 (a) GC–DMS topographic plot showing the resolution of sulfur-containing natural gas odorants (MES, THT, TBM) from hydrocarbon background. (b) Traditional chromatographic presentation of GC–DMS data from Fig. 10.31a.

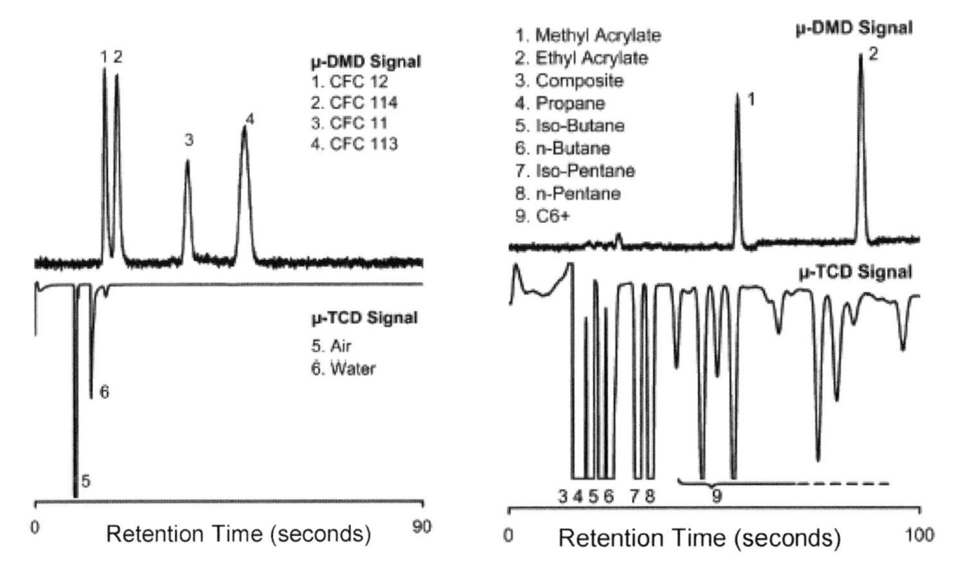

Fig. 10.32 (a) CP-4900 DMD application to detection of CFCs in air. (b) CP-4900 DMD application to detection of sulfur-free odorants in natural gas. Reprinted with permission from Jos Curvers, Varian Inc., Copyright 2002.

The resultant CP-4900 DMD product is an information-rich detector designed for trace level, multi-component analysis of gaseous samples. It offers sensitivity in the 100 ppb range and is virtually matrix independent. It is factory tuned according to specific application requirements and can be used as an online, at-line and laboratory analyzer. Varian, Inc.'s CP-4900 with DMD is the first system of its kind that can accurately select multiple target components out of co-eluting peaks in a single analysis run, and eliminates the usual requirements for multiple sample injections, repetitive calibrations, and cumbersome sample preparation [40, 42]. Other GC–DMS applications that have been explored include detection of anesthetic gasses, Freon's in ambient air (Fig. 10.32a) and sulfur-free odorants in natural gas (Fig. 10.32b) [41]. The top plots in Fig. 10.32 are the DMS response, while the bottom plots are the TCD response. The TCD is connected in series with the DMS in the CP-4900 system.

10.6.1.2 Security applications of DMS (explosives detection)

A second example illustrating the benefit of coupling DMS to GC is demonstrated for explosives detection. Detection of explosives at airports and security checkpoints is of ever growing importance. The challenge is detecting the explosive compounds without false negatives and with minimum false positives. While false positives are not nearly as catastrophic as false negatives, they adversely affect passenger throughput. The explosives detection problem is further compounded by the introduction of large amounts of contaminants and potential interferents into the analyzer as part of the sampling process. Most current detectors deployed at airports for hand baggage screening are based on ion mobility spectrometry. While highly sensitive for explosives compounds they tend to have high false positive levels due to matrix effects from interferences. The combination of GC with DMS is understood to have value in several facets. The pre-separation step adds retention information to compensation voltage for enhanced specificity, and the introduction of constituents as single compounds or an elution peak with a few compounds which increases the reliability of the APCI reactions. The combination of the GC and DMS can lead to a reduction of at least an order of magnitude in the level of false alarms. Results from GC–DMS detection of a mixture of NG, DNB, DNT, TNT, PETN explosives are shown in Figs. 10.33 and 10.34. Fig. 10.33 shows a total ion current GC chromatogram, where the DMS compensation voltage separation is not available. This one-dimensional chromatographic trace shows only four apparent peaks. Fig. 10.34 demonstrates the improved resolution

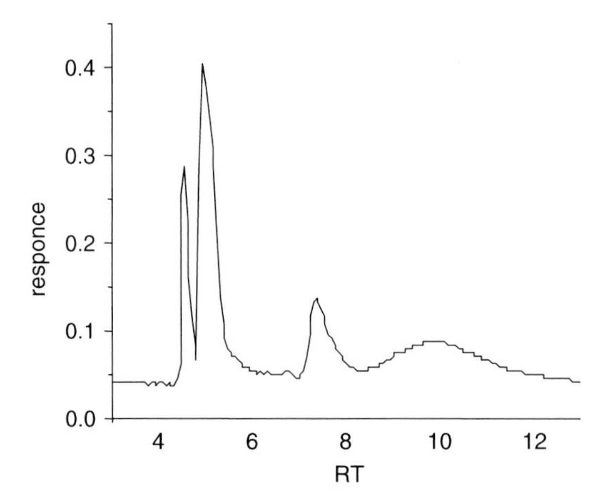

Fig. 10.33. Total ion current, GC–DMS chromatogram for a 2 μl injection of a mixture of 0.078 mg/ml NG, 0.156 mg/ml DNB, 0.047 mg/ml DNT, 0.047 mg/ml TNT, 0.125 mg/ml PETN [43].

achievable with the Sionex microDMx™ when the second dimension of separation is provided [43]. Furthermore, the differential mobility-based separation of DMS enables the resolution of non-nitrogen containing explosives such as triacetone triperoxide (TATP). The data presented in Figs. 10.33 and 10.34 were collected with a conventional gas chromatograph, so separation times were relatively long. However, in an implementation by Thermo Electron Corporation, the EGIS™ Defender, the GC separation time is reduced to less than 10 s.

The EGIS™ Defender is an explosives detection product which combines Thermo Electron's high-speed gas chromatography technology with the Sionex microDMx detector. It is designed to simultaneously detect plastic, commercial and military explosives, TATP, HMTD, and enhanced AN-FO as well as international civil aviation organization (ICAO) marker compounds with high sensitivity. Other trace detection systems for explosives, such as Ion Mobility Spectrometers (IMS) rely on one technology for identification and therefore must sacrifice either accuracy or sensitivity to achieve marginal results. IMS field units typically produce two to three percent false positive results while the EGIS™ Defender's advanced technology is claimed to produce less than 1.5 percent false positives [44].

10.6.1.3 Future trends

Smaller, handheld, analytical instruments with performance comparable to much larger laboratory instruments, but which require much less user intervention and are more rugged, are being sought and developed for field applications. Toward that end, miniaturized and microfabricated instruments for GC are under development in several laboratories [45–47]. These instruments are of interest because of their very small size, weight, and reduced utility requirements which makes them suitable for on-site analysis. Instruments with two-way wireless communications, and batteries with remote recharging capabilities are also being explored

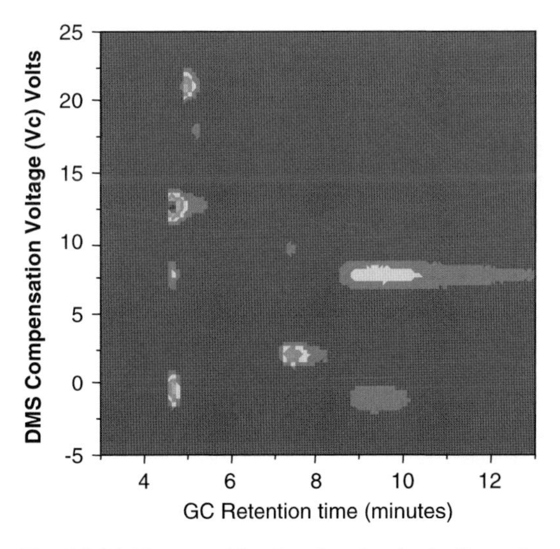

Fig. 10.34. Topographic plot showing both dimensions of GC and DMS separation. NG elutes at 4.2 min, DNB at 4.7 min, DNT at 4.8 min, TNT at 7.2 min, and PETN at 10 min [43].

for remote operation. Fig. 10.36 shows a series of different length microfabricated GC columns ranging from 3 m down to 10 cm columns being developed at the University of Michigan [58]. The performance of these microfabricated columns is improving with time, but detectors which are suitable themselves to miniaturization and which enhance or compliment

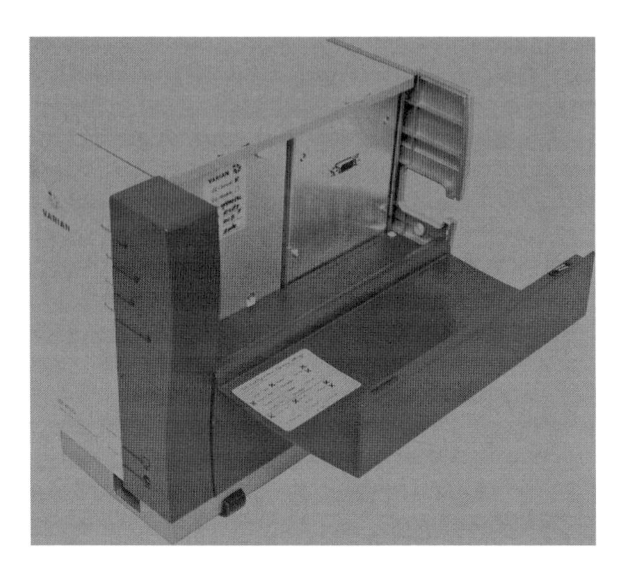

Fig. 10.35. Varian CP4900 DMD incorporating the Sionex DMS detector, the MicroDMx^{TM}. Reprinted with permission from Jos Curvers, Varian Inc., Copyright 2002.

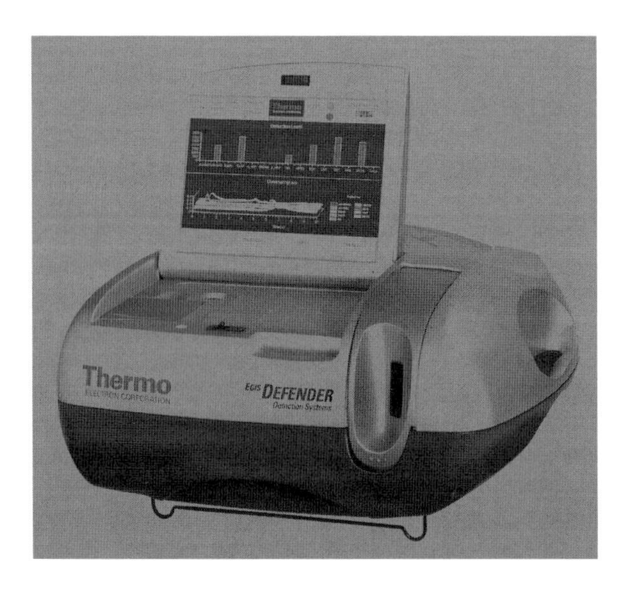

Fig. 10.36. Thermo electron's EGIS defender incorporating Sionex's microDMx^{TM} detector. Reprinted with permission from Thermo Electron Corporation.

the selectivity of these columns are desired. Recently, work to couple and characterize microfabricated GC columns with differential mobility detectors has been performed [58,59]. The ultimate objective of this work is the integration of DMS with micro-GC onto the same substrate, or at least packaging the systems in a further miniaturized platform.

The micromachined GC columns were first characterized with a standard FID detector to assess their performance (Fig. 10.37). Fig. 10.38 shows a chromatogram obtained from a 45-component test mixture injected into a 3.0-m-long microfabricated GC column. The composition of the mixture is described in Table 10.1. Peak numbers correspond to component numbers in Table 10.1. The portion of the chromatogram in the dotted-line box is shown on an expanded timescale above the main chromatogram. The test mixture elutes over

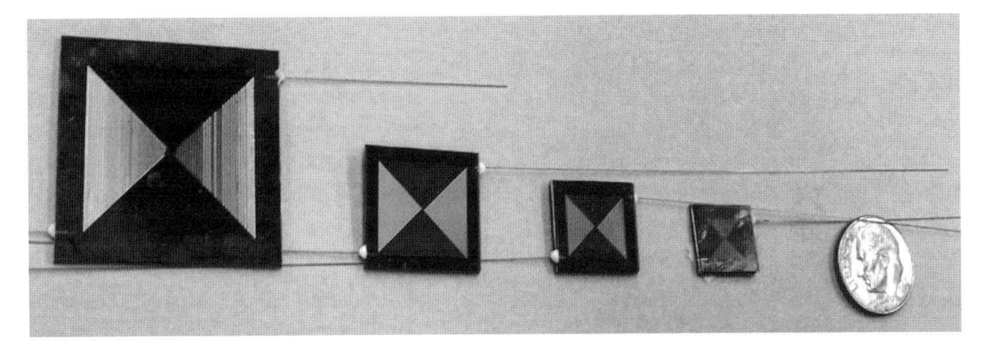

Fig. 10.37. Microfabricated silicon–glass GC Chip Columns. Reprinted with permission from Lambertus *et al.* [58]. Copyright 2005 American Chemical Society.

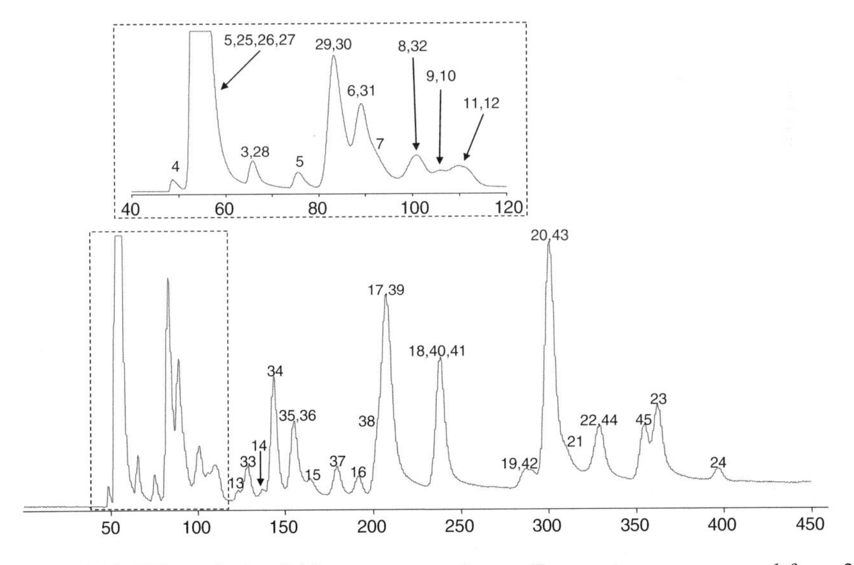

Fig. 10.38. FID analysis of 45 component mixture. Temperature programmed from 30–120°C @ 10°C/min. Reprinted with permission from Lambertus *et al.* [58]. Copyright 2005 American Chemical Society.

TABLE 10.1

TEST MIXTURE COMPONENT LIST

No.	Compound	No.	Compound
(1)	Pentane	(24)	Propyl benzene
(2)	Methyl Sulfide	(25)	Iodomethane
(3)	Isopropyl Alcohol	(26)	Ethyl Bromide
(4)	Acetonitrile	(27)	Dichloromethane
(5)	Propionitrile	(28)	Carbon disulfide
(6)	Ethyl acetate	(29)	Iodoethane
(7)	Tert-butyl ethyl ether	(30)	Chloroform
(8)	Tetrahydrofuran	(31)	Bromopropane
(9)	3-Methyl-2-butanone	(32)	1,2-Dichloroethane
(10)	Benzene	(33)	1,1,1-Trichloroethane
(11)	Thiophene	(34)	Iodopropane
(12)	Ethylene glycol dimethyl ether	(35)	Bromobutane
(13)	2-Pentanone	(36)	1,3-Dichloropropene
(14)	2-Ethylfuran	(37)	1-Chloropentane
(15)	4-Methyl-2-pentanone	(38)	1-Iodo-2-methyl propane
(16)	Toluene	(39)	Chlorodibromomethane
(17)	2-Hexanone	(40)	Tetrachloroethylene
(18)	Butyl acetate	(41)	Iodobutane
(19)	Ethyl benzene	(42)	1-Chlorohexane
(20)	*m*-Xylene	(43)	Bromoform
(21)	Cyclohexanone	(44)	1-Chloro-2-methyl propane
(22)	*O*-Xylene	(45)	Iodopropane
(23)	Bromobenzene		

the volatility range from *n*-C5 to *n*-C10 with a nonpolar stationary phase. Only a few components are adequately resolved for quantitative measurements. The column generates 5500 theoretical plates using air carrier gas at an average linear velocity of 10.8 cm/s. The pentane solvent peak is very intense and masks peaks from several target compounds. In contrast Fig. 10.39a shows the topographic plots for the positive ion channel and the negative ion channel (Fig. 10.39b) obtained using the DMS as a detector for the same column and separation conditions as for Fig. 10.38. The vertical axis is the time after sample injection, and the horizontal axis is the DMS compensation voltage (Vc). The display intensity indicates peak amplitude. The enhancement in information content in the two-dimensional data in Fig. 10.39 relative to the one-dimensional chromatogram in Fig. 10.38 is remarkable. The peak capacity of the microfabricated column combined with DMS detection is dramatically increased, and a distinct if not completely separated peak is observed for each of the 45 components in the test mixture [58].

The positive ion channel, Fig. 10.39a, shows peaks for 28 of the target compounds and the pentane solvent (peak 1). Note that *n*-pentane and other low-molecular-weight saturated

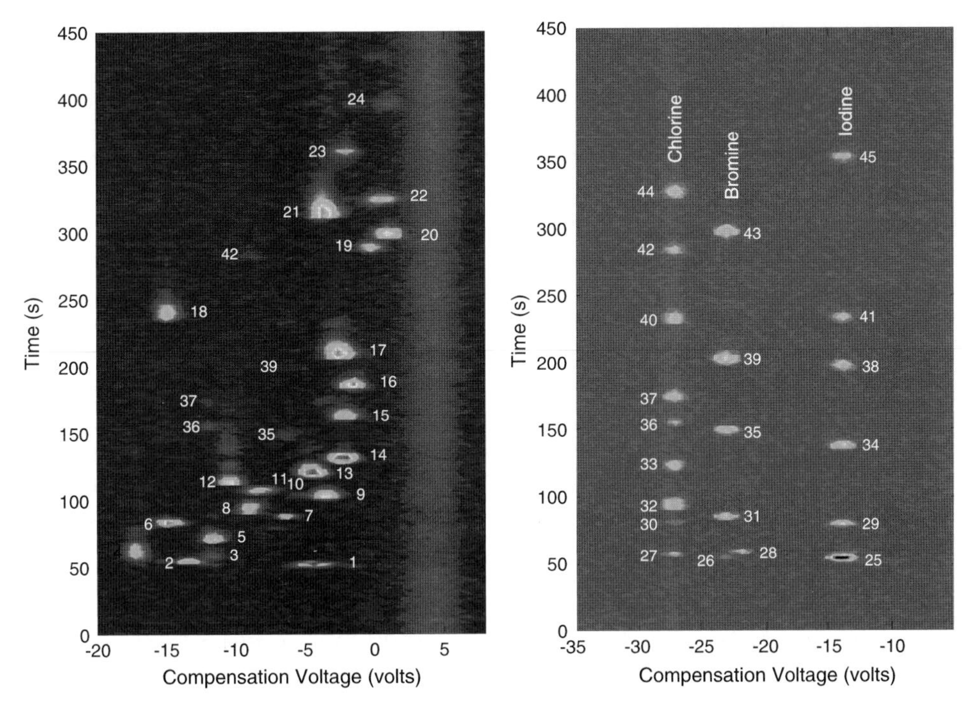

Fig. 10.39. μGC-microDMx analysis of 45 component mixture. The RF voltage for the DMS was 1300 V, and the detector temperature was 100°C. The compensation voltage was scanned from +10 to −35 V at 1 Hz. Reprinted with permission from Lambertus *et al.* [58]. Copyright 2005 American Chemical Society.

hydrocarbon compounds are not detected at low concentrations with APCI. Thus, the solvent does not produce a large and very wide peak as it did in the FID chromatogram shown in Fig. 10.38. Positive ion peaks are distributed over a wide range of compensation voltages. In general, higher-molecular-weight compounds show a smaller range of compensation voltages than the lower-molecular-weight compounds. This is useful for GC detection, since volatile compounds often are more difficult to separate due to inadequate interaction with the stationary phase in the column. The appearance of the negative ion channel plot contrasts with the positive channel plot in that peaks tend to form vertical patterns. That is, there are three groups of peaks with different retention times but similar compensation voltages. Peaks 27, 30, 32, 33, 36, 37, 40, 42, and 44 are all chloride compounds with a chlorine attached to an aliphatic carbon. Peaks 26, 31, 35, 39, and 43 are all from aliphatic brominated compounds, and all have peaks with the central compensation voltage of about −23 V. Peaks 25, 29, 34, 38, 41, and 45 are all aliphatic iodides, and all produce peaks with compensation voltage centered near −14 V. The only other major peak observed in the negative ion channel is peak 28 from carbon disulfide, which is a frequently used extraction solvent. Eliminating this as a possible interference in the analysis of volatile compounds will be useful [58]. While preliminary, the results obtained from coupling these two technologies are extremely promising. They promise a path to the realization of truly miniaturized, portable, high-performance detection systems.

10.6.2 Application to MS

As with gas chromatography, the use of DMS with MS has been shown to enhance the quality of mass spectrometric analysis. In this case, the DMS serves as an atmospheric pressure ion pre-separator between the atmospheric pressure ion source and the inlet to the mass spectrometer. Fig. 10.40 shows a conceptual schematic of a planar DMS chip coupled to a mass spectrometer inlet orifice. One advantage of the planar configuration is that only ions are introduced into the mass spectrometer inlet, while sample neutrals are exhausted out of the analyzer. Another advantage that was discussed previously is that of lack of ion focusing. This enables very rapid and quantitative DMS ion separation.

Even though differential mobility ion separation is not entirely orthogonal to mass based separation [60], the DMS separation is still powerful for resolving between many compounds which cannot be or are difficult to resolve by MS alone.

The degree to which the DMS can be utilized as an analyte separator in conjunction with mass spectrometric detection depends on the type of analysis being performed, and the sample complexity. For some sample analyses the DMS–MS has the potential to replace some conventional LC–MS and GC–MS methods.

DMS–MS separation was not necessary for the APCI analysis of acetone since the mass spectrum for acetone is relatively simple (Fig. 10.25). In contrast, the added dimension of orthogonality can be very useful for quantitation and detection of other compounds which produce more complex APCI mass spectrums. Fig. 10.41 shows a three-dimensional APCI plot of DMS–MS spectral information for a trace amount of butanone in a transport gas of clean air. The DMS and mass spectrometric information are projected onto corresponding axes. This three-dimensional plot illustrates how the DMS provides orthogonality to MS and enhances compound analysis. There are two ways to look at the DMS–MS plot. For a particular DMS spectral peak (e.g., compensation voltage, approx. $-6\,V$) the corresponding mass spectrometric composition of the DMS ion peak can be observed. Or at a particular

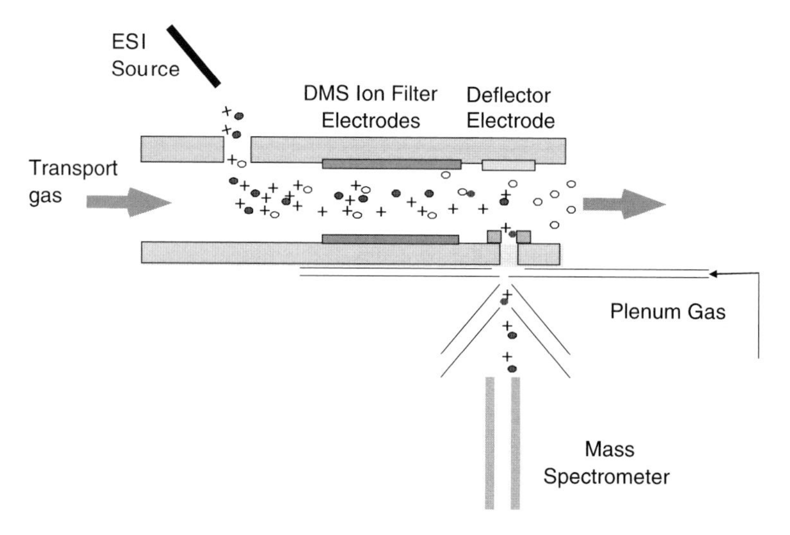

Fig. 10.40. Schematic of a planar parallel plate DMS mass spectrometer pre-filter.

mass value, the corresponding DMS spectrum can be observed. The protonated butanone monomer should have an *m/z* value of 73, and one would expect to see only a single DMS peak corresponding to the butanone monomer. In actual fact, one finds that in addition to the butanone monomer peak, the dominant ion peak in the DMS, there are two additional peaks. One DMS peak has a compensation voltage that corresponds to the reactant ion peak RIP (a water cluster peak $(H_2O)_nH^+$), and a second peak corresponding to heavier cluster ions (Cluster 2). The mass spectrometer is not capable of resolving between these ion species and would provide a single intensity response at an *m/z* value of 73. By combining the DMS and MS separation abilities, a truly quantitative determination of the concentration of butanone alone can be obtained.

The combination of DMS with ESI–MS has proven beneficial to the analysis of various biological, pharmaceutical, and environmental compounds [61–66]. Early work by Buryakov *et al.* illustrated the possibility to couple DMS to MS [67]. This DMS–MS system was utilized to provide insight into the nature of DMS ion separation and ion chemistry. Since then, a number of groups have explored the use of the DMS primarily as a pre-filter to a mass spectrometer to enhance the quality of the mass spectrometric analysis [68–70]. Guevremont *et al.* were the first to commercialize a DMS-based product, the Ionalytics Selectra, which served as a front-end cylindrical ion filter for MS.

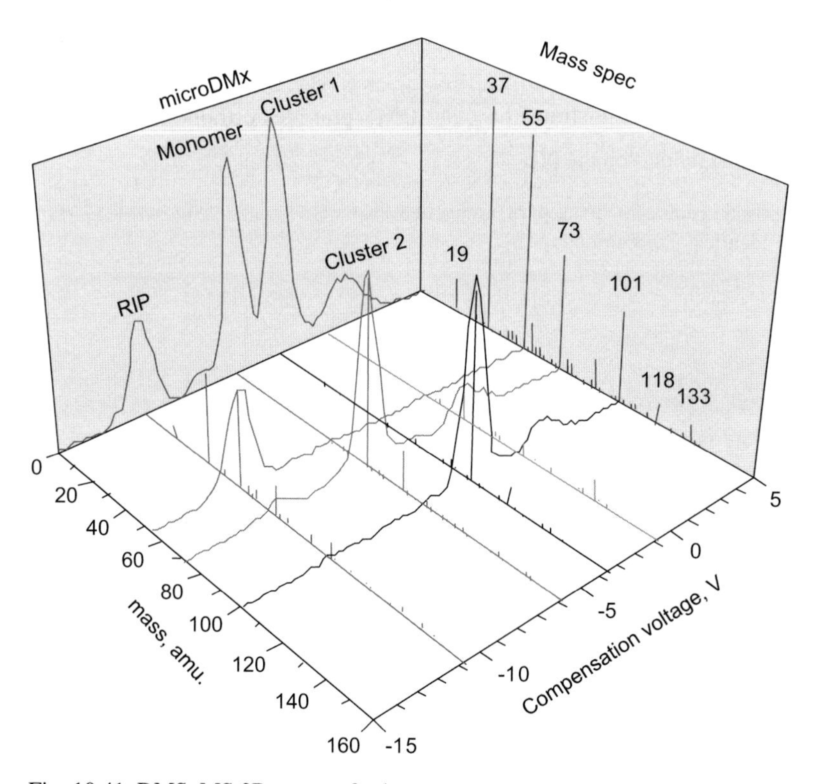

Fig. 10.41. DMS–MS 3D spectra for butanone.

Another illustration of the utility of coupling the DMS to a mass spectrometer is illustrated by the ability of the DMS–MS system to resolve between structural, and isobaric, isomers. In many cases, MS has difficulty resolving these compounds, or the separation technique is very complex or time consuming. Ion separation in the DMS can be achieved in the gas phase in a matter of seconds. Fig. 10.42 illustrates the DMS ability to resolve between iso-, meta-, and paraxylene isomers [9]. Fig. 10.43 shows a plot of DMS peak position as a function of RF voltage (V_{RF}) for isomers of $C_{12}H_{18}$ [71]. Roger Geuvremont and colleagues demonstrated the ability to separate isomers such as leucine and isoleucine [72] and to resolve between ortho-, meta-, and paraphthalic acids [20], a separation impossible with MS alone. In 2003 W. Gabreyelski and K.L. Froese demonstrated the ability of the DMS

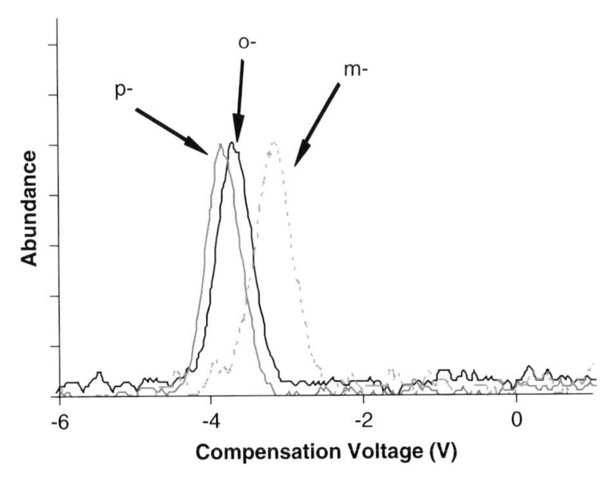

Fig. 10.42. Superimposed spectra for meta, para, and orthoxylene, showing varying levels of separation between these identical mass compounds.

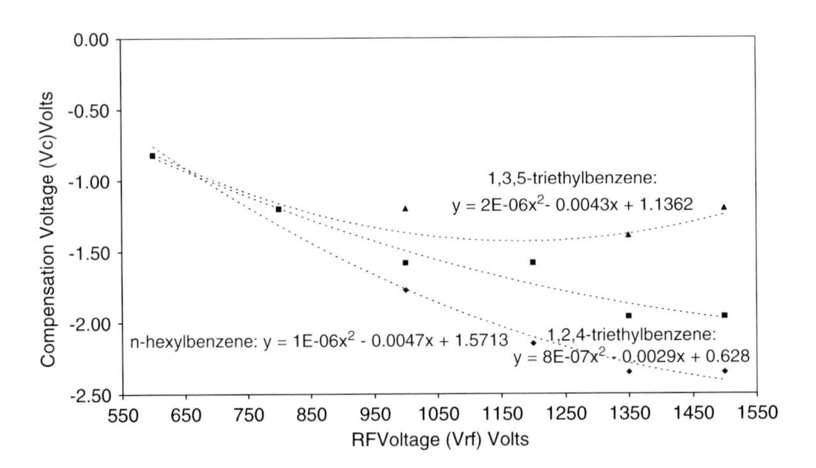

Fig. 10.43. Dependence of compensation voltages on RF voltage for isomers of $C_{12}H_{18}$. Reprinted with permission Helko Borsdorf, Copyright 2005.

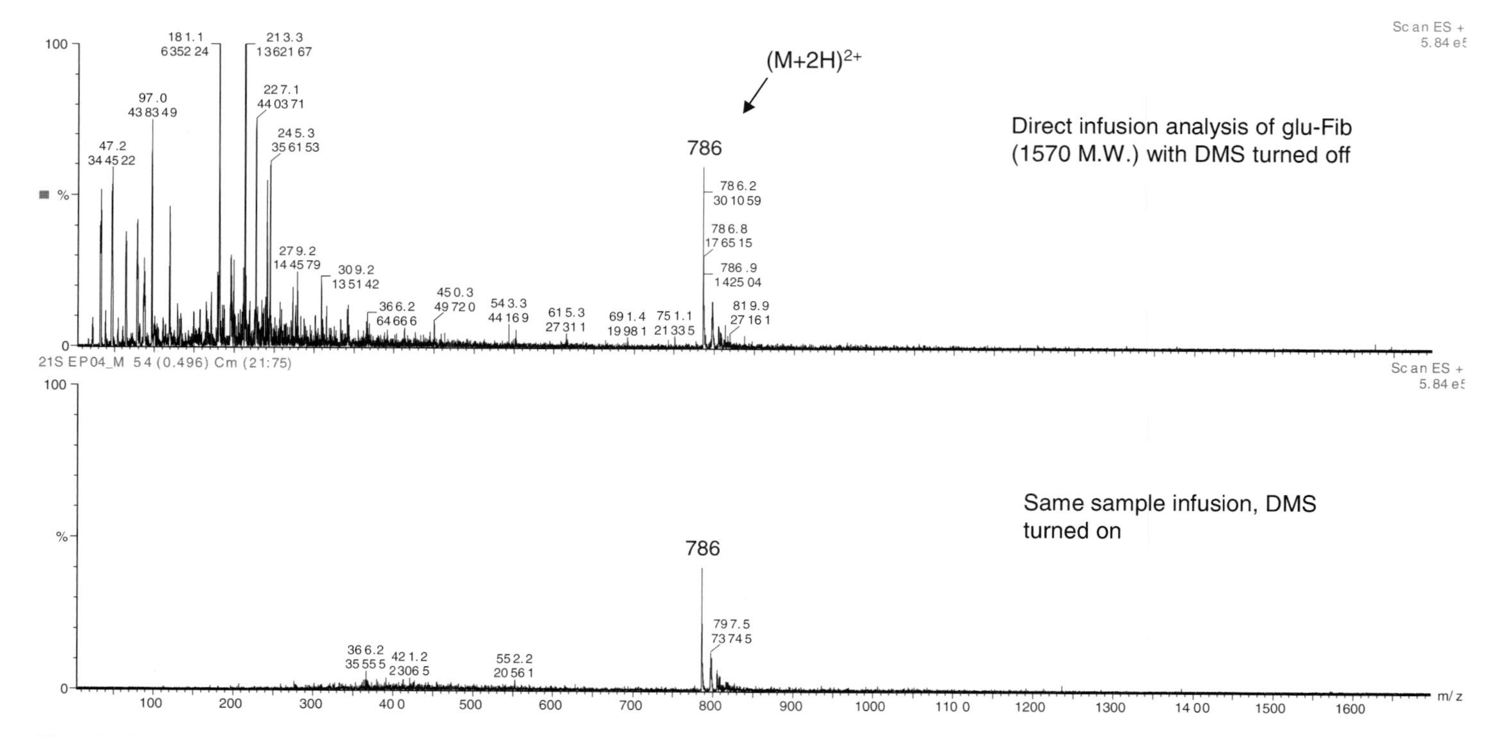

Fig. 10.44. ESI–DMS–MS analysis of Glu-Fibrinose B.

to resolve and differentiate anomers, linkage, and position isomers of disaccharides at low nM levels in a matter of seconds without sample purification or fractionation [73].

The DMS can also be used to reduce background noise and improve the signal-to-noise ratio of an analysis. Fig. 10.44 illustrates how the DMS can be tuned to suppress an undesired background and improve the signal to noise, in this case for the ESI–DMS–MS analysis of a directly infused Glu-Fibrinose B (M.W. 1570) peptide sample. The top spectrum shows the response for the Glu-Fibrinose B buried in the background with the DMS filter off (i.e. all ions pass through the DMS filter). The bottom spectrum shows the result with the filter on and tuned to a compensation voltage optimized to allow the Glu-Fibrinose B to pass through the filter while reducing the background signals. Meng Cui *et al.* demonstrated a 30-fold improvement in signal-to-noise ratio for Cisplatin, for a study which identified the modes of action and toxicity of Cisplatin. Cisplatin is a principal chemotherapeutic drug for cancer treatment for solid tumors (ovarian, testicular). Detection limits of 0.7 ng/ml with no derivatization or liquid chromatographic separation prior to analysis were obtained [74].

In a series of ESI–DMS–MS experiments using a triple quadrupole, MS Barnett *et al.* [72] demonstrated that the negative ions of the leucine and isoleucine amino acids could be detected separately, and that each isomer could be analyzed in the presence of over 600 times excess of the other isomer. In an additional study, a signal-to-noise improvement of over 50-fold was observed in the ESI–DMS–MS experiments compared with conventional ESI–MS. In yet other experiments with the DMS filter, the low background intensity allowed detection limits of 20 ng/ml of morphine and 60 ng/ml of codeine in urine samples to be detected [75].

DMS–MS has been applied to the analysis of peptides for environmental as well as proteomics applications. In one application the DMS was used as a pre-filter for MS analysis of microcystins. Microcystins are a class of peptides formed by blooms of cyanobacteria in fresh and brackish waters. Some of these species release toxic compounds. An ESI–DMS–MS system was used to obtain qualitative and quantitative measurements of the microcystin compounds in water. In a comparison between ESI–MS and ESI–DMS–MS response, an over 30-fold improvement in signal-to-background ratio was observed with the DMS filter on, and detection limits of 1–4 nM of microcystin were obtained [11,76]. A number of studies on simple peptides and complex mixtures of peptides resulting from the digestion of pure proteins and mixtures of proteins have been performed with DMS as a pre-filter [77,78,79]. For peptide applications the DMS separation was shown to remove many background ions with the same mass to charge of the peptide of interest, resulting in a higher proportion of structurally informative ions with the use of DMS than without the DMS present. Barnett *et al.* did a more systematic study of DMS for tryptic digests looking at mixtures from 14 proteins including two variants of hemoglobin and five variants of albumin [79]. More recently a nanoelectrospray LC–DMS–MS/MS system for the analysis of tryptic digests was used to demonstrate a ten times improvement in signal-to-noise compared with conventional nano LC–MS. Detection limits were observed in the high attomole range [80]. R. Geuvremont *et al.* were able to demonstrate that the intact protein, bovine ubiquitin, could be transmitted through a DMS at a defined compensation voltage that was non-zero (indicating resolution in the DMS). They further observed that different conformers of the protein could be resolved from one another. They extended their study to bradykinin demonstrating that the DMS is sensitive to the three-dimensional structure of the protein ion. In 2004 A.E. Ashcroft investigated the use of DMS coupled to ESI–MS for the differentiation of co-populated protein conformers on the amyloidogenic protein β2-microglobulin.

The data showed that different protein conformers could be detected and analyzed by ESI–DMS–MS. This approach offers significant opportunities for the study of the conformational properties of proteins, potentially enabling an understanding of the roles that different conformers play in diseases related to protein folding [81].

10.6.2.1 Enhancements to ESI–DMS–MS separation

As discussed in Section 10.3.3, modifying the transport gas composition affects the differential mobility behavior of the ions and the resultant spectral response. D.A. Barnett showed that a 95:5 mixture of N_2 and CO_2 could resolve the isomers of phthalic acids [20]. Levin *et al.* investigated the role of incorporating drift gas modifiers or dopants to the DMS sensor to reduce peptide clustering or aggregation from the ESI process. The drift gas modifiers provided the capability for selective enhancement of peptide ion separation by nano ESI–DMS–MS [82]. For this study, the angiotensin fragment 11–14 (M.W. 481) peptide was selected as a test compound to establish the optimum parameters for peptide ion "selection" in the DMS sensor. Three different drift gas modifiers were investigated for their effects on the DMS ion selection of the angiotensin fragment peptide. DMS dispersion plots and selected compensation voltage, Vc, point mass spectra were collected to determine the ideal Vc point corresponding to the maximum MH^+ ion signal of the angiotensin fragment. The extent to which the angiotensin singly charged monomer ion, *m*/*z* 482, was shifted to a larger negative compensation voltage (Vc) for a given Rf voltage was monitored for each drift gas modifier condition. It is presumed that the greater the shift away from a Vc of zero to a larger negative Vc value, the greater the decrease in clustering of the peptide aggregate ions, resulting ultimately in a reduction in effective cross-sectional area and an improved drift gas medium for separation between individual peptides.

Fig. 10.45 (a–d) provides the DMS dispersion plots and selected Vc point mass spectra (at Rf = 1000 V) for the maximum angiotensinogen fragment monomer ion signal of a 0.05 mg/ml sample, without (a) and in the presence of approximately 8000 ppm each of three different drift gas modifiers, methanol (b), 2-propanol (c), and 2-butanol (d). The dispersion plots on the left side of Fig. 10.45 are plotted as Rf voltage in the *y* axis (500–1500 V), compensation voltage in the *x* axis (−20 to +5 V) and ion intensity as reflected in the plot color (red equaling no signal and blue the most intense signal). The dispersion plots provide a comprehensive view of DMS ion separation for a given sample. The observed bands (tracts) in the dispersion plots demonstrate unique differences in DMS ion separation for the different drift gas modifiers. As demonstrated in Fig. 10.45a, without any modifier the dispersion plot contains only one low resolved tract, reflecting limited ion separation. The maximum angiotensinogen fragment 482 *m*/*z* ion signal for the DMS spectrum of Fig. 10.45a is found at a Vc of −0.5 V. Fig. 10.45 (b–d) demonstrate that the introduction of the drift gas modifiers at 8000 ppm produce new tracks in the dispersion plots. Each of the drift gas modifiers induced an analyte peak shift toward a greater negative Vc value: with 2-butanol providing the greatest shift to −6.8 V, while 2-propanol provided a shift to −6 V and methanol to −5 V.

Levin *et al.* introduced a new platform which has demonstrated the capability for ultra rapid quantitative analysis with direct sample infusion and nano ESI–DMS–MS [82]. After demonstrating DMS separation of individual peptide ions, the nano ESI–DMS–MS system was utilized to create an ultra rapid quantitative analysis platform with sample times of 10 s. It is anticipated that sample analysis times can be reduced down to 0.5–1 s. The platform is created

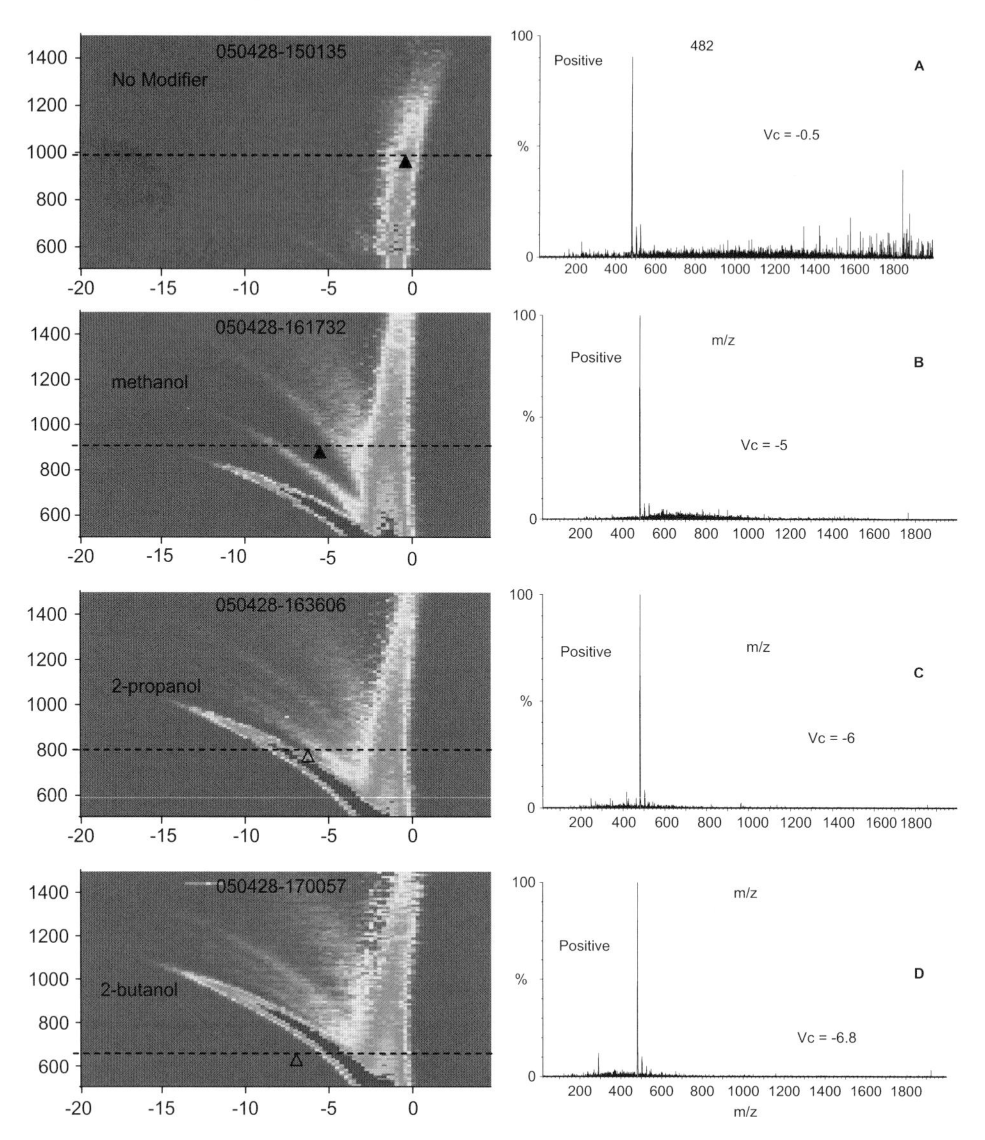

Fig. 10.45. 0.05 mg/ml angiotensin fragment 11–14 sample DMS dispersion plot and selected Vc point mass spectra (at an Rf = 1000 V) for the maximum MH^+ *m/z* 482 ion signal without any drift gas modifier (A), 8000 ppm methanol modifier (B), 8000 ppm 2-propanol modifier (C), and 8000 ppm 2-butanol modifier (D).

by rapidly scanning a compensation voltage range (10 s) and monitoring a selected ion (SIM) signal from the mass spectrometer for a directly infused nanospray sample. The analyte-specific peak in the DMS spectrum is then integrated for peak area. The reproducibility of generating the peak area ranged from 2.5 to 4.6% RSD. A minimum linear dynamic range of 2500 going down to 20 ng/ml was demonstrated for the angiotensin fragment 11–14 peptide

(M.W. 481). Linked scans of the DMS and MS data enables the generation of DMS spectra where the compensation voltage value corresponding to the analyte peak apex can be accurately identified. Determination of peak apex compensation voltage is used to provide an increased measure of analyte specificity in the same manner that retention time is used for chromatography. The future direction of research surrounding this platform is to combine a fully automated sample handling/nanoelectrospray system with DMS–MS, improving issues commonly associated with the use of nanoelectrospray for quantitative work. Such improvements are stable nanospray, no carryover, and consistent spray position. In addition, such automation would provide increased sample throughput times and low sample consumption.

10.7 CONCLUSION

The DMS method is a powerful tool for rapid gas phase ion separation and detection. Coupled with gas chromatography it offers a highly sensitive detector that provides a second dimension of information to enhance compound identification and the overall accuracy of the analysis.

When interfaced with MS, the DMS has been demonstrated to enhance the quality of mass spectrometric analysis, by providing additional information related to ion species conformation at atmospheric pressure. As greater understanding of the DMS operation mechanisms develops, further enhancements to the quality of ion separation and compound analysis will be realized. The DMS should prove a valuable tool to solve important chemical analysis problems.

ACKNOWLEDGMENTS

The authors would like to acknowledge the following people for their contributions. Dr. Paul Vouros from Northeastern University, Dr. Gary Eiceman from New Mexico State University, Scott J. Robbins, and Mark Bowers.

REFERENCES

1 I.A. Buryakov, E.V. Krylov, A.L. Makas, E.G. Nazarov, V.V. Pervukhin and U.Kh. Rasulev, *Pis'mav Zhurnal Tekhnicheskoi Fiziki*, 17 (1991) 60–65.
2 I.A. Buryakov, E.V. Krylov, E.G. Nazarov and U.Kh. Rasulev, *International Journal of Mass Spectrometry and Ion Processes*, 128 (1993) 143–148.
3 B. Carnahan, S. Day, V. Kouznetsov, M. Matyjaszczyk and A. Tarassov, Proceedings of the 41st annual ISA analysis division symposium, Framingham, MA, 21–24 April, 1996.
4 R.A. Miller, G.A. Eiceman, E.G. Nazarov and A.T. King, *Sensors and Actuators B*, 67 (2000) 300–306.
5 M.P. Balogh, Emerging technologies in the mass spectrometry arsenal, *Spectroscopy*, 20(2) (2005) 54–59.
6 E.W. McDaniel and E.A. Mason, The mobility and diffusion of ions in gases, John Wiley, 1973.
7 R.A. Miller, E.G. Nazarov, G.A. Eiceman and A.T. King, *Sensors and Actuators A*, 91 (2001) 307–318.

8 B. Carnahan, S. Day, V. Kouznetcosv and A. Tarassov, Proceedings on the 4th international workshop on ion mobility spectrometry, Cambridge, UK, Aug 6–9, 1995.
9 M. Groshkov, USSR Inventors Certificate no. 966583 (1982).
10 R. Guevremont, D.A. Barnett, R.W. Purves and L.A. Viehland, *J. Chem. Phys.*, 14 (2001) 10270–10277.
11 R. Guevremont, *J. Chrom. A*, 1058 (2004) 3–19.
12 G.A. Eiceman and Z. Karpas, *Ion Mobility Spectrometry*, CRC Press, Boca Raton, FL, 1994.
13 E.A. Mason and E.W. McDaniel, *Transport Properties of Ions in Gases*, Wiley, New York, 1988.
14 T.W. Carr, *Plasma Chromatography*, Plenum Press, New York, 1984.
15 Y. Raiser, *Fizika Gazovogo Razryada*, Moscow, Nauka, 1987.
16 G.A. Eiceman and Z. Karpas, *Ion Mobility Spectrometry*, Second Edition, CRC Press, Boca Raton, FL, 2005.
17 N. Krylova, E. Krylov and G.A. Eiceman, *J. Phys. Chem. A*, 107 (2003) 3648–3654.
18 G.A. Eiceman, N.S. Krylov and N.S. Kylova, *Anal. Chem.*, 76(17) (2004) 4937–4944.
19 D.A. Barnett, B. Ells, R. Guevremont, R.W. Purves and L.A. Viehland, *J. Am. Soc. Mass Spectrom.*, 11 (2000) 1125–1133.
20 D.A. Barnett, R.W. Purves, B. Ells and R. Guevremont, *J. Mass Spectrom.*, 35 (2000) 976–980.
21 A.A. Shvartsburg, K. Tang and R.D. Smith, *Anal. Chem.*, 76 (2004) 7366–7374.
22 R. Guevremont and R.W. Purves, *Rev. Sci. Instrum.*, 70 (2) (1999) 1370–1383.
23 E.V. Krylov, *Tech. Phys.*, 44 (1999) 113–116.
24 R. Guevremont, L. Ding, B. Ells, R.W. Purves, G.U. Thekkadath and L.A. Viehland, Proceedings of the 50th ASMS conference on mass spectrometry and allied topics, Orlando, FL, 2–6 June 2002.
25 M. Dole, L.L. Mack, R.L. Hines, R.C. Mobley, L.D. Ferguson and M.B. Alice, *J. Chem. Phys.*, 49 (1968) 2240–2249.
26 M. Yamashita and J.B. Fenn, *J. Phys. Chem.*, 88 (1984) 4451–4459.
27 K. Benkestock, C.K. Van Pelt, T. Åkerud, A. Sterling P.-O. Edlund and J. Roeraade, *J. Biomol. Screen.*, 8 (2003) 247–256.
28 S. Zhang, C.K. Van Pelt and D.B. Wilson, *Anal. Chem.*, 75 (2003) 3010–3018.
29 J.A. Loo, *Mass Spectrom. Rev.*, 16 (1997) 1–23.
30 http://www.chm.bris.ac.uk/ms/theory/esi-ionisation.html.
31 A.T. Blades, M.G. Ikonomou and P. Kebarle, *Anal. Chem.,* 63 (1991) 2109–2114.
32 P. Kebarle and L. Tang, *Anal. Chem.*, 65 (1993) 972A–986A.
33 C.A.J. Evans and C.D. Hendricks, *Rev. Sci. Instrum.*, 43 (1972) 1527–1530.
34 C.M. Whitehouse, R.N. Dreyer, M. Yamashita and J.B. Fenn, *Anal. Chem.*, 57 (1985) 675–679.
35 www.msterms.com.
36 L. Rayleigh, *Philos. Mag.,* 14 (1882) 184–186.
37 A. Gomez and K. Tang, *Phys. Fluids*, 6 (1994) 404–414.
38 V.A. Williamsburg, 2nd joint conference on point detection for chemical and biological Defense, March 1–5, 2004.
39 R.A. Miller, A. Zapata, E.G. Nazarov, E.V. Krylov and G.A. Eiceman, Proceedings of 2002 MRS conference, San Francisco, CA, 2002.
40 R.F. Firor and B.D. Quimby, A comparison of sulfur selective detectors for low level analysis in gaseous streams, Publication Number 5988-2426, April 2001.
41 J. Curvers and H.V. Schaik, *American Laboratory*, June (2004) 18–23.
42 Varian Inc. Press release, March 8, 2004.

43 G.A. Eiceman, E.V. Krylov, N. Krylova, E.G. Nazarov and R.A. Miller, *Anal. Chem.*, 76 (2004) 4937–4944.
44 Thermo Electron Press Release, New EGIS™ defender explosives trace detection desktop system launched at AVSEC World 2004, January 14, 2005.
45 E.B. Overton, K.R. Carney, N. Roques and H.P. Dharmasena, *Field Anal. Chem. Technol.*, 5 (2001) 1–2.
46 C.M. Yu, M. Lucas, C. Koo, P. Stratton, T. DeLima and E. Behymer, *Micro-Electro-Mechanical System (MEMS)* DSC, Vol. 66 (1998) 481.
47 H. Noh, P.J. Hesketh and G.C. Frye-Mason, *J. Microelectromech. Syst.*, 11 (2002) 718–725.
48 I.A. Buryakov, E.V. Krylov, V.B. Luppu and V.P. Soldatov, USSR certificate of invention SU 1627984 A2, 1980.
49 E.V. Krylov, PhD. Dissertation, published St. Petersburg, Russia, 1995.
50 USSR inventors certificate, SU 1485808A1, G01N27/62, 1987.
51 E.G. Nazarov, doctoral thesis, "Dissocative surface ionization of molecules in steady and non steady state conditions", St. Petersburg, Russia, 1992.
52 B.L. Carnahan and A.S. Tarassv, US Patent 5,420,424 1995.
53 K. Tan, A.A. Shvartzburg, 14th international symposium on ion mobility spectrometry, Chateau de Maffliers, France, July 24–28, 2005.
54 A.A. Shvartzburg, F. Li, K. Tang and R.D. Smith, *Anal. Chem.*, 78 (11) (2006) 3706–3714.
55 NIST Chemistry WebBook (http://webbook.nist.gove/chemistry).
56 G.A. Eiceman, E.G. Nazarov and R.A. Miller, *IJIMS* 3, 1 (2000) 15–27.
57 R.F. Firor and B.D. Quimby, Agilent publication number 5988-2426, April 2001.
58 G.R. Lambertus, C.S. Fix, S.M. Reidy, R.A. Miller, D. Wheeler, E. Nazarov and R. Sacks, *Anal. Chem.*, 77 (2005) 7563–7571.
59 R.A. Miller, 14th international symposium on ion mobility spectrometry, Chateau de Maffliers, France, July 24–28, 2005.
60 G.A. Eiceman, B. Tadjikov, E. Krylov, E.G. Nazarov, R.A. Miller, J. Westbrook and P. Funk, *J. Chromatogr. A*, 917 (2001) 205–217.
61 K. Venne, E. Bonneil, K. Eng and P. Thibault, *Anal. Chem.*, 77 (2005) 2176–2186.
62 J. Li, R.W. Purves and J.C. Richards, *Anal. Chem.*, 76 (2004) 4676–4683.
63 W. Gabryelski and K.L. Froese, *J. Am. Soc. Mass. Spectrom.*, 14 (2003) 265–277.
64 D.A. Barnett, B. Ells, R. Guevremont and R.W. Purves, *J. Am. Soc. Mass Spectrom.*, 13 (2002) 1282–1291.
65 R.W. Purves, D.A. Barnett, B. Ells and R. Guevremont, *Rapid Commun. Mass Spectrom.,* 15 (2001) 1453–1456.
66 J.T. Kapron, M. Jemal, G. Duncan, B. Kolakowski and R. Purves, *Rapid Commun. Mass Spectrom.*, 19 (2005) 1979–1983.
67 I.A. Buryakov, E.V. Krylov, A.L. Makas, E.G. Nazarov and V.V. Pervukhin, *U. Kh. Rasulev, Zhurnal of Analiticheskoi Khimii*, 48 (1993) 156–165.
68 R. Guevremont, *J. Chromatogr. A*, 1058 (2004) 3–19.
69 A.A. Shvartsburg, K. Tand and R.D. Smith, *J. Am. Soc. Mass Spectrom.*, 15 (2004) 1487–1498.
70 A.A. Shvartsburg, K. Tang and R.D. Smith, *J. Am. Soc. Mass Spectrom.*, 16 (2005) 1447–1455.
71 H. Borsdorf, E.G. Nazarov and R.A. Miller, Atmospheric pressure ionization studies and field dependence of ion mobilities of isomeric hydrocarbons using a miniature differential mobility spectrometer, *Anal. Chim. Acta*, Submitted for publication.
72 D.A. Barnett, B. Ells, R.W. Purves and R. Guevremont, *J. Am. Soc. Mass Spectrom.*, 10 (1999) 1279–1284.
73 W. Gabreyelski and K.L. Froese, *J. Am. Soc. Mass Spectrom.*, 14 (2003) 265–277.

74 M. Cui, L. Ding and Z. Mester, *Anal. Chem.*, 75 (2003) 5847–5853.
75 M.A. McCooeye, B. Ells, D.A. Barnett, R.W. Purves and R. Guevremont, *J. Anal. Toxicol.*, 25 (2001) 81–87.
76 B. Ells, K. Froese, S.E. Hrudey, R.W. Purves, R. Guevremont and D.A. Barnett, *Rapid Commun. Mass Spectrom.*, 14 (2000) 1538–1542.
77 R. Guevremont and R.W. Purves, *J. Am. Soc. Mass Spectrom.*, 10 (1999) 492–501.
78 R. Guevremont, D.A. Barnett, R.W. Purves and J. Vandermey, *Anal. Chem.*, 72 (2000) 4577–4584.
79 D.A. Barnett, B. Ells, R. Guevremont and R.W. Purves, *J. Am. Soc. Mass Spectrom.*, 13 (2002) 1282–1291.
80 K. Venne, E. Bonneil, K. Eng and P. Thibault, presentation to *Pharmagenomics* (2004).
81 A.J.H. Broysik, P. Read, D.R. Little, R.H. Bateman, S.E. Radford and A.E. Ashcroft, *Rapid Commun. Mass Spectrom.*, 18 (2004) 2229–2234.
82 D.S. Levin, R.A. Miller, E.G. Nazarov and P. Vouros, *Anal. Chem.*, 78 (15) (2006) 5443–5452.

Subject Index

S

T

U

V